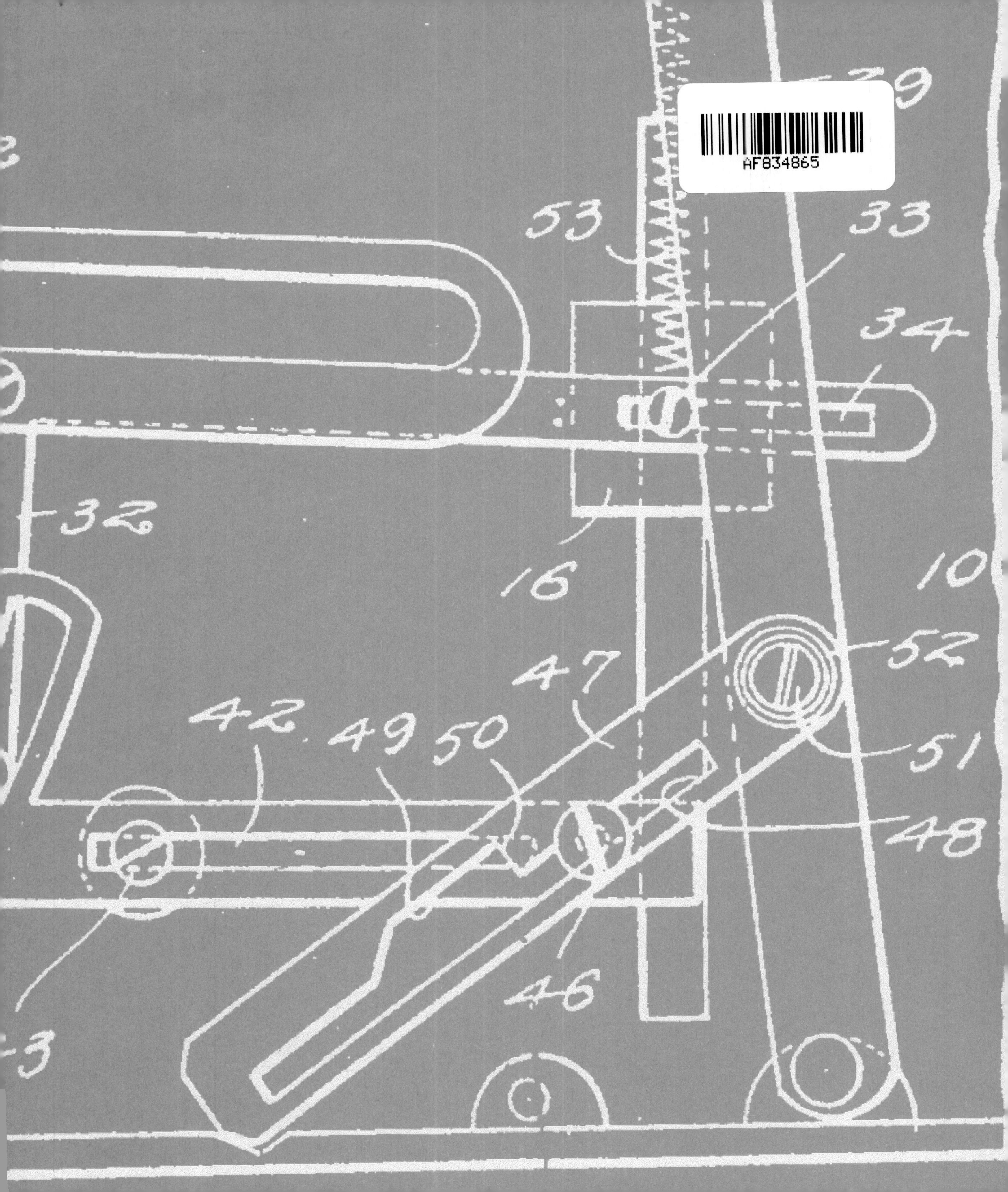
39
53
33
34
32
16
10
52
47
42
49
50
51
48
46

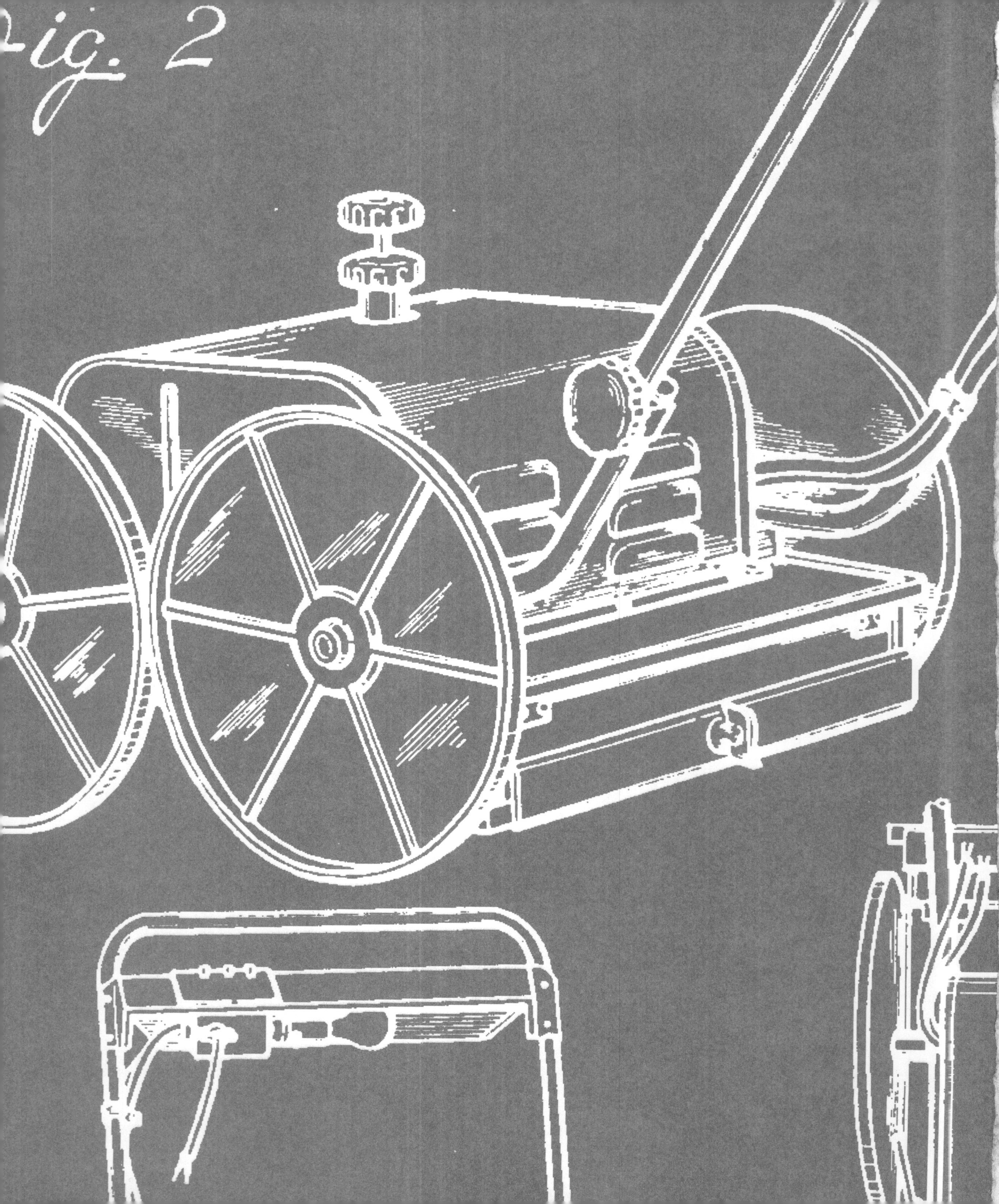

Fig. 2

INDUSTRIAL DESIGN IN THE MODERN AGE

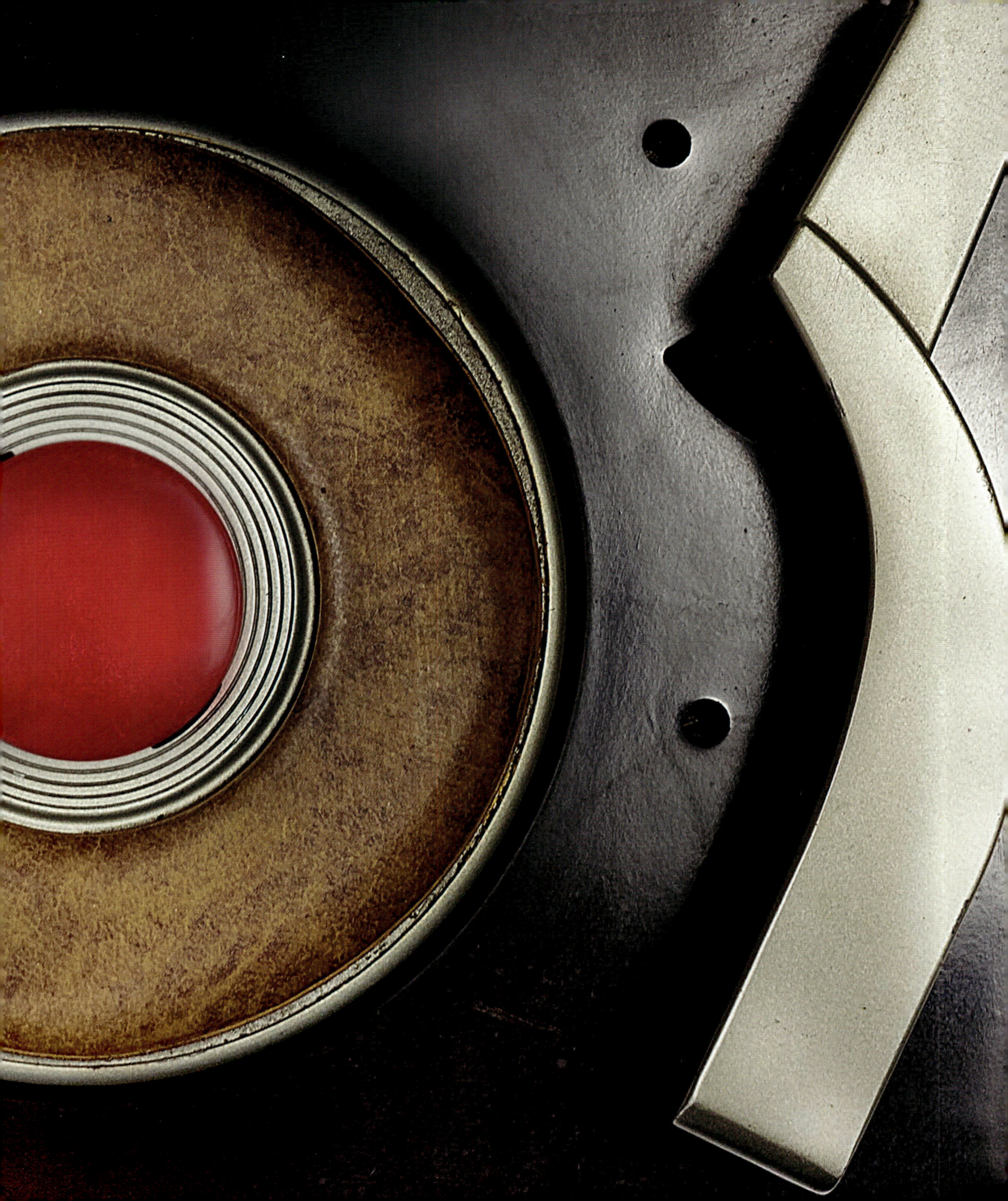

INDUSTRIAL DESIGN IN THE MODERN AGE

INTRODUCTION BY **PENNY SPARKE**

KRAVIS DESIGN CENTER | Rizzoli Electa

CRAFTSMAN

CONTENTS

PREFACE

George R. Kravis II

IT HAS LONG BEEN MY BELIEF THAT GOOD INDUSTRIAL DESIGN CONTRIBUTES TO A BETTER life. I also believe that we can actually improve our prospects for the future with our understanding and recognition of the importance of design. This publication presents a selection of design objects I have been collecting over the last fifty years. The look of a designed object is as important to me as the beauty of a painting is for others. I am especially drawn to the machine aesthetic, which was influenced by the simple geometric designs of the Bauhaus.

The objects in my collection now number more than four thousand. They can be dated from the late 1800s to the present and encompass all design media, including furniture, drawings, ceramics, metalwork, textiles, and graphic design, including record covers. My idea from the beginning has been to build a collection that could be used by others for education and research purposes, and toward that end I have sought out archival items relating to the objects—such as period advertisements, manufacturers' brochures, and patent drawings—which provide context. Whenever possible, I acquired the range of colors in which a piece was produced, as well as variations in production.

Though the core of my collection comprises work by American industrial designers who emerged after World War I, as well as that of European designers, many of the objects were anonymously designed. Work by unknown industrial designers has always intrigued me, and through the years I've had little competition from museums or other collectors, who have primarily sought well-known masterpieces. To this day, I notice a prejudice against anonymous objects, and even shock from those who do not see how a refrigerator or a typewriter might be worthy of an art museum. Given the lack of interest in anonymous objects from auction houses and dealers, I have had access to the best. I enjoy not only the detective work required to find great pieces, but also the risk involved in purchasing objects that I find beautiful but for which the designer and manufacturer are largely unknown.

PAGE 2
Worker polishing chemical laboratory kettle, Aluminum Company of America, 1930.

PAGE 3
Weaver at American Woolen Company, 1931.

PAGE 4
Detail of *RCA Victor* record player by John Vassos (page 232), c. 1949.

PAGE 6
Detail of *Craftsman* portable air compressor by John R. Morgan (page 291), 1935–40.

above left View of Russel Wright *American Modern* dinnerware pitchers at the Kravis Design Center, Tulsa, Oklahoma.

above right View of collection, Kravis Design Center.

Given that industrial design is such a large field, I came to believe that a comprehensive collection would be nearly impossible to achieve, so over time I began to focus on the areas I am passionate about. One area is audio-visual equipment, particularly radios and record players from the 1930s through the 1970s. This was a natural outgrowth of my work in the broadcasting business, and I grew up with some of these wonderful products. Bakelite and Catalin table radios appealed to me with their vibrant colors, which were made possible by new technologies and materials. Certain radios, such as the *Patriot* designed by Norman Bel Geddes, were of exceptional aesthetic interest, and I acquired multiples of some of these important designs to show the range of colors that were available. I particularly like designer and artist John Vassos's work for Radio Corporation of America (RCA) and am interested in the development of that extraordinary company under his design leadership.

I am fascinated by the endless array of electronic equipment that has developed in my lifetime, and I have sought to learn all I could about each design. The broad range of my collection illustrates the evolution of technological advancements, demonstrating the progress of certain forms. Apple products from the late 1970s onward have been exciting to me due to the combination of remarkable technology and outstanding aesthetics. Beginning in the 1980s, I acquired new Apple models as they were introduced to the marketplace.

I have also had a lifelong interest in automobiles, and I have been fortunate to acquire some great examples, such as the 2008 Smart Car. One of my favorites is the silver 2008 Audi R8 with Tuscan brown leather interiors. I saw this car when it was first unveiled at Design Miami in 2007, and I was subsequently invited to Las Vegas to test drive it on the speedway there.

I grew up during a time when the cocktail hour became an institution and American drinks became as popular in Europe as they were here. Perhaps that explains my attraction to designs for the cocktail hour, especially those made between the two World Wars, reflecting Jazz Age sophistication and glamour. Back then the world's problems seemed less threatening and, with the progress of modern design, life became simpler, brighter, and easier. In these cocktail objects, I sense the optimism of their designers and their dreams for the future. The cocktail shaker was the most important and enduring form of this period, and I have many of them from different countries. I have collected all the accoutrements of mixing and serving cocktails—ice buckets, trays, glassware, bottle openers—many in complete sets and some still in the eye-catching original packaging.

One American designer in particular has captured my attention—Russel Wright. In collaboration with his wife, Mary, he created *American Modern,* the bestselling ceramic tableware line in history. His ergonomic forms and earthy color palette greatly appealed to me, as they did so many others at mid-century. They were intended for informal living as formal dining gave way to casual gatherings—Wright's designs were modern and stylish, perfect for the new American lifestyle that emerged in the postwar years. I wanted every form and color ever produced, which has led to the large assemblage of Wright's dinnerware, machine-age aluminum ware, and textiles now in my collection.

My various collections, which had grown to over three thousand objects by the year 2013, were scattered among my house and several warehouses. In 2015, I consolidated and moved them to a central location, a 14,000-square-foot storage warehouse with museum-quality shelving, high-level security, and controlled environmental conditions. Shortly thereafter, I inaugurated the Kravis Design Center to accommodate and care for the collection and to make it available to others for study and enjoyment. Also in 2015, I gave more than one hundred objects to the Cooper Hewitt, Smithsonian Design Museum, in New York, which were presented in the museum's exhibition *Energizing the Everyday* in 2017. Selected design objects from my collection were also included in *Oklahoma and Beyond,* an exhibition organized by the Oklahoma State University Museum of Art. I now think in terms of the needs of museums and the future. Most of all, I look at the educational potential of my collection to help others understand and appreciate design in their daily lives.

right View of collection, including Godley-Schwan's *Checkers* cabinet, Kravis Design Center, Tulsa, Oklahoma, 2017.

Black+Blum

BY DAN BLACK

An idea for a new design often comes about when we're looking for a new product but find that the marketplace doesn't offer what we want. It is always a good justification for a project and an exciting starting point when we discover a need for a product. As industrial designers, we are creating products of our own choosing and hopefully the passion that we put into everything we make shows.

In the case of the *Venti* fan, we were looking for a large floor fan but found nothing that (excuse the pun) was cool. As a general rule, we do not have a set recognizable design aesthetic. We tend to design each product on an individual basis and try to imagine a design that is not trendy in form or color, that is undateable and yet familiar. This is often a combination of modern and vintage aesthetics, but always underpinned with minimalism and functionality. For the *Venti*, we were inspired by the vintage fans of the 1930s that are now so fashionable. They harken back to the glamour of old Hollywood, where it might have sat on the messy desk of a trilby wearing, smoking detective. We wanted the vintage appeal, but also a utilitarian modern retro look. Even the remote control is bespoke and takes inspiration from the tips of old airplane propeller blades.

Venti floor fan designed by Black+Blum, 2006. In 2017 George Kravis acquired a floor fan, identical to the one illustrated, as an example of contemporary design inspired by streamlined objects in his collection. It is a prized object, and Dan Black of Black+Blum is one of Kravis's favorite industrial designers.

INTRODUCTION

Penny Sparke

> The ultimate form of any object—natural or man-made—is inherent in the object itself: it must be evolved naturally to suit the function which the object is intended to perform, the materials out of which it is made and the methods used in its making.
>
> **WALTER DORWIN TEAGUE**[1]

FIG 1
opposite Florence Knoll at her desk, Knoll Associates, New York, 1956.

THE EMERGENCE OF THE CONCEPT OF INDUSTRIAL DESIGN WAS A DIRECT CONSEQUENCE of the division of labor in the factory, a defining feature of industrial manufacturing that was first developed in the United Kingdom in the mid-eighteenth century and continues to the present day. Industrialization took the work of the craftsman and broke it down into an organized set of new tasks. From the outset, industrial designers had a two-fold responsibility. While conceiving new products and planning their manufacture, in order to ensure that their products contained and communicated appropriate meanings, they also had to understand the principles underpinning consumption. Designers were uniquely placed, therefore, to straddle both production and use (fig. 1).

The term *industrial design* evokes an era in the late nineteenth and early twentieth centuries when the factory was the epicenter of production (fig. 2), and designers, working alongside engineers and others, transformed base materials into highly desirable objects. In turn, their sewing machines, bicycles, and automobiles—as well as, a little later, radios, hair dryers, and hi-fi systems—entered the lives of mass populations in the Western industrialized world, carrying with them the utopian message of modernity. The machines of the factory produced another set of machines, ones that quickly came to inhabit everyday life. These seemingly banal objects told, and continue to tell, fascinating stories about their own pasts as well as about the world in which they were created and from which they evolved. These products of industry were active agents of change, rather than passive onlookers. They were key symbols of the modern age and indicators of the future.

Yet most of us know very little about these objects. We don't know where they came from before we purchased them. We don't think about the fact that someone was responsible for giving them their forms and their rich meanings. And we don't contemplate that someone had to decide which materials would work best. This book sets out to show the results of that creative decision-making process through a vast range of objects designed and produced since the late 1800s.

FIG 2
Production line, Singer Manufacturing Company, c. 1929.

Although industrial design established itself as the prime mover of economies and livelihoods well over a hundred years ago, it is as relevant today as it ever was, if not more so. It continues to play a part in our complex contemporary world, in which mass manufacturing has declined in some places and has only just begun in others. Today, industrial production is accompanied by other forms of making, including the work of artisans employing local materials in their handmade products and the labor of fine arts craftspeople producing one-of-a-kind and limited edition pieces, which reign supreme in the highly sophisticated luxury goods arena. The technologies of rapid prototyping and 3D printing offer new variants of the design process while echoing many features of the craft process.

Although handicraft preceded industrialization, it did not disappear from sight in the era of industrial design. Rather, it continued along a parallel path, one that moved to a rural from an urban setting. It has recently reared its head again with the rise of the maker, DIY, or handmade movement. However, with the arrival of new manufacturing techniques, the distinction between craft (the making of one-offs) and industrial design (mechanized mass production) has become blurred.

In a world in which both the digital and the virtual have such a powerful presence, the strong link between industrial design and materiality might seem to be under threat, but it still has an important role to play. Indeed, the materials and 3D forms of mass-produced objects can be seen as powerful antidotes to the digital. Also, in the 2010s, the thrill of consumption that characterized the 1940s and 1950s has somewhat abated, and design has attempted to break its links with the market economy and transform itself to serve society in new and different ways, both socially and environmentally.

THE HISTORY OF INDUSTRIAL DESIGN
TWO APPROACHES

Two distinct approaches toward the history of industrial design are outlined in this introduction and are represented by the objects in this book. First, there are the well-trodden stories of the heroic alliance between art and industry; the efforts made by designers, in collaboration with national agencies and manufacturing companies, to raise standards; and the inevitable absorption of modern design into the international market economy. Second, there is the story of the ways in which anonymous industrial design has long played a role in our everyday lives, humbly and unostentatiously enriching them in a myriad of ways through the seemingly banal objects with which we surround ourselves. Although these stories touch on different phenomena, and approach the subject of industrial design from different angles, they ultimately come together in this publication to offer a more multi-dimensional and nuanced story than has hitherto been told. They are two sides of the same coin. While the heroic aspect of industrial design is allied to the finer arts and the idealistic histories of art and architecture, the other side of the coin, anonymous design, is embedded in popular culture, that aspect of everyday life that makes us who we are and defines many of the spaces we inhabit. It is less overtly idealistic, perhaps, than its high-culture equivalent, but it nonetheless affects us all on a daily basis.

PART ONE
ART AND INDUSTRY

The United Kingdom was the first country in the world to industrialize. In the eighteenth century, a few of the early mass manufacturers—notably the ceramics firm Wedgwood—commissioned artists to make their goods more attractive for particular markets. The marriage between art and industry continued into the nineteenth century and is still going strong today. It was, and remains, an avant-garde, utopian movement. Pushing boundaries, it responded to the new challenges of materials and production technologies and stimulated consumer tastes and desires. Following artists' practice of signing their work, the movement gave birth to what we now call the signature designer, which eventually became a desirable commodity in itself.

From the mid-nineteenth century onward, first in the UK, subsequently across Europe, later in the United States, and eventually in many other global manufacturing centers, designers responded to what was widely believed to be a mindless use of the machine and the production of objects that pandered to lowest-common-denominator taste, as well as to the social aspirations of the expanding and increasingly affluent classes. Indeed, the emergence and rise to fame of the designer for industry, whose aim was to add aesthetic excitement and raise production standards, is one of the key stories of the modern age. The leaders in the field—from Christian Barman, Christopher Dresser, Zaha Hadid, and Jasper Morrison in the UK; to Peter Behrens, Dieter Rams, Timo Sarpaneva, Ettore Sottsass, Philippe Starck, Wilhelm Wagenfeld, and Tapio Wirkkala in Europe; Norman Bel Geddes, Henry Dreyfuss, Charles and Ray Eames, Walter Dorwin Teague, and Russel Wright in the US; and, later, to Shiro Kuramata in Japan and Marc Newson in Australia, all of whom have designs illustrated in this book—have been celebrated as heroes of modern and postmodern design.

The UK and Europe, 1900–1930

In late-eighteenth-century Britain, developments in steam power inspired the invention of new machine tools and production techniques, which in turn facilitated the manufacture of new consumer goods. New materials, cast iron among them, brought about dramatic changes in the design of products, made new forms of ornamentation possible, and transformed the production of traditional decorative arts. In the textile industry, for example, the Spinning Jenny and the Jacquard loom (fig. 3) revolutionized how fabrics were both conceived and made. Designers were required to make decisions in advance of manufacture, unlike craft workers, who for centuries had made aesthetic and material choices as they went along in the creation process. That seemingly simple change represented an important shift in working processes: the craftsperson depended on tacit skills, whereas the new designers engaged in rational planning in advance of producing an object.

From the late nineteenth century onwards, technological innovations such as electricity allowed for vast numbers of newly conceived products to be manufactured and enter the marketplace. Their imaginations challenged by the latest goods, designers developed new objects, among them electrical appliances and advanced modes of transportation, which provided a means through which consumers could negotiate their entry into modern life.

As mass production in the UK increasingly became the primary means of manufacture, a design reform movement emerged in response to what were seen as its excesses, such as lavish ornamentation and too-bright synthetic-dye colors. First, the British government took steps to improve the standards of design, including the formation of a number of design schools in key manufacturing areas, such as the Norman School in London, which later became the Royal College of Art, and the Birmingham School. A little later on, led by William Morris (fig. 4) and John Ruskin, the protagonists of the arts and crafts movement—including C.F.A. Voysey, C. R. Ashbee, and W. A. S. Benson—sought to re-instill traditional

FIG 3
Henderson and Co.'s carpet power-loom, from the 1862 International Exhibition, published in *The Illustrated London News,* October 11, 1862.

FIG 4
above left William Morris, 1874.

FIG 5
above right The Wiener Werkstätte silver workshop, Vienna, 1906.

craft values into design, thereby temporarily halting the emergence of a modern industrial design movement in the UK. Nevertheless, a number of Britain's industrial designers rose to prominence during that time—such as Christopher Dresser in the late nineteenth century and Wells Coates, Serge Chermayeff, and Christian Barman by the 1930s—who became pioneering figures of international repute.

Continental Europe also rose to the challenge of industrialization. Artists and architects linked to the Viennese workshops, or Wiener Werkstätte (fig. 5), for example, began to work with manufacturers. Germany was also quick to embrace the union of art and industry. In the first decade of the twentieth century, the German architect and graphic designer Peter Behrens created a complete visual identity—from sales brochures to kettles to fans to the factory building itself—for the Allgemeine Elektricitäts-Gesellschaft (AEG) company (fig. 6), while Deutscher Werkbund (fig. 7) provided a support body for the country's new, rational approach toward the design of everyday industrial products. The Werkbund was formed in 1907 as an association of artists, architects, designers, and industrialists that aimed to link designers with industry and thereby improve the quality of German goods. Key figures involved in the group include Behrens, Richard Riemerschmid, Bruno Paul, and Theodor Fischer. At the Werkbund Exhibition of 1914 in Cologne, a debate broke out between Werkbund member Hermann Muthesius, who supported standardization, and the Belgian art nouveau architect and designer Henry van de Velde, who defended a more individualistic approach.

The Bauhaus

In 1919, architect Walter Gropius founded a design school named the Bauhaus in Weimar, Germany. It was responsible for consolidating a new and modern industrial design aesthetic and for engaging a group of designers who were to prove enormously influential. It was

FIG 6
above left *AEG Preisliste* ("AEG Price List"), April 1924.

FIG 7
above right Herbert Bayer, Deutscher Werkbund exhibition, Paris, 1930.

also important to design pedagogy, which will be explored later. The Bauhaus was created through a merger of the Grand Ducal School of Arts and Crafts and the Weimar Academy of Fine Arts, and was housed in the building of the Grand Ducal School, which was designed by the Belgian artist and architect Henry van de Velde.

Gropius's vision was to create a school that brought art, craft, and design into a single place, with architecture leading the way (although it wasn't introduced into the curriculum until later). In the early years, German expressionism was influential: a woodcut of a cathedral with a strong expressionist flavor by artist Lyonel Feininger was featured on the cover of the first Bauhaus manifesto and the first staff members to be recruited by Gropius included Feininger and artists Gerhard Marcks and Johannes Itten. They were soon followed by Oskar Schlemmer, Paul Klee, and Wassily Kandinsky. In 1922, Theo van Doesburg arrived from the Netherlands, where he had played a part in the Dutch De Stijl artistic movement. De Stijl ideas were quickly integrated into the thinking at the Bauhaus, as were those of the Russian constructivists, which were brought to Germany by graphic designer El Lissitzky.

The Bauhaus teachers encouraged their students to work with abstract forms and a variety of materials—wood, ceramics, metal, and textiles—in order to create a wide range of functional products, from chairs to teapots, lamps, and rugs. Perhaps the most radical of all the Bauhaus's pedagogical innovations was the preliminary course (*Vorkurs*) led by Johannes Itten, who belonged to a kind of cult called Mazdaznan and encouraged students to wear a certain style of dress and be on a particular diet. Itten's extreme approach soon fell from favor, however, and he was dismissed in 1922. The Hungarian László Moholy-Nagy, who was less spiritual in his approach, replaced Itten. Moholy-Nagy emphasized aesthetic ideas and theories of form that had origins in constructivism. In the preliminary course, students engaged with a broad curriculum, encountering many different principles of art and design. Paul Klee and Wassily Kandinsky gave classes about color theory and composition, light, and shade.

The early years of the Bauhaus proved financially difficult, and there was little support from the government in Weimar. In 1923, an exhibition was held to put what was being achieved at the school on display. Called *Art and Technology: A New Unity*, it featured the Haus am Horn, a prototypical house created by Bauhaus students and staff in which a number of novel ideas were embraced, including different colors to distinguish the internal spaces. The kitchen, designed by Benita Koch-Otte and Ernst Gebhardt (fig. 8), was especially innovative, containing, as it did, a built-in workbench and a stool on which a housewife could sit while she worked. The idea was borrowed from the American writer and household advisor Christine Frederick, who penned a book titled *Household Engineering* that was published in 1923. Koch-Otte and Gebhardt were more radical than Frederick; they added eye-level cupboards with doors on them to avoid any contamination of their contents by dust. Most memorably, the Bauhaus kitchen featured a set of standardized and labeled ceramic containers, designed by Theodor Bogler, which were the precursors of containers labeled "flour" or "sugar" that are still found in kitchens today.

Although the Bauhaus curriculum was craft-based in 1923, Gropius was committed to the notion that the skills acquired by students should be applied, ultimately, to the world of industry. A year or so later, however, the Weimar Academy was closed, and the Bauhaus had to move to a new purpose-built building, designed by Gropius, in Dessau. The new building featured impressive workshops and a residential block for the students, in which each room had its own balcony.

FIG 8
Kitchen in Haus am Horn experimental house, designed by Benita Koch-Otte and Ernst Gebhardt for the Bauhaus exhibition, Weimar, Germany, 1923.

After completing the preliminary course, Bauhaus students moved on to work in one of the materials workshops. There they learned craft skills from specialists in ceramics, metalwork, weaving, and woodwork, while simultaneously studying form and applying their knowledge to practical work. The objects created in the Bauhaus workshops were understood as prototypes for factory production, although only a few tentative links with industry were forged in the 1920s. The Bauhaus's best-known objects were manufactured after the school was shut down by the Nazis in 1933.

Some of the most memorable Bauhaus creations were produced in the metal workshops. They included Wilhelm Wagenfeld's little glass table lamp and Marianne Brandt's small metal teapot. Each was the result of its designer's deep interrogation into how the object was to be used, as well as the user's needs and desires. Along with others, Wagenfeld and Brandt created a brand new look for the modern home and its contents, which proved highly influential in future decades. Their initial aim was not visually focused, however; instead, they had set out to revisit the form of the object to change people's behaviors and to create a new vision of contemporary living.

Two main teachers at the Bauhaus's ceramics workshop were Otto Lindig and Theodor Bogler. At the furniture workshop, which had a focus on wood, Marcel Breuer emerged as a star pupil. Breuer went on to become a staff member at the institution, and his work proved greatly influential throughout the twentieth century. Breuer developed ideas from De Stijl, promoted by Gerrit Rietveld, in which furniture pieces were conceptualized as a set of geometric elements. He established an original language for modern furniture, especially through his work in tubular steel, and he helped develop a fresh aesthetic for the home that focused on space rather than mass. Breuer's designs have stood the test of time and are still popular in the twenty-first century.

Anni Albers and Gunta Stölzl were among the designers who introduced innovations in the textiles workshop, which was largely peopled by women, unlike the rest of the workshops. There was a strong male bias at the Bauhaus; nearly all the female applicants were directed to the weaving workshop, as Gropius believed that women did not have excellent spatial skills but had a strong sense of color.

Graphic design and typography were also modernized at the Bauhaus. When Moholy-Nagy was appointed to teach the preliminary course in 1923, he brought with him the New Typography, a radical approach to this Russian-developed discipline. New Typography emphasized clarity of communication, sans serif fonts, and asymmetry. The Bauhaus typography workshop was led by Herbert Bayer, who designed his Universal typeface there. Bayer later moved to the United States and became one of the twentieth century's most influential graphic designers.

Although Gropius viewed architecture as the ultimate medium, the subject wasn't taught at the Bauhaus until 1927. It was finally introduced into the curriculum by Hannes Meyer, who rejected the role of aesthetics in design and advocated an approach that was entirely based on rational principles. This signaled a retreat from the less rational work of Itten and others, as well as from the visually oriented work of the fine artists, toward a more systematic approach in artistic production.

Meyer took over as head of the school when Gropius resigned in 1928, and the school inevitably moved in a different direction. Meyer was replaced in 1930 by German architect

FIG 9
Baker House dormitory designed by Alvar Aalto, 1949, Massachusetts Institute of Technology, Cambridge, Massachusetts.

Ludwig Mies van der Rohe, who moved the Bauhaus to Berlin in 1932, when it was under attack by the National Socialist Party for its perceived radical ideals. The school lasted only ten months before it was closed by the Nazis the following year.

Importantly, in addition to completely radicalizing art and design education, the Bauhaus established a modern industrial design aesthetic that has remained firmly in place for many a decade. The combination of abstract art and craft skills resulted in a cutting-edge approach toward industrial design. Although the application of this formula to industrial manufacture was not extensive during the school's lifetime, some industrial projects were produced while the school existed. An income was earned by the factory production of some wallpaper designs, for example.

The true legacy of the Bauhaus was not felt until after 1933, when Gropius, Breuer, and Moholy-Nagy all settled for a short time in Britain. A number of other Bauhaus protagonists, including Mies van der Rohe, transferred to the United States where they continued their important work.

Scandinavia

Following in the wake of the Bauhaus Germany, the Scandinavian countries Denmark, Sweden, and Finland developed applied art practices, a softer model of industrial design, in the 1930s. They focused on the craft-based manufacturers of glass, ceramics, and furniture. The Finnish glassworks Iittala, for example, brought in artists Simon Gate and Edward Hald, and the Swedish ceramics manufacturer Gustavsberg worked with William Kåge to create a range of influential contemporary designs. The Finnish architect Alvar Aalto designed several modern furniture items that were made of bent plywood rather than Germanic tubular steel (fig. 9).

Danish Modern

Of all the Scandinavian countries, Denmark had the greatest influence on international design in the mid-twentieth century. Although Danish modern design had a huge impact on Europe and America in the years immediately following World War II, it originated in the 1930s, when a new generation of innovative Danish designers came to the fore in the furniture, textiles, metalwork, ceramics, and glass industries. They offered an alternative to the highly rational objects made from industrial materials that were emerging from Germany at that time by focusing on traditional materials and working with small-scale manufacturers. Denmark's modern design movement was more organic and more natural—some would say more human—than Germany's, and it was exclusively domestic in orientation. It was also ideologically driven, as it sought to offer everyone an opportunity to bring high quality design into their everyday lives by making it affordable to a larger audience.

Among the first design areas to be modernized in Denmark was furniture. Kaare Klint was the head of the School of Architecture at the Royal Danish Academy of Fine Arts, and under his leadership furniture forms were taken back to basics (fig. 10). Espousing a philosophy of fitness for purpose, Klint spent considerable time in the 1930s investigating the requirements of the human body in relation to furniture. In the process, he looked back at traditional pieces of functional furniture, such as those used by the military, and he developed a set of designs that could be described as more like equipment than furnishings. They reflected modern Danish life and the decreasing size of the average home by, for example, including fewer pieces of bulky upholstered furniture.

FIG 10
Deck chair designed by Kaare Klint, 1933, shown at Copenhagen Cabinetmakers' Guild exhibition, 1933. Design Museum Denmark.

Danish craftsmen also embraced the modern style in the area of silver at this time. In the early twentieth century, the Georg Jensen company had established a quality standard for Danish silver goods; Kay Bojesen, one of Denmark's leading silver designers, was trained there in the 1930s. He is best known for his flatware set of 1938, the Grand Prix, which was made of polished steel. It was one of the simplest and most elegant examples of Danish modern in that era. Bojesen is also known for his wooden toys, including a monkey that was exhibited at London's Victoria and Albert Museum in the 1950s. He also helped set up the artist cooperative Den Permanente in 1931, which featured an exhibition center that showed the best examples of Danish modern design and played the role of arbiter of exemplary Danish design until 1981.

Denmark also produced some remarkable modern ceramics. The Royal Copenhagen porcelain manufactory was founded in the late eighteenth century, and although it was best known for its traditional designs, the interwar years saw some strikingly modern looking pieces made in a muted, natural stoneware that transformed the company's public image. The porcelain manufacturer Bing and Grøndahl also embraced the modern style alongside its more traditional ware. The Holmegaard Glassworks took the lead in the area of glass: its radical art director, Jacob E. Bang, pushed the company to embrace the new aesthetic in the 1920s. Textiles were also given a face-lift by designers like Marie Gudme Leth, who set up her own studio in 1941 where she undertook screen printing.

Lighting was another area in which Denmark excelled during the interwar years. Poul Henningsen produced a number of designs for the Louis Poulsen company that were novel both in their appearance and in the way they illuminated spaces. His famous PH light, produced in 1925, remains a design classic to this day. He continued to create models throughout the middle years of the century.

It wasn't until after 1945 that Danish modern design gained its international reputation. After the trauma of the war, a new generation of innovative designers emerged whose aim was to modernize the design of the home and improve the quality of people's lives. They emphasized affordability and, toward that end, worked closely with industry.

After the war, furniture became important once again. Hans Wegner, who worked at the craft end of the spectrum, was a carpenter and a sculptor in wood. His simple designs, such as his Chinese Chair of 1944, Peacock Chair of 1947, and "The Chair" of 1949, quickly became iconic. Today, his designs, which were created for different manufacturers including Carl Hansen & Son and Fritz Hansen, represent Danish modern at its best. Børge Mogensen is the other father of postwar Danish modern furniture. He worked for Klint for a while and was trained as both a cabinetmaker and an architect, and formed his own studio in 1959. Mogensen was interested in the way furniture functioned in a domestic setting, and his 1945 sofa with leather ties that allow the sides to drop down is among his most successful items. Finn Juhl, another highly respected Danish furniture designer as well as an architect and interior designer, took a more overtly sculptural approach to his work.

Many of the most notable furniture designers to emerge in Denmark in the 1950s rejected the craftsman aesthetic and introduced metals and plastics into their pieces. Poul Kjaerholm, a craftsman by training, produced some strikingly minimal pieces, including the well-known PK series. Arne Jacobsen, an architect, used molded plywood and steel rods in his Ant Chair of 1952; the simple chair had initially been created for the canteen of a Danish pharmaceutical company, but it soon found its way into the home. Jacobsen designed a number of iconic furniture pieces, among them the Swan and the Egg chairs, in the later 1950s.

Nanna Ditzel was another key member of the postwar Danish modern movement. In 1959, she produced her wicker egg-shape chair, which was frequently featured in the pages of fashion and modern interior magazines, suspended from a ceiling by a chain. Ditzel adopted a broad approach to design; in addition to creating exhibitions, she produced jewelry, textiles, and other objects in the Danish modern idiom.

Furniture designer Verner Panton broke the mold of Danish modern. His best-known design, the S Chair, was created in 1960 and was the very first injection-molded all-plastic chair. Panton rejected the emphasis on domesticity and everyday life, as well as the craft and natural materials that had characterized Danish design up to that point. His was a far more industrial approach, and he succeeded in projecting Danish modern design into the international arena.

Danish textiles also became more innovative in the postwar years through the work of Lis Ahlmann and others. The firm Kvadrat played a leading role in this category. Henning Koppel and Erik Herlow, among others, played a similar role with regard to designs in silver.

Danish modern made its greatest international impact after it became popular in America. An important moment was when Edgar Kaufmann Jr., director of the Industrial Design Department at the Museum of Modern Art (MoMA) in the 1950s, purchased some Danish modern items for Fallingwater, the famous Pennsylvania home that was designed by Frank Lloyd Wright. Fallingwater inspired some American manufacturers to mass produce Danish designs under license. Finn Juhl's designs, such as the Pelican and Chieftains chairs, were particularly popular. Soon, Danish modern became an international phenomenon. By the late 1960s it had been replaced by Italian design, but later, by the early twenty-first century, mid-century Danish modern had come back in vogue, relevant once again for its strong craft roots and democratic ideology.

Art and Commerce in the United States, 1920–1940

Up until 1914 in the United States, no single model or school for design had presented itself. American design tended to evolve in an ad hoc, pragmatic manner, dependent on the past functions of the decorative arts industries, the requirements of industrial production, and the vagaries of the marketplace. In short, it remained the task of a wide range of diverse individuals, from established fine artists to teams of lowly and anonymous art workers to architects, engineers, craftsmen, decorative artists, and a new breed of creative individuals, referred to as commercial artists, to contribute to the visual face of the emerging modern material world. In addition, different areas of production—including traditional decorative arts, new consumer products, fashion, two-dimensional designs of various kinds, and interior spaces and environments within the private and the public spheres—developed models of design practice that suited both their industries and their markets. In general terms, however, art and industry increasingly developed strategic alliances with the aim of creating products, imagery such as posters and advertising, and spaces that would appeal to the ever expanding population of consumers and embody the complex messages that were required of them.

In the 1920s and 1930s, the concept of the consultant designer for industry came of age, partly as a solution to economic challenges and partly as a means of bringing together the

FIG 11
Assembly line, Ford Motor Company plant, Highland Park, Michigan, c. 1913.

expanding numbers of tasks that were required to link production and consumption in a way that met the needs of industry, the commercial sector, and the consumer. The consultant designer required not only a high level of specialist skills in the areas of visualizing and conceptualizing, but also a broad approach, in order to enable the new professionals to consider a wide range of industries and design fields.

While Britain witnessed transformations in manufacturing in the traditional decorative arts industries, as well as in textiles, ceramics, glass, and furniture, the United States made huge advances in the efficient organization of labor and production in the new, technologically oriented industries, such as electrical goods, transportation, and office machines; in the development of highly specialized machine tools; and in the manufacture of new goods with a technological bias.

The interwar years in the United States were a crucial period in the story of international industrial design. During this time, it became clear that, in the context of the mass manufacture of consumer goods, technologically and otherwise, the design of an object or product constituted a form of added value and had the power to dictate commercial success or failure. The groundwork for this rise in appreciation for design was laid with the success in mass production of arms in the nineteenth century, and continued through the 1920s with automobile manufacturing. The clocks, sewing machines, bicycles, agricultural machinery, and automobiles that were produced over time are markers of that manufacturing success. Henry Ford's Model T automobile was the result of the first moving assembly line to be put into practice in mass production (fig. 11); however, it lasted only twelve years. General Motors's vice president, Alfred P. Sloan, brought in a more flexible means of mass production in response to the demands of the market—which, it turned out, required product differentiation.

FIG 12
above Duplicating machine by Sigmund Gestetner, 1929.
below Raymond Loewy's redesign of the duplicating machine for Gestetner Ltd., London, England, 1933.

Drawing on earlier nineteenth-century industrial models, Ford's system of mass production depended, above all, on the concept of standardization. His famous dictum that people could have his car in any color "as long as it's black" led to the idea of a widely available basic car made from standardized components that could be easily manufactured, assembled, and repaired. The status that a Model T imbued on its buyers was linked solely to ownership, to the simple fact that they had exchanged their horse for an automobile. It didn't matter, in the context of a first purchase, what the object looked like; it only mattered that it was affordable. When it came to the acquisition of a second car, however, the principle of outdoing one's neighbor in order to gain a level of social status came into play, and it became necessary to provide a form of added value through design.

It was at this point that General Motors's principle of flexible mass production came to the fore, and Ford's philosophy of manufacture had to be revisited. The result was that the latter had to close its plant for a year in order to re-adjust its production process and to add design to the mix. The result was the Model A Ford, first released in 1927.

The industrial designer—better known as a stylist at that time—originally emerged at General Motors, out of the need to create a set of design standards that established differentiation among its automobiles. This represented an important moment in the story of modern industrial design, one in which technology and culture came into direct conflict. In this case, culture won the day, and design provided the means by which the two forces could achieve a workable compromise. From that point onward, the new mass production industries became sensitive to the vagaries of fashionable taste and trends. The role of designer became highly valued by the industries that understood the advantages of accommodating art within their rational operations.

The Emergence of the Industrial Design Profession

After recognizing the need for variation and added value through design, Sloan hired a coach builder, Harley Earl, to inject a certain style into his automobiles to make them more appealing to consumers.[2] Other industries were watching, and they quickly adopted this strategy for their own technological goods, including refrigerators and other domestic appliances, office machines, telephones, and radio sets. Challenged by the economic depression of the 1930s and faced with a new level of competition, manufacturers began to hire designers who imbued an artistic style into their products. Raymond Loewy's 1929 design for the Gestetner Company's duplicator, a precursor to the copy machine, is a case in point (fig. 12). He modernized a rather crude-looking piece of office machinery by simply adding a smooth body-shell, which minimized its resemblance to a piece of factory machinery and made it more of-the-moment for an office setting.

It was no coincidence that several of the leading industrial designers of the day had begun their careers in graphic advertising and department stores, which, emulating French models, had been quick to embrace the modern style. Loewy, for example, had worked in the New York department store Saks Fifth Avenue, designing, among other things, the lift attendants' uniforms.

The best-known American industrial designers of the day—Loewy, Norman Bel Geddes (fig. 13), Walter Dorwin Teague, and Henry Dreyfuss among them—penned their own accounts of their careers in which they all stressed their perceptions of themselves as

FIG 13
Norman Bel Geddes in a Chrysler *Airflow* automobile, c. 1933.

idealistic modernists. Linking themselves to the modern European art and architectural movement provided them with a means of elevating the status of their contributions and of distinguishing themselves from the commercial context that had given birth to their roles. In his 1940 text *Design This Day: The Technique of Order in the Machine Age*, Teague made repeated references to the European modernist architects Le Corbusier, Walter Gropius, and Mies van der Rohe, hinting, perhaps, that his name should be added to the list.[3] In his autobiography, *Horizons*, published in 1932, Norman Bel Geddes also referenced European modernism, in particular the abstract paintings of Pablo Picasso and Paul Cézanne, claiming them as his artistic roots.[4] Their pretensions served to reinforce their claims of adding artistic value to the products of industry.

Most accounts of the work by American consultant designers of the 1930s have concentrated on their advocacy of dramatic streamlined forms depicting a concept of the future. The majority of their time was, in reality, devoted to redesign projects. And they worked for conservative markets as well as more adventurous ones. Histories of their work have tended to play down their market-oriented pragmatism and presented them as celebrities, as they were heralded by the press. *Time* and *LIFE* magazines did pieces on their work on several occasions and reported on the details of their daily lives as if they were Hollywood stars. While their fame was undoubtedly important to them as individuals, it was even more crucial for the manufacturers who employed them; designers bestowed an instant added value on their products. They became branded entities, and their names were used as a form of product endorsement. In turn, the products became their own advertisements.

Consultant designers quickly became intrinsic components of the American commercial system that had engendered them. Their arrival marked the moment when designer culture—the perception of value being added to an object, image, or environment as a result of a well-known designer's name attached to it—came into being. The simple application of a designer's name to a product, whether it be a railroad train or a biscuit, bestowed instant distinction on those artifacts.

A number of explanations have been offered for that phenomenon. The most persuasive suggested that it represented the need for consumers to construct their identities through the acquisition of commodities that were associated with the name of an individual known to possess a high level of taste. That rationalization builds on the fact that, in the years before industrialization and the rupture between production and consumption, the upper classes had been guaranteed social status either by birth or through the acquisition of custom-made, crafted artifacts. The association of a designer name with otherwise anonymous mass-produced goods could be seen to restore individualism to the products by personalizing them.

As the number of standardized mass-produced goods increased, the need to inject them with a kind of distinctiveness became paramount. The consultant designers were acutely aware of the diversity of consumer tastes in the marketplace. Bel Geddes acquired his knowledge of the market by undertaking extensive consumer questionnaires, and in response to one of them, he proposed four different radio designs to the Philco electronics company—the Highboy, the Lowboy, the Lazy-boy, and a radio-phonograph combination—each of which was styled with a different market in mind.

While the automotive industry preferred to keep its stylists in-house, maintaining their anonymity for the most part, many other new industries engaged in making consumer products such as domestic appliances and office machinery that benefited greatly from consultant designers who had gained a unique generalized overview of the industry as a whole. The model of consultant design that emerged in the United States in the late 1920s was rapidly adopted by many other countries, as they began to look for ways to embed design into their emerging industries and to inspire and enhance consumer desire. In the years after 1945, the dissemination of this industrial design paradigm constituted one of America's key strategies of cultural influence that stretched across the globe.

International Industrial Design, 1945–1980

The economic explosion that transpired in many European countries after 1945 encouraged manufacturers to embrace the lessons learned from the growth of industrial design in the United States and to use them to develop their own products. Across the globe, the rise to fame of the designer for industry continued to be one of the crucial stories of the era. From the 1950s onward, the streamlined aesthetic of interwar products was replaced by a new visual vocabulary, including what became known as the mid-century modern look, visible in decorative arts and in domestic interiors. The Germanic neo-modernist minimal look, from the 1960s onward, heavily influenced the "high-tech" appearance of many machines, from hi-fi equipment to television sets. An aesthetic that emerged later in this period was inspired by post-modernist stylistic pluralism, a sense of fun, and nostalgia. It was evident in the pink goods produced in Japan in the 1980s, in Michael Graves's Whistling Bird teakettle for Alessi in 1985, in James Dyson's vacuum cleaners of the 1990s, and in Jonathan Ive's jewel-like iMac computer for Apple in the same decade.

The mid-century modern style emerged in the United States, where it was first presented to the public in a series of *Good Design* exhibitions held at MoMA in the 1950s. The style was driven by designers such as Charles and Ray Eames (fig. 14), Alexander Girard, Eero

FIG 14
Installation view, *New Furniture Designed by Charles Eames*, the Museum of Modern Art, New York, 1946.

Saarinen, Harry Bertoia, and George Nelson, who were all working for Herman Miller and Knoll Associates, among other companies. The Grand Rapids, Michigan–based Herman Miller furniture company, which had employed Gilbert Rohde and subsequently George Nelson to advise on design, had been working in the modern style since the 1930s. Through Nelson, a connection was established between Herman Miller and the architect Charles Eames. Together they went on to create one of the best-known and most influential of all postwar relationships between manufacturer and consultant designer. By the 1950s, Eames had become a leading international figure. He and his wife, Ray, became a talented and innovative couple who created a highly desirable modern lifestyle for themselves. They even made a film about their Santa Monica house, which was filled with their own designs and objects they had collected on their world travels. In all their work, they were guided by the sincere belief that design could improve people's lives. They became role models for a progressive way of living that reflected their time. Importantly, their modern idyll was accessible to other people through the purchase of an Eames chair or table or the many other pieces they designed for Herman Miller.

Under the curatorship of Edgar Kaufmann Jr. in the 1950s, MoMA took on the role of important arbiter of taste and good design, and its seal of approval became internationally recognized. Manufacturers around the world looked to it for guidance and new talent. The British furniture company Hille first encountered the work of its long-term collaborator, Robin Day, at the prestigious New York museum. Kaufmann's rhetorical description of the concept of good design stressed the qualities of "integrity, clarity, and harmony."[5] Significantly, the examples he cited were all chairs created by early modernists—Marcel Breuer and Le Corbusier among them—as well as by a number of later adherents of the same school of design, including the Scandinavian designers Finn Juhl and Bruno Mathsson and the Americans Eames and Nelson. Kaufmann's work at MoMA helped establish an elite group of second generation, innovative designers who became internationally renowned. The agenda of MoMA was clear: modernism was alive and well in postwar Scandinavia and the United States, as epitomized by the work of the designers they supported. Above all, in Kaufmann's resounding words, "Streamlining is not good design."[6]

The United States clearly sought to define its postwar design strategy by emulating Europe, rather than by acknowledging its own indigenous modern design movement of the interwar years, which it considered vulgar by comparison.

In Britain, the contemporary style, as seen in the work of Ernest Race and Robin Day, for example, owed much to the USA. It was also influenced by developments in Scandinavia, especially Danish modern, where the emphasis was also on the home and on mass-produced goods with a decorative arts lineage.

Germany took a slightly different route, focusing on a revival of the prewar Bauhaus approach and applying that aesthetic to machines and to products of the decorative arts industries. The notion of *gute form* ("good form") grew out of the minimalistic work of Dieter Rams for the Braun company. The University of Ulm also looked back to Bauhaus ideas.

In Italy, many furniture and product manufacturers promoted their designs through the names of designers, similar to the situation in the United States. In the 1950s, the furniture manufacturer Cassina abandoned its production of mass-produced, standardized

FIG 15
above left *Superleggera* chair designed by Gio Ponti, 1957, produced by Cassina, Meda, Italy. Cooper Hewitt, Smithsonian Design Museum, gift of George R. Kravis II.

FIG 16
above right Living room, Residential Design Standards exhibition, Milan Triennale, 1954.

furniture for the Italian navy and moved into the realm of designer furniture—first through a collaboration with Franco Albini and subsequently with Gio Ponti (fig. 15) and Vico Magistretti. Many modestly sized Italian furniture companies realized that the way forward was to produce limited edition, high quality, designer-led, and modern furnishings for an international, taste-conscious, elite market. The connection to a designer worked in two ways: it ensured innovative, modern work, and it also allowed the product in question to be sold in the name of art. That strategy was reinforced by developments in the infrastructure of design, which included the continuation of the Triennale exhibitions in Milan (the 1954 exhibition was subtitled *The Production of Art*) (fig. 16); the instigation by the Rinascente department store of an annual design prize, the Compasso d'Oro (the Golden Compass); and the formation, in 1956, of the ADI, an association of industrial designers.[7] Many important manufacturer/consultant designer partnerships were established in the 1950s, including Achille Castiglioni with Flos and Brionvega and Marco Zanuso with Arflex and Kartell. The large-scale manufacturer Olivetti adhered to the same model: working extensively with Marcello Nizzoli and subsequently with Ettore Sottsass (fig. 17) on a consultancy basis, the company understood the benefits of collaborating with an external designer in tandem with its in-house team. While the former provided a high level of innovation and cultural capital, the latter ensured that the designs could be effectively realized.

By the 1960s, the luxurious nature of much of Italy's industrial design output, which incorporated a great deal of marble, stainless steel, and high quality plastics, led Sottsass and others to initiate the anti-design movement, which, by the 1980s, had merged with postmodernism to undermine the international dominance of Germanic "good form" and what many considered the industry's excessive control over the work of the creative designer. The 1981 Memphis exhibit in Milan represented a key moment in that dramatic volte-face, bringing designers from many countries together. Its irreverent attitude

FIG 17
above left *Synthesis 45* side chair designed by Ettore Sottsass, 1972, produced by Olivetti, Milan, Italy.
The Metropolitan Museum of Art, Anonymous gift, 1987 (1987.386).

FIG 18
above center *Bubu 1er* stool designed by Philippe Starck, 1986, produced by XO, Servon, France.

FIG 19
above right *Voido* rocking chair designed by Ron Arad, 2006, produced by Magis, Milan, Italy.

surfaced again later, in the work of Philippe Starck (fig. 18) in France, of Ron Arad (fig. 19) and Tom Dixon in Britain, of Frank Gehry in the United States, and of the members of the Dutch design group Droog.

The designers in Japan, unlike their Italian counterparts, who mostly worked as independent consultants, were primarily in-house employees of large-scale manufacturing companies, such as Sony, Sharp, and Toyota, whose names were unknown to the public at the time. In the 1970s, Japan became a contender for leadership in the field of international industrial design when a group of designers established a new product design movement that emphasized the impact of high, or more advanced, technology, especially in electronics and miniaturization, as well as the nature and appearance of a range of sophisticated products, from cameras to calculators to high-fidelity systems. The Japanese high-tech product designers combined German minimalist style with visual displays of virtuosity, as seen in their complex control systems.

These Japanese consumer goods began to pour into Western markets in the 1970s. Sophisticated and complex electronics—like hi-fi audio equipment and cameras—manufactured by companies such as Sony, Sharp, Canon, Toshiba, and Hitachi, were created by automated production systems (fig. 20). Not surprisingly, their appearance reflected the high-tech environment in Japan. In spite of the fact that the new hi-fi equipment was destined for the living room, their complex control panels and multiple knobs and switches—redolent of the interiors of space shuttles—suggested an emphasis on high performance levels and technological virtuosity, rather than comfortable domesticity. Yet these products created a new style for domestic machines that proved extremely popular.

The degree of technological utopianism represented by those overtly masculine artifacts, colored mostly black and silver, was expressed not only through their appearance, but also by their multiple and novel functionalities. No longer was a watch simply a timepiece; it

FIG 20
Production line at the Toyota Tusutsumi car factory near Nagoya, Japan, 2004.

was also a stopwatch and an alarm clock. That important move away from a single object performing a single function, and therefore a move away from form being determined by function, accelerated dramatically in the last decades of the twentieth century and the first decade of the twenty-first century. The fact that the objects emanating from Japan expressed a faith in the power of technology was due to interventions by the industrial designers who created a look for them. The products' high-tech aesthetic rapidly expanded beyond the workplace and transformed leisure environments and the private sphere.

Japan was the only country to witness the emergence of a postmodern design movement that impacted on mass market goods. By the late 1970s, the rapidly developing country had begun to reject its earlier marketing strategy of selling its products on the basis of technological virtuosity; instead it adopted a more culturally oriented view and a more strategic use of design to capture its home market, especially its wealthy young consumers. That led, in the 1980s, to a range of high-tech goods produced in a variety of highly decorative color schemes that were clearly directed at female and youth markets. The products were chosen for their carefully selected aesthetic and lifestyle potential rather than their technological attributes. They were not expensive designer objects but were available in the general marketplace. It has been suggested that postmodernism was closely aligned to the Japanese love of contradiction and was therefore easily assimilated into Japanese culture.[8] Japan certainly led the way in injecting a consumption-led postmodern aesthetic into its mass-produced goods.

In spite of the principle of designer anonymity in Japan, a small group of designers—Sori Yanagi, Shiro Kuramata, Shigeru Uchida, and Masanori Umeda among them—became well known in international design circles. They focused on furniture and brought indigenous craft traditions together with progressive approaches to the discipline.

International Industrial Design, 1980 to the Present

Design highlights of the period after 1980 include the emergence of postmodernism on a global scale—a consequence of the burgeoning influence of mass culture on everyday life and a growing disillusionment with modernism—the development of a minimalist style epitomized in the work of the British designer Jasper Morrison (fig. 21), and the arrival of a new, sleek style that was led by Apple and its designer Jonathan Ive.

By 1980, industrial design and the mass production that accompanied it had become well integrated in global economies and cultures. As a result, it became increasingly hard to isolate unique national visual characteristics in products. So strong were the forces of pluralism and diversity that it also became hard to identify single trends. Designers felt free to look to the past and incorporate historical references into their work.

By the turn of the twenty-first century, a growing disillusionment with the idea of design as a marriage of art and industry, as a form of added value—the icing on the cake—and as the key to consumer desire within the market economy, led to a radical reassessment of design's social, cultural, and economic roles. The electronics revolution of the 1970s broke the centuries-old bond that linked design to the decorative arts, forcing a redefinition of design as a feature of the new post-industrial world. In that new context, the question of style, which had become synonymous with lifestyle, became urgent for many individual consumers in an increasingly globalized world, and designers continued to respond to that imperative. Simultaneously, however, designer culture began to lose its prominence, and designers were replaced by celebrities, young entrepreneurs, chefs, computer wizards, and others who were widely visible in the media. Arguably, the extension of good design into areas such as food, gardening, and web design—arenas in which the amateur could become increasingly confident and active—rendered it ubiquitous and, ironically, invisible.

In the 1990s, design theorists and practitioners pulled back from the dilution of design as it had been in its heroic decades—the 1920s, 1930s, 1960s, and 1980s in particular—and sought both to reform it and to revitalize its potential to bring about change. The phenomenon called "design thinking," or a method used by designers to solve intricate problems and find desirable solutions, grew out of that moment of rebirth. Design thinking affected both practicing designers and design education. One of its earliest manifestations in an educational institution occurred at Stanford University's design school, d.school,

FIG 21
Sofa from the *Three Sofa Deluxe* series designed by Jasper Morrison, 1991, produced by Cappellini, Milan, Italy.

also known as the Hasso Plattner Institute of Design, which was established in the mid-1980s by brothers Tom and David Kelley. By the early twenty-first century, the d.school aimed to inspire multidisciplinary teams to solve problems in ways that did not necessarily result in new products or services, but that had the potential to produce new business, implementation, and strategic plans for entrepreneurs and companies. While the majority of design thinking projects are business focused, some are more altruistic in nature, such as developing strategies for managing the aftermath of a disaster.

Also in the 1990s, Jonathan Ive, the designer responsible for Apple's key products, achieved celebrity status, while James Dyson continued to represent the successful designer-entrepreneur and even opened his own design school. An important distinction between these two individuals is that Ive was on the board of Apple Inc. and reported directly to its CEO, Steve Jobs, while Dyson owned his design-led company; yet both men had the power and authority not only to create design icons but to be closely involved with key business decisions, and therefore to integrate design into the hearts of their respective consumer product–based businesses.

Starting in the 1980s, the sophistication of modern production technologies, which by the early twenty-first century included rapid prototyping and 3D printing, introduced new freedoms into the scale of manufacturing and to the visual possibilities of products (fig. 22). The period also saw designers becoming increasingly preoccupied with environmental issues, the service economy, and virtual design, resulting, in certain contexts, in the object becoming secondary. In the twenty-first century this has been reinforced by many designers such as IDEO and Frog Design who are attempting to confront the multiple challenges of the era. They are becoming interested in new ways of working, whether co-design, participatory design, or social design. As a result, the celebrity designer and the iconic object have become less and less dominant. Instead, today's designers such as Anthony Dunne and Fiona Raby of Dunne & Raby (fig. 23) are increasingly conscious of the broad role they can play in the world, rather than simply continuing to act as the glue between art and industry.

FIG 22
below left *Fractal MGX* coffee table designed by Wertel Oberfell, Jan Werteland, and Matthias Bär, 2009, produced by Materialise, Leuven, Belgium. The Montreal Museum of Fine Arts, Liliane and David M. Stewart Collection.

FIG 23
below right *Risk* watch prototype from the *Do You Want to Replace the Existing Normal?* series designed by Anthony Dunne, Fiona Raby, and Michael Anastassiades, 2007, produced by Dunne & Raby, London, England.

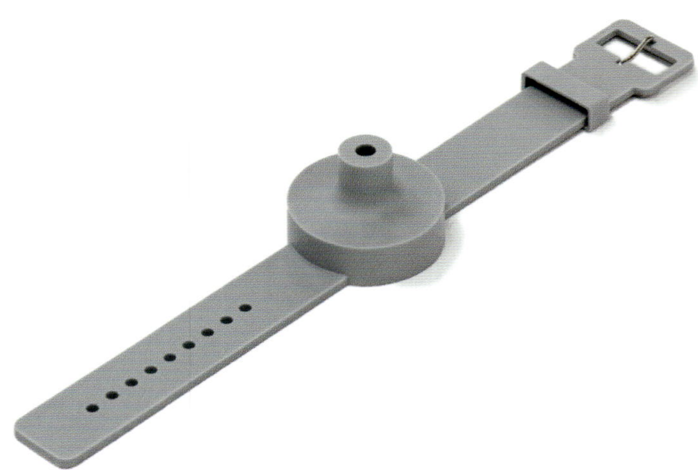

Industrial Design Beyond Europe and the United States

Design played a role in the branding of nations in the years after 1980. The opening up of Eastern Europe to market economies provided an opportunity for Hungary, Poland, the re-united Germany, the Czech Republic, and Slovakia, among others, to use design to reinforce and communicate their craft traditions and to develop a modern post-Communism face for the world.[9] Following Japan's lead, several Far Eastern countries including Korea, Singapore, and Taiwan also began to inject modern design into their programs of modernization. The Western modernist model was widely used. Bauhaus principles were imported into Latin America, for example, while modern design in India was based on the ideas of American designers Charles and Ray Eames, who wrote a report for the Indian government just after World War II.

Former British colonies such as Canada, Australia, and South Africa also began to develop their own design traditions as a way of moving beyond their imperial pasts. Canada made huge efforts to develop a modern design identity, epitomized in such artifacts as the Skidoo, a kind of motorized sledge (fig. 24). By the early twenty-first century, the previous domination of the field by the United States–Europe axis since the first half of the twentieth century had diminished, as many other countries across the globe began to embrace design as a means of demonstrating their aspirations to become part of the modern developed world.

Industrial Design in China

By the first decade of the twenty-first century, it was apparent that China had become an industrial power to reckon with on the global stage, and was also beginning to think about the role that design could play in its modern economy and culture. The country's rapid growth and competitive manufacturing costs encouraged many product-related companies in Europe and America, as well as others in the Far East, to outsource their production to China. Vast manufacturing plants producing high-tech components and other products for global firms were built. The largest of the plants, Foxconn, in Shenzhen, formed by the Hon Hai Precision Industry Co. Ltd in 1974, was soon producing electronic components for Apple's iPods and iPads, among numerous other technology-based products.

As China developed as an industrial power, and as its urban population became increasingly wealthy, a new middle class emerged—mirroring the societal developments in nineteenth-century Britain—that was keen to demonstrate its new social status through the acquisition of goods. Given the enormous size of its population, the home market became significant for Chinese manufacturers. In order to provide their customers with the added value they were seeking, Chinese producers began to think not only about the quantity of their manufactured goods, but also about the quality. As consumption expanded, so consumer taste began to drive the appearance of Chinese products, despite the continuing control of the ruling Communist Party. Once again echoing developments in mid-nineteenth-century Britain, as well as what happened in the United States in the early twentieth century, China transformed into a consumer society in which the taste of its middle class was reflected in the emergence of a highly conspicuous aesthetic. Theirs emulated styles in the West, from streamlining to Philippe Starck.

The rapid shift from production to consumption that occurred in China in the early twenty-first century closely paralleled the ways in which General Motors's market-oriented strategies replaced Ford's principles of mass production in the 1920s. The transformation in China was made easy with the help of high-tech flexible manufacturing. Under increasing pressure from its home market to compete with the price and the quality of goods produced elsewhere, China began to evolve its own model of modern design. At first, it relied on copying Western models. Sponsored by indigenous manufacturing companies, especially in the high-tech and automotive industries, Chinese design students were sent in significant numbers to study in European and American design schools. Most importantly, however, they were obliged to return to China to apply their new knowledge. One manifestation of their version of modern design can be seen in the Geely Meirenbao sports car, which debuted in 2003. The country's educational institutions also experienced a rapid expansion.

Once Chinese producers realized that they had to embrace design, they moved fast to learn from models developed elsewhere. Importantly, however, they sought to retain a sense of national identity and developed a concept of Chinese modern design (fig. 25). Given the strengths of the country's decorative arts traditions, it made sense to ensure that longstanding craft skills and ritualistic artifacts could inform a national modern design movement. While on one level that approach resulted in some interesting hybrids—examples include a General Motors–produced concept sedan that looked like a pagoda when viewed from the rear; a speaker phone based on traditional hot pot; and a corporate identity scheme for Pizza Hut China's website that made extensive use of China's traditional colors, red and yellow—on another level it stimulated programs of research that focused on reviving traditional Chinese applied arts—such as garden design, calligraphy, and ceramics—as a means of offsetting the bland Westernized products that were increasingly filling their retail centers. The principles underpinning Daoism and Confucianism were also widely discussed in that context, alongside efforts to ensure that Chinese designers understood how to innovate.

By 2010, the Chinese home market had become so important to international companies that they began not only to manufacture their products in the country, but also to target Chinese consumers.

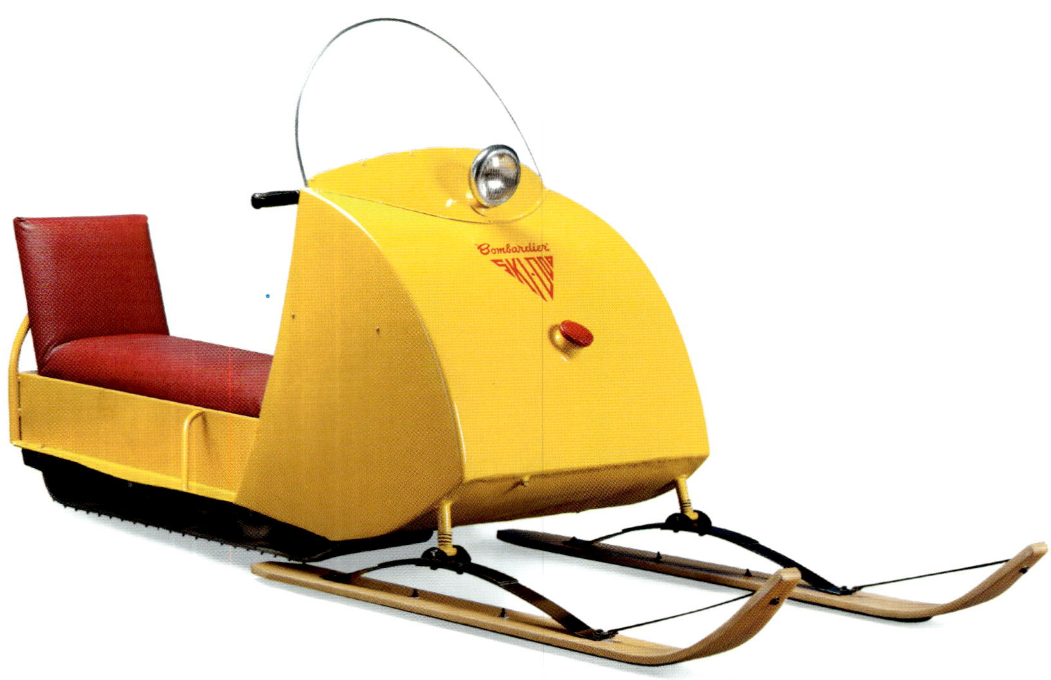

FIG 24
below *Ski-Doo* snowmobile designed by Joseph-Armand Bombardier, 1959, produced by Bombardier Snowmobile Ltd., Valcourt, Quebec. The Montreal Museum of Fine Arts, Liliane and David M. Stewart Collection, gift of Mrs. Claire Bombardier Beaudoin.

FIG 25
opposite Armchair designed by Shao Fan, 2000. The Metropolitan Museum of Art, gift of Pearl Lam.

PART TWO
INDUSTRIAL DESIGN AND EVERYDAY LIFE

The history of designers and manufacturers producing progressive objects that aspire to the status of fine art is a familiar and heroic story that is well represented by the Kravis collection, but is only part of the tale. The collection also represents many other aspects of the world of industrial design, which is, ultimately, far more complex than a timeline of designers and their products.

Designers and manufacturers cannot work in isolation. Without consumers of those products, the cycle would be incomplete. It is often only in the context of use that objects acquire their meanings, many of which were unintended by the designer. The Kravis collection features innumerable objects that not only express the significant role each played individually, but also enhance and illuminate the larger story of industrial design since 1900.

Fashion and Taste

Just as style is produced by designers, taste derives from the consumers and users of objects in their everyday lives. As a modern concept, design developed in part as direct result of the expansion of consumer goods markets and the democratization of taste. For centuries, handmade furnishings, ceramics, glass, metalwork, clothing, printed artifacts, textiles, and vehicles for transport had provided comfort; were the markers of propriety; provided the glue of social, family, and gender relations; and were the visible signs of social status and taste. Starting in the nineteenth century, as progress and modernity influenced the lives of ever larger numbers of people, design—as defined as the visual and conceptual components of the growing set of mass-produced goods and images—became the primary means through which goods became attractive to a mass market and helped to give meaning to people's lives. In both Europe and the United States, as increasing numbers of consumers embraced the new goods made possible by the industrial age, new classes emerged and the link between design and taste was forged. Gradually, as industrially produced goods became more accessible, designed goods and images began to replace decorative arts as the primary indicator of social class, becoming the key messengers of fashionableness and modernity.

In the nineteenth century, the British middle class grew, and its capacity to consume continued to increase. The expansion of the British market for manufactured goods was a direct result of the population's growth, increase in spending power, and shifts in taste and behavior, which could encourage people to transfer their spending from one set of goods to another. For example, in the 1860s, more gas cookers became available as a result of a drop in gas prices; by 1914, new household objects, such as the Bissell carpet sweeper from America, came on to the British market; and by the first decade of the new century, the bicycle and the automobile had become common sights on the urban scene.

Technological novelty played a key role in most consumer choices in the early twentieth century. But given that many consumers saw certain technological advancements as a threat, visual strategies were sometimes employed to conceal the perceived threat. On some of the new electrical products destined for the home, surface patterns were added to enhance familiarity and relieve anxiety.

The rise of branded goods in the United States signified a major shift in the way in which goods were bought and sold and created a new link between design and marketing that was to become increasingly complex over the nineteenth and twentieth centuries. Companies established names for their products that evoked particular lifestyles. The Pittsburgh Reduction Company, a manufacturer of aluminum pots, pans, and other cookware in the late 1880s, coined the name Wearever for its range of products, building on the assumption that their potential consumers were rational beings who wanted value for their money. The company employed artists to develop magazine advertisements to promote the Wearever products, and shops were encouraged to place banners with the Wearever name emblazoned on it in their windows. A sophisticated visual promotional framework was thus developed around the campaign to sell their aluminum pots and pans. Importantly, it required the input of visually skilled specialists in the areas of graphics, packaging, and window display.

In the years before 1914, however, there were few attempts to make goods themselves—aluminum saucepans, for example—look attractive. Rather, objects retained a utilitarian appearance, and their desirability was communicated through the visual marketing information that accompanied them, as embodied in the Wearever campaign. That early means of conveying the added value of mass-produced items was a first step in the shift that eventually led, in the interwar years, to the emergence of the concept of industrial design. By that time, manufacturers had realized that the visual impact of goods themselves gave a competitive edge that advertisements, packaging, and shop window displays could not achieve on their own.

During the first decade of the twentieth century, in its attempt to stimulate and satisfy consumer desire, the commercial world had developed a set of sophisticated marketing strategies that involved a wide range of visualizing and conceptualizing practices. The expanded market, the phenomenon of these new consumers using visual, material, and spatial culture to define and communicate their identities and social aspirations, and the availability of more and more goods, images, and services, all combined to create the necessary foundation upon which the concept of modern industrial design could begin to build.

A prime example that illustrates this evolution is the vacuum cleaner. The basic shift in appearance of the vacuum cleaner over time, from a set of visually distinct elements to unified products that performed their tasks on trust, also occurred in most mass-manufactured products from the first forty years of the twentieth century across the industrialized world. Craft-based objects—furniture, ceramic, glass, and textiles—followed the lead of the technological products as they entered into mass production.

Anonymous Industrial Design

An equally significant thread in the story of industrial design was the development of products that were not created for private consumption, such as tools, instruments, and machines, as well as many utility goods that are not necessarily destined for the domestic arena, but rather play a role in public spaces. The invisible components of complex objects also fall into this category. Never designed to be seen, they exist only as functional items, exhibiting form but not style. In the worlds of fine and decorative arts, craftsmen seek stylistic categories and authorship for their objects, in an attempt to imbue them with status and add them to the canon; but industrial design does not always fit neatly into this model. In fact, many objects have no designers' names attached to them, and there is no need to seek them. Nor were they intended to exhibit a particular style. They exist, simply, as functional items, created anonymously by the hands of nameless engineers and standardized machines. Their meanings are expressed through their materials, their utilitarian forms, and their natural fit within daily life. Their very mundaneness can be seen as a positive, rather than a negative, attribute in the role they play in our everyday lives.

In times of war, the power of design is inevitably harnessed to different ends. In this situation, style, self-expression, and fulfillment of desire are luxuries that become secondary considerations. In wartime, the boundary between engineering, innovation, and design is very fine.

From Functional Form to Functional Style

Yet these anonymous utilitarian objects exhibit a kind of beauty of their own. Indeed, it was their "natural" engineered beauty that inspired modernist architects and designers to develop the theory of functionalism in the early twentieth century. The aesthetic shift in products from the 1920s, from surface decoration to pure and unelaborated form—known as the functional style—is one of the most important visual transformations in the story of modern design. Mechanized production did not lead immediately to machine-styled products, as one might expect. Rather, a couple of generations of forward-looking designers took their cue from contemporary architects and fine artists, who had been quicker to respond to the cultural imperatives of the machine age, and applied their knowledge to the factory-made products that had inspired the architects and artists in the first place.

The idea that pure form derives from function, and that products should not display surface decoration, does not necessarily mean that the use of an object determines its appearance. The idea that "form follows function," as promulgated by the modernists, was partly based on the architecturally focused notion that inner structure should influence outer form. The phrase was coined by Chicago-based architect Louis Sullivan, who believed that just as a flower grows out of the root and stem of a plant, so the outer appearance of a building should be determined by its inner structure. Sullivan advocated for the use of a steel frame as the supporting structure for a modern building, and believed its outer form should reflect that simple geometry. Although he did not always practice what he preached—several of his buildings displayed decorative details on their exteriors, especially at the entrance—his progressive idea had a great impact on early modernist architects and designers. Sullivan's ideas were picked up by the American architect Frank Lloyd Wright, who became highly influential on later modernist thinking.

On mainland Europe, one of the most powerful manifestos in this context was *Ornament and Crime* by the Austrian Adolf Loos, published in 1913. Loos proposed that, through its rejection of ornament, mankind had become increasingly civilized. The most mature society, he maintained, was represented by the simplest of forms, and it was the responsibility of the modern architect-designer to develop that principle.

As the twentieth century progressed, architects and designers sought an objective formula based on the precepts espoused by Louis Sullivan and Adolf Loos. The collision of nineteenth-century individualism with twentieth-century functionalism reached a head in 1914 in a confrontation between the Belgian architect-designer Henry van de Velde and the

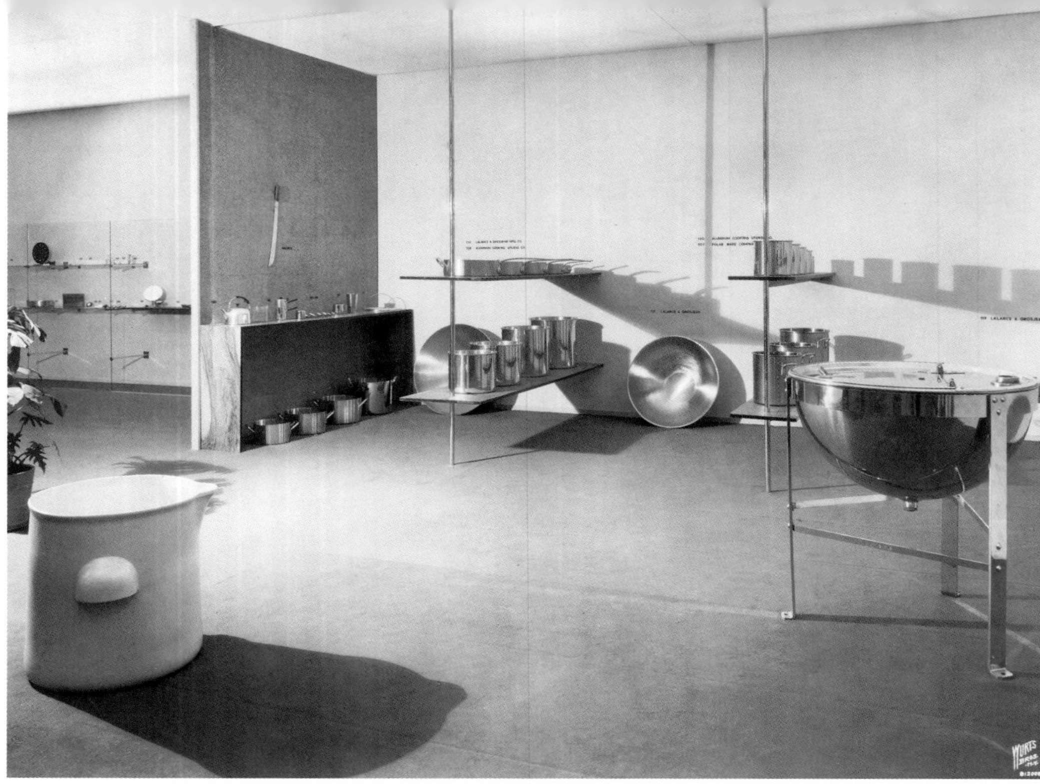

FIG 26
Installation view, *Machine Art*, Museum of Modern Art, New York, 1934.

German diplomat Hermann Muthesius. The latter was strongly committed to an abstract aesthetic and, unlike van de Velde, he believed in the principles underpinning industrial standardization and the sacrifice of the individual to the masses. The lively debate between the two men pinpointed what was to remain a fundamental dilemma for the supporters of modernist design for some time, which was whether modernism represented an attempt to develop an aesthetic using the idea of the machine as a metaphor or to design for machine production and facilitate the widespread availability of goods for the masses, giving everybody direct access to modernity.

By 1914, modernism's architecturally derived ideas were fully formed, and their implications for design clearly expressed. "Form follows function" had become widespread. The Bauhaus was rooted in craft-based thinking that, in line with Sullivan's earlier philosophy, proposed that a product is a result of the process that created it. Thus, the appearance of a ceramic pot derives from the fact that clay has been thrown on a wheel, and there is no need for surface treatments. Similarly, the tubular steel chair exposed its inner structure, which in turn became its aesthetic. The desire to cover it up had fled.

By the 1930s, this idea was embedded in modern design, and surface decoration had become a thing of the past. In its place, simplicity and geometric form had become the aesthetic for the modern age. Ironically, the functional forms that Le Corbusier and other early modernists developed aspired to the status of art, and their designers came to be lauded as heroes of modern design.

The functional style did not only have architectural origins; it was also inspired by the non-styled objects described above. The idea that anonymous engineered industrial design is beautiful has been widely recognized by writers, curators, and designers over the last hundred years. In early European modernist writings, a strong sense of awe was expressed about the simple products of early American mass production. Clocks, keys, bicycles, farm machinery, and standardized bookcases, for example, were widely revered as functional and

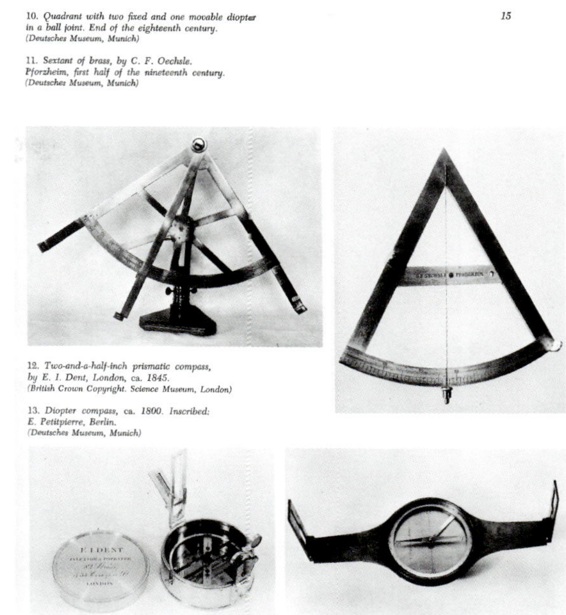
15

10. *Quadrant with two fixed and one movable diopter in a ball joint. End of the eighteenth century.* (*Deutsches Museum, Munich*)

11. *Sextant of brass, by C. F. Oechsle. Pforzheim, first half of the nineteenth century.* (*Deutsches Museum, Munich*)

12. *Two-and-a-half-inch prismatic compass, by E. I. Dent, London, ca. 1845.* (*British Crown Copyright. Science Museum, London*)

13. *Diopter compass, ca. 1800. Inscribed: E. Petitpierre, Berlin.* (*Deutsches Museum, Munich*)

FIG 27
right Cover of *Mechanization Takes Command* by Sigfried Giedion, published by Oxford University Press, New York, 1948.

FIG 28
far right Illustrations of measuring devices and compass from *The Roots of Modern Design: Functional Tradition in the 19th Century* by Herwin Schaefer, published by Studio Vista, Great Britain, 1970.

visually unself-conscious products of the machine. The modernists' shared nostalgic longing for that age of innocence was a response to their deeply felt anxieties about the commercial context of design.

The exhibition *Machine Art*, held at MoMA in 1934 and curated by Philip Johnson, took the beauty of anonymous, functional design as its subject (fig. 26). It featured a wide range of objects, including machine parts, pots and pans, scientific equipment, airplane propellers, and ball bearings, defining them as modern art because of their simple, abstract forms. The show was instrumental in turning functional form into functional style. In 1948, Sigfried Giedion penned his famous book *Mechanization Takes Command: A Contribution to Anonymous History*, in which he, too, lauded the quality of simple utility items that played a role within everyday life (fig. 27). Giedion saw a clear link between the well-known Thonet bentwood chair of the mid-1800s (frequently used by Le Corbusier in his interiors) and the products of the Bauhaus.

Herwin Schaefer's 1970 publication *The Roots of Modern Design: Functional Tradition in the 19th Century* looked back to the anonymous products of nineteenth-century factories, explaining how a wide range of goods—from instruments and machines to guns to ceramics, glass, and furniture—were manufactured anonymously to perform a range of functions (fig. 28). In his introduction, he explained how the heroic, stylistic story of modern design that was set out by Nikolaus Pevsner in his well-known 1936 publication *Pioneers of Modern Design: From William Morris to Walter Gropius* had become *the* story of modern design, marginalizing the functional tradition in the process. In Schaefer's view, "The efforts of reformers and pioneers, be they piously medievalizing or aesthetically artistic, pale to insignificance beside the strength and consistency of this [functional] tradition which furnished the solid core of modern design."[10]

Experiencing Industrial Design

Looking at industrial design artifacts from the perspective of the user, rather than viewing them as iconic art objects, helps illuminate the different ways in which they can be experienced. On an everyday basis, users interact with design sensorially—whether visually, in response to their shiny, matte, or colored surfaces; tactilely, through the feel of the surfaces and textures; or aurally, through the sounds that are sometimes emitted. We see, hear, and touch designed objects, eat and drink from them, sit and lie on them, drive in them, walk on them, and so on. During all of these activities, an emotional relationship develops between the user and the design of the object, sometimes experienced consciously but more often subliminally.

Ergonomics

The physical relationship between design and the body is a particularly strong one. Many design objects—typewriters, keyboards, control systems, and knobs—are controlled by users' hands, while masks and goggles adorn the face. Chairs and bicycles have a relationship with whole bodies. Over the years, studies of the design/body relationship have helped designers ensure that their objects meet the physical needs of users.

The science of ergonomics advanced significantly during World War II. Building on early twentieth-century scientific management studies, which sought to increase efficiency in the workplace, work on human/machine interaction became increasingly important in the context of military equipment and flying fighter aircraft. Although on one level there was a sense that people on the battleground were ultimately expendable, on another, fitting machines to the size of their operators and having effective control panels and buttons were crucial to successful warfare.

After the war, the new knowledge acquired in this field fed back into a number of different areas. In Sweden, a neutral country during the war, this knowledge was rigorously applied to the interior spaces of postwar buildings created to house the expanding population. Because space was at a premium, it was important that minimum sizes were established, that storage was measured properly, and that beds and sofas were the "right" size. The Svenska Slöjdföreningen, the equivalent of Britain's Council of Industrial Design but with deeper roots, undertook a number of scientific measurement studies in the postwar period and developed sets of guidelines laying out the ideal sizes for objects destined for the domestic interior. In the United States, Henry Dreyfuss pursued a similar direction and invented a standardized human couple—Joe and Josephine—as the "users" of design that took into account the measurements, and the psychological needs, of people. Today, the relationship between design and the body has taken a new direction with the advent of wearable computing and smart textiles.

The Meanings of Industrial Design

Industrial design not only engages users' senses in order to stimulate their emotions, it also helps form their personal and collective identities. An essential component of the cultural backdrop of everyday life, it contains, communicates, and helps us form values and behaviors that in turn enable us to know who we are and where we are in the world. Those values are encoded within design products not only by designers, but also by the contexts in which they find themselves. Brightly colored toys, for example, suggest the age of their intended users, while wearing headphones on a train discourages fellow travelers from interacting with the user. Industrial design objects express their meanings in many ways, including through the use of symbols, metaphors, or visual stereotypes. Anthropomorphism can also be used to create an empathetic relationship between object and use. In short, designed objects articulate a form of language that is "spoken" to users on a daily basis.

The meanings of industrial design are many and varied. It can play a part in the formation of geographical identities—whether global, national, or local—and of other identities according to gender, race, ethnicity, age, religious persuasion, consumer group, or taste. In its capacity as a cultural force, design plays a fundamental role in everyday life of the early twenty-first century. As a tool to convey the power of nations, it has come a long way from the nineteenth-century exhibitions filled with huge halls displaying early production machinery. Now, national identities are on display in the global marketplace and embedded in the mass media. As individuals, consumers have learned to negotiate their identities and their sense of belonging, aided by designed goods, services, spaces, and images.

The meanings of industrial design are also formed and reinforced by the activities they support and the physical spaces in which they are found. Products facilitate human activities at home, at work, and in a range of other spaces, whether private or public. Many objects can move between spaces, while others are designed specifically for particular environments. There is a difference, say, between an industrial washing machine and a domestic one; a difference not only of performance and specification but also of appearance. Particular materials are felt to be appropriate in specific spaces. While, for example, one would expect to find a lot of metal in an airport, in a country cottage one would expect less metal and more natural materials.

The objects that are featured in this publication have been organized according to the activities they enable and the spaces they inhabit. They are drawn from the collection of George R. Kravis and are all deemed important examples of industrial design from the late 1800s to the present. The Kravis collection comprises over four thousand designed objects and is growing, with a special emphasis on industrial design. These objects are presented here in sections defined by activities that take place in the home, the office, and workplace, leisure spaces, the street, and the spaces of transportation. With the exception of traveling, all of the other activities—homemaking, housekeeping, controlling the climate, measuring, food preparation and storage, eating and drinking, listening, viewing, working, and playing—can take place in the home. Only homemaking and housekeeping are exclusive to their location. The objects that facilitate these activities represent the iconic and the anonymous, the visible and the invisible. All of them contribute to life in the everyday world. Whether they are aspirational art objects created by signature designers and given pride of place in living rooms, or mundane tools that we use every day at home or at work, without noticing or acknowledging their existence; whether they are anonymously made invisible components of functional objects that we use regularly, or something we pass on our way to work—they are all important to us. Hopefully this publication will allow us to see them all with new eyes.

ENDNOTES

1 Walter Dorwin Teague, *Design This Day: The Technique of Order in the Machine Age* (London: Studio Publications, 1940), 49.

2 Alfred P. Sloan, *My Years with General Motors* (New York: Mcfadden-Bartell, 1965).

3 Teague, *Design This Day*, 215.

4 Norman Bel Geddes, *Horizons* (Boston: Little, Brown, 1932).

5 Edgar Kaufmann Jr., *Introductions to Modern Design* (New York: Museum of Modern Art, 1950), 9.

6 Kaufmann, *Introductions to Modern Design*, 8.

7 Penny Sparke, *Italian Design: 1870 to the Present* (London: Thames and Hudson, 1988).

8 See Penny Sparke, *Japanese Design* (London: Michael Joseph, 1987).

9 David Crowley, *National Style and Nation-State: Design in Poland from the Vernacular Revival to the International Style* (Manchester: Manchester University Press, 1992).

10 Herwin Schaefer, *The Roots of Modern Design: Functional Tradition in the 19th Century* (London: Studio Vista, 1970) 5.

01
objects

HOMEMAKING

THE CONCEPT OF HOME IS FUNDAMENTAL TO OUR MENTAL AND physical well-being. From a modest caravan to a sumptuous palace, home is where our identities are formed and are most easily expressed. The art of homemaking involves turning that personal space into a sanctuary, a safe haven from the crowds and noise outside. It is where material items are combined to create a space that is psychologically and physically comfortable and relaxed. That sense of restfulness is represented by a wide range of objects designed in all shapes and sizes and materials, and the Kravis collection contains many of them, from chairs to lamps to vases, all demonstrating our desire to beautify and personalize our domestic spaces.

Verner Panton (1926–1998)

Heart Cone (Chair K3) Chair

Designed 1959 (utility patent 3,086,818)

Wool upholstery, foam rubber, steel,
stainless steel, plastic

36 x 40 x 28¾ in. (91.4 x 101.6 x 73 cm)

Produced by Plus-Linje, Copenhagen, Denmark

K2009.55

Cooper Hewitt, Smithsonian Design Museum,
New York, promised gift of George R. Kravis II

Verner Panton sitting in his *Heart Cone* chair, c. 1960.

THE INDUSTRIAL DESIGN FIELD WAS NOT IMMUNE TO THE INFLUENCE OF POP ART. DANISH designer Verner Panton was one of its most powerful advocates in the 1950s and 1960s. Panton was trained as an architect and worked in the office of Arne Jacobsen from 1950 to 1952, opening his own studio in 1955. He quickly established himself as a major designer with works such as the *Cone* series of chairs, which have no conventional legs or seating shapes. His designs evolved rapidly with his use of radical forms, bright colors, and geometric patterns that were almost psychedelic in intensity; his total environments from the 1960s are the epitome of the pop aesthetic. His *Heart Cone* chair for Plus-linje was one of his first successes, and the series included other chairs, a bar stool, an ottoman, and a dining table. The earliest examples for Plus-linje incorporated traditional thin sheets of steel, rather than molded plywood or fiberglass, to create the heart-shaped shell. Panton moved to Basel in 1963 to begin his collaboration with Vitra to produce his famous *Panton* side chair (1967). The rights to the *Heart Cone* chair were transferred to Vitra in 2001. Current versions of the chair have shells made of glass-fiber-reinforced plastic laminate (GRP). After more than a half century, it remains an icon of modern design. –RCM

Alvar Aalto (1898–1976)
Armchair 41 Paimio Lounge Chair
Designed 1931–32
Birch plywood, lacquered wood
25½ x 33½ x 23½ in. (64.8 x 85.1 x 59.7 cm)
Produced by Artek, Helsinki, Finland,
for ICF, New York, New York
K2009.64

ALVAR AALTO IS PERHAPS THE MOST ACCLAIMED FINNISH DESIGNER OF THE twentieth century. This chair was designed for the Paimio Sanatorium (built 1929–33), one of Aalto's first celebrated buildings in the international style. In the 1920s, modernist furniture was characterized by simple geometric forms and often used tubular steel. Aalto was among the leaders who pushed modernism in a new direction during the 1930s by using more organic forms made of molded plywood, which was thought of as warmer and more humanistic. The *Paimio* chair is Aalto's reinterpretation of Marcel Breuer's *Wassily* (Model No. B3) lounge chair (1925). The frame is made of one continuous piece of undulating bent plywood; the seating panel is also a continuous but more sinuous scroll-like form made of molded plywood, in this example lacquered black. The bicolor scheme emphasizes the lightness and beautiful interplay of curves. Aalto devised a number of variations of the chair; the seating panel could be made in black, white, or natural wood, and he also designed an upholstered version, *Armchair 44,* for the same frame. The *Paimio* lounge chair (renamed *Armchair 41 Paimio*) was originally produced by Huonekalu-ja Rakennustyötehdas in Oy Ab, Turku, Finland, and later put into mass production when Artek was established in 1935 and became one of his most famous designs. This example was produced in c. 1970. –RCM

Charles Eames (1907–1978)
Ray Eames (1912–1988)

LCW Chair

Designed c. 1946 (D150,683)

Dyed ash plywood, rubber

26⅜ x 24¼ x 22 in. (67 x 61.6 x 55.9 cm)

Produced by Molded Plywood Division, Evans Products Company, Venice, California, and by Herman Miller Furniture Company, Zeeland, Michigan

K2008.335

THE EAMES LOUNGE CHAIR—ALSO KNOWN AS THE *LCW* CHAIR—IS A LOW-SEATED chair designed by Charles and Ray Eames, who used innovative technology that they developed for molding plywood before and during World War II. After the war, they applied their technology to mass-produced chairs. Through many experiments, they devised ways to bend wood furniture in new directions and give hard materials a soft look. The seat and back of the *LCW* were connected to the supporting spine with dense rubber "shock mounts" inserted between the wood pieces to provide flexibility and comfort in the absence of padding and upholstery. It was through George Nelson that the Eameses were hired to design furniture for the Herman Miller Company in 1946, which then took over production of their plywood creations. *Time* magazine recognized the plywood chairs as "The Best Design of the 20th Century." —PM

Gerald Summers (1899–1967)
Bent Plywood Armchair (BPAC)
Designed 1934
Laminated beech with birch veneer
29⅜ x 35¼ x 24¼ in. (74.6 x 89.5 x 61.6 cm)
Produced by Alivar, Ltd., Italy
K2012.100

UNLIKE GERMANY, FRANCE, OR FINLAND, ENGLAND WAS NOT IN THE VANGUARD of modern design in the 1930s, as illustrated by Gerald Summers's now classic *Bent Plywood Armchair*. Summers was fascinated by molded plywood and established his own firm, Makers of Simple Furniture, which produced a wide range of products in the 1930s, including the original production of this chair (this example produced c. 1990s). The chair was innovative in that it was one of the first examples of a continuous chair made from one piece of material. However, his company did not have the sophisticated technical ability of Artek to bend or mold plywood in two dimensions with heat or steam (page 48). Rather, the chair is in many respects a hand-crafted object in which layers of thin veneer are glued together in a mold with the grain of each sheet running at right angles to the next. Summers's chair is thus an instance when the concept is perhaps as important as the resultant design. By the 1940s, plywood furniture would move in radical new directions when designers such as Eero Saarinen and Charles and Ray Eames were able to produce highly sculptural three-dimensional designs. –RCM

Frank Gehry (born 1929)

Chair and Ottoman

Designed c. 1970 (utility patent 4,067,615)

Laminated corrugated cardboard, Masonite

Chair: 29⅝ x 37½ x 23⅝ in. (75.3 x 95.3 x 60 cm)

Ottoman: 15 x 29½ x 17½ in. (38.1 x 74.9 x 44.5 cm)

Produced by Easy Edges, Inc., New York

K2012.14

Publicity photo of a Volkswagen *Beetle* supported by test versions of Frank Gehry's chairs, c. 1972.

THE *EASY EDGES* FURNITURE SERIES, INTRODUCED IN 1972, BROUGHT FAME TO Frank Gehry long before his monumental masterpiece, the Guggenheim Museum in Bilbao. Inexpensive materials and assembly-line production made the seventeen pieces of the line affordable, costing only $35 to $100 each. First sold at Bloomingdales, the line was presented in a simple fold-out paper catalogue. Layers stamped out of single-ply cardboard sheets were stacked together and glued, like plywood, with their corrugations running perpendicular to one another. As proof of the strength of the resulting chairs, a marketing photo shows a Volkswagen Beetle balanced on a few of them. Cardboard furniture had first been designed in the 1960s, when plastic still dominated the market. However, with the energy crisis in the 1970s, cardboard became an attractive alternative to the environmentally destructive plastics. Although Gehry resisted his fame as a furniture designer in favor of his wish to be considered foremost as an architect, he continued to design furniture through the decades. –DAH

Charles Eames (1907–1978)
Eero Saarinen (1910–1961)

Side Chair

Designed 1940

Honduran mahogany, upholstery

33 x 17¾ x 21 in. (83.2 x 45.1 x 53.3 cm)

Produced by Haskelite Manufacturing Corporation and Heywood-Wakefield, Grand Rapids, Michigan

K2012.105

Cooper Hewitt, Smithsonian Design Museum, promised gift of George R. Kravis II

Installation view, *Organic Design in Home Furnishings*, Museum of Modern Art, New York, 1941.

IN 1940, THE MUSEUM OF MODERN ART (MOMA) IN NEW YORK SPONSORED THE GROUNDBREAKING Organic Design in Home Furnishings competition, which, in collaboration with department stores and manufacturers, showcased low-cost furniture and other household items for the American public and discovered "a group of designers capable of contributing to the erection of a useful and beautiful environment for today's living."[1] Eames and Saarinen's *Side* chair was the winning entry for their Seating and Other Living Room Furniture collection, which included a suite of various armchairs, a sectional sofa, a variety of modular case units, a desk, tables, and this side chair.

Eames and Saarinen designed the chair with clean curving lines that were strikingly different from anything else on the market at the time. It is one of the first pieces of furniture produced by either designer and became the foundation for their future practice. To achieve the three-dimensional compound curves of the chair back, Eames and Saarinen worked with the Haskelite Manufacturing Corporation, a company that had developed a process to mold a three-dimensional bent plywood form or shape, using a cast-iron mold that had been used for shells for boats and planes. The technological achievement of bent plywood was applied to furniture in the United States for the first time, producing this lightweight chair—made comfortable with padding and woven upholstery designed by Marli Ehrman—that laid the groundwork for further innovations in modern seating design. –PM

Alexander Girard (1907–1993)

Armchair

Designed 1967

Cast aluminum, Naugahyde, upholstery

27⅝ x 23¾ x 21 in.
(70.2 x 60.3 x 53.3 cm)

Produced by Herman Miller Furniture Company, Zeeland, Michigan

K2012.119

Alexander Girard at his desk with his armchair in the foreground, 1967.

IN 1965, ALEXANDER GIRARD RECEIVED A MAJOR COMMISSION TO UPDATE ALL DESIGN aspects of Braniff International Airlines, including furniture for airport lounge areas. Two years later, Herman Miller introduced a line of Girard furniture, including this chair, based on the Braniff designs. Girard compared his furniture with his varied upholstery designs to a chameleon: "its character remains intact, yet its skin texture and color change to suit the environment. . . . The permutations are infinite." Other features of his aesthetic were a low sight line, suited to "contemporary low-ceilinged spaces," and a small scale to "create a feeling of spaciousness and repose." Girard maintained that "the components are clearly articulated and visually expressed," and stated that such details as soft rounded corners, wide choice of fabrics, and simple attachments between structural elements were basic to his concept.[2] Herman Miller discontinued the line after one year.[3] –DAH

Poul Kjaerholm (1929–1980)

PK 24 Chaise Lounge

Designed 1965

Mirror-finished stainless steel, cane, leather

34¼ x 62 x 26¼ in. (87 x 157.5 x 66.7 cm)

Produced by Fritz Hansen,
Copenhagen, Denmark

K2012.73

UNTIL HIS DEATH IN 1980, POUL KJAERHOLM MADE A CONSIDERABLE contribution to the international reputation of Danish furniture design by combining both crafted and industrially finished elements and adapting aspects of the classical European modern style to his simple, purist designs. *PK 24* was first put into production in 1967 by E. Kold Christensen; production was taken over by Fritz Hansen in 1982. The chair is an example of stripped down minimalism, where Kjaerholm eliminated the physical connections between the main parts of the chair and instead utilized gravity and friction to stabilize the pieces. All three parts of the chair—base, seat, and neck rest—are loose, allowing the occupant to easily adjust according to preference. For the reclining surface, the designer chose a rattan cover, and the only padding is a neck rest, which hangs loosely over the back section and is held in place by a counterweight. Because of the way the seat is suspended from front to back, Kjaerholm called it the "hammockchair." Its metal elements were mirror-polished stainless steel; after 2005, the frame was altered to satin-brushed stainless steel. –PM

Florence Knoll (born 1917)
Credenza
Designed c. 1955
Walnut, marble, chromium-plated steel
25¼ x 74⅝ x 17¾ in. (64.1 x 189.6 x 45.1 cm)
Produced by Knoll International,
New York, New York
K2012.230

AS AN ARCHITECT, INTERIOR DESIGNER, FURNITURE DESIGNER, AND businessperson, Florence Knoll is one of the most influential designers of postwar America. Along with her husband, Hans Knoll, she founded Knoll Associates in 1946, bringing on many architects and designers, including Eero Saarinen, Harry Bertoia, and Ralph Rapson, to design furniture for the company. When the famous-designer objects didn't fulfill the needs of a particular interiors project, Knoll designed the pieces herself, making them, as she said, "the meat and potatoes" of the collection. "Eero and Bertoia did the stars," she said, "and I did the fill-in." She was director of the Knoll Planning Unit, the interior design department responsible for the company's showroom and interior projects; it was also the engine of the company's corporate success and aesthetic achievement by bringing architecture, furniture, fabric, and spatial planning together, probably for the first time in American design culture. This credenza is an example of her work. With its clean, geometric forms that display their structural supports, it was used not only as storage, but also as a "gallery" on which to display art, sculpture, or plants. –PM

Promotional illustration for Knoll Associates' *Desk Designs* by Florence Knoll, c. 1962.

Pierre Paulin (1927–2009)

Tongue Chair, Model No. 577

Designed 1963

Tubular-steel, nylon-jersey-covered latex

24¼ x 33⅞ x 35¾ in. (61.6 x 86 x 90.8 cm)

Produced by Artifort, Maastricht, the Netherlands

K2013.155

Cooper Hewitt, Smithsonian Design Museum, promised gift of George R. Kravis II

PIERRE PAULIN FIRST BEGAN DESIGNING FURNITURE IN THE LATE 1950S FOR the Dutch producer Artifort. In 1963, supposedly inspired by the tight, stretchy effect of the fabric used in a new style of swimsuit that emerged at the time, Paulin designed the 577—the stackable *Tongue* chair—which was a kind of salon chaise lounge, or fireside chair, with a tubular steel frame and brightly colored upholstery. The chair, put into production by Artifort in 1967, was clad with inexpensive polyurethane foam. The frame is at once the body and the base of a supple, uncluttered monolithic structure. The simplicity of the removable nylon stretch-jersey fabric, which fit snugly and seamlessly around the chair's complex three-dimensional volumes, reduced production costs. In 1969, Paulin was commissioned to provide furniture for rooms in the Élysée Palace in Paris. Paulin included the *Tongue* chair with his designs for the French pavilion at the 1970 Osaka World's Fair. –PM

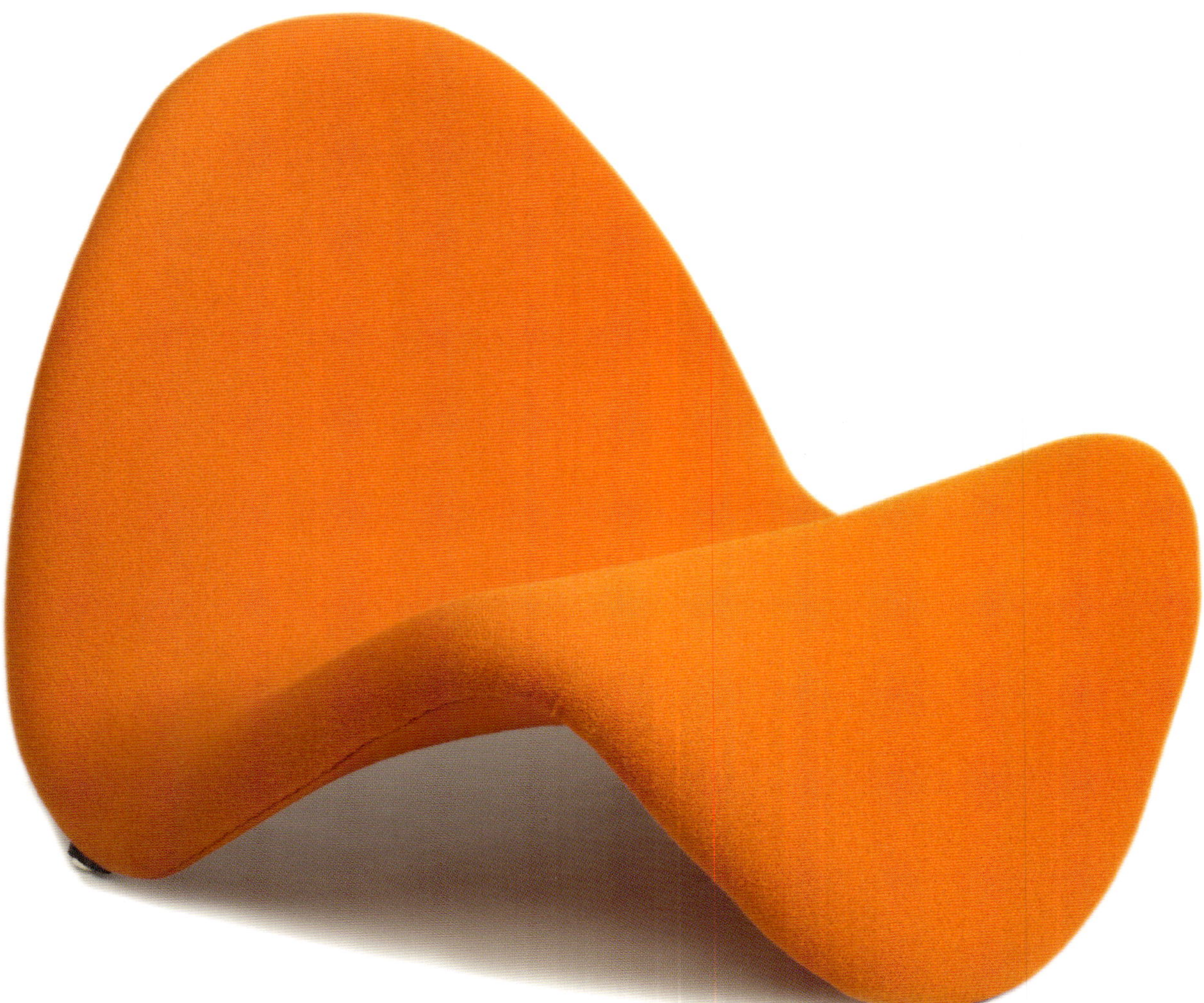

DDL Studio (active 1966–91):
Gionatan "Jonathan" de Pas (1932–1991)
Donato d'Urbino (born 1935)
Paolo Lomazzi (born 1936)

Blow Armchair
Designed 1967
Transparent PVC plastic
32⅝ x 43½ x 31½ in. (83 x 110 x 80 cm)
Produced by Zanotta, Nova Milanese, Italy
K2016.23

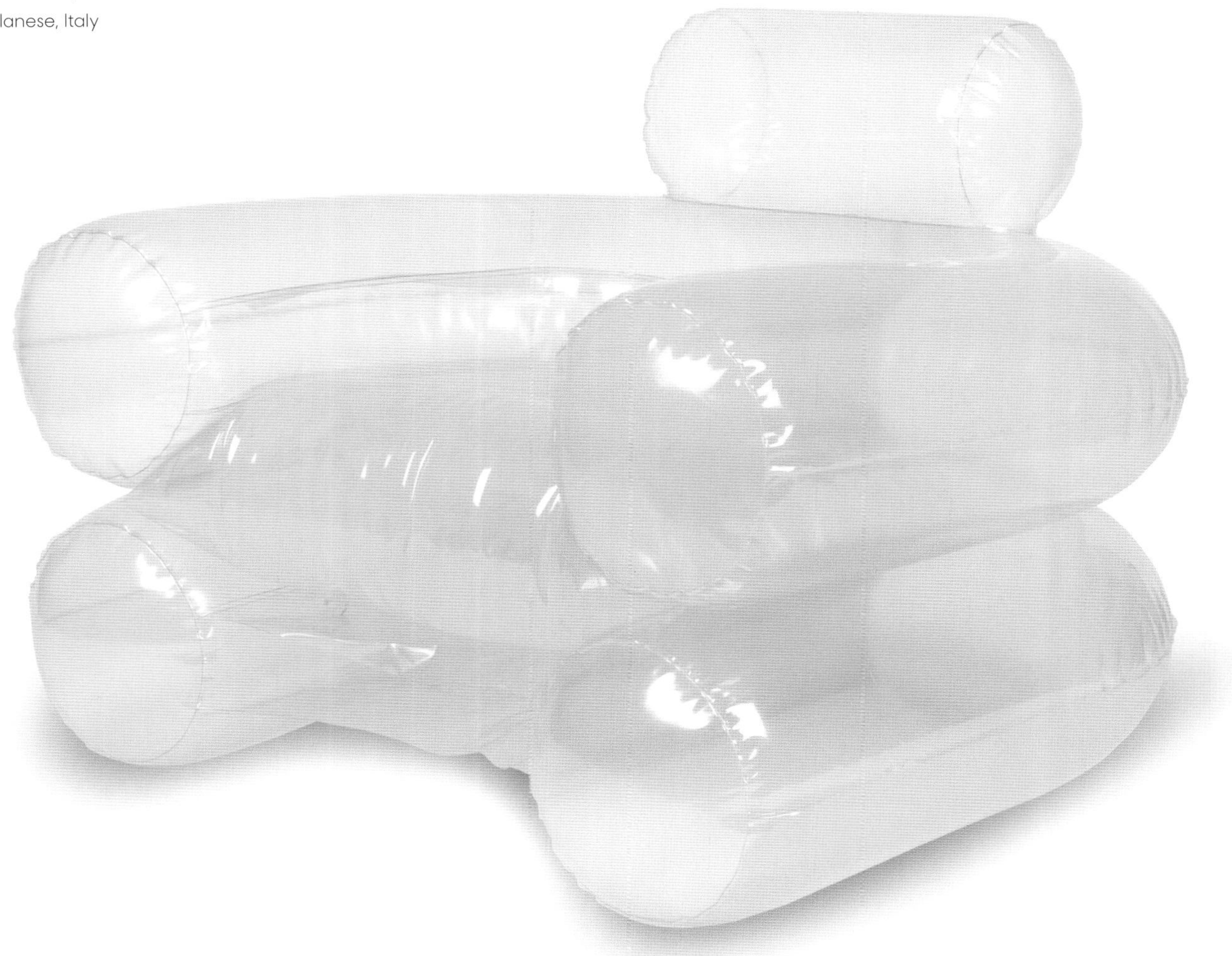

THE *BLOW* ARMCHAIR, AN ICON OF POP FURNITURE, WAS THE FIRST INFLATABLE CHAIR TO SUCCEED in commercial production. The chic design had no structural supports; without air, it was an empty plastic skin, easy to produce and easy to transport. Innovative though it was, *Blow* was not intended for long-term endurance—this example is accompanied by its original box and a repair kit to fix the anticipated leaks. Although it was marketed for indoor and outdoor use, the chair was somewhat impractical to have outside—too hot to sit on in the summer sun, and too light to hold its ground in a strong wind. The design suggests the appeal of informal, casual living that this chair exemplified in contrast to traditional upright wood furniture. The design was the first collaboration of DDL Studio in Milan, a firm that started in 1966 and aimed to design simple, inexpensive objects for a younger generation. This chair was included in MoMA's 1972 landmark exhibition *Italy: The New Domestic Landscape*, which introduced contemporary Italian design to the American public. –DAH

József Bottlik (1897–1984)

Orion Poster

Designed c. 1930

Offset lithograph

37¼ x 24⅛ in. (94.6 x 61.3 cm)

Produced by Bruchsteiner es Fia,
Budapest, Hungary

K2011.97

Cooper Hewitt, Smithsonian Design Museum,
promised gift of George R. Kravis II

JÓZSEF BOTTLIK WAS ONE OF THE MOST important poster designers in Hungary. His 1927 poster advertising the film *Metropolis* was his most famous and iconic design. From 1924 to 1965, he was the lead designer for the Hungarian electronics company Orion, which produced light and radio bulbs.[4] This Orion light bulb advertisement features the product box with orange rays of light radiating from each side of the bold-type Orion logo with a white light bulb above. A larger transparent light bulb containing the word ORION hovers at the top of the poster, superimposed over a large *S* curve. Orion was founded in 1913 in Budapest, and at some point in 1924 the firm was associated with the General Electric company in the United States.[5] –DAH

Ingo Maurer (born 1932)
Bulb Table Lamp
Designed 1966
Enameled steel, glass, metal
11⅜ x 7½ x 7½ in. (28.9 x 19.1 x 19.1 cm)
Produced by Design M (Ingo Maurer GmbH), Munich, Germany
K2013.20

IN 1966, INGO MAURER, A TYPOGRAPHER AND graphic designer who trained in Switzerland and Germany, designed this simple lighting fixture for an installation at the Herman Miller showroom in Munich. "I have always been fascinated by the light bulb because it is the perfect meeting of industry and poetry," says Maurer. "The bulb is my inspiration."[6] He designed this lamp to take a mirrored bulb to avoid glare. Maurer called his pop-art fixture *Bulb*, and he started his company, Design M (it later became Ingo Maurer GmbH), which put the lamp into production. His continuing work in lighting design for events and spaces explores the poetic underpinnings of form and materials, whether through the use of industrial wire mesh, low-halogen lighting, or crafted materials such as transparent rice paper or bird feathers. –PM

Timo Sarpaneva (1926–2006)
Claritas Vase
Designed 1983 (utility patent 4,367,086)
Glass
8⅛ x 7⅛ x 7⅛ in. (20.6 x 18.1 x 18.1 cm)
Produced by Iittala Lasitehdas, Iittala, Finland
K2012.130

IN 1950, FINNISH GLASS AND TEXTILE DESIGNER Timo Sarpaneva joined Iittala glassworks, where he initially worked in publicity and exhibition design and created Iittala's iconic logo. At the time, he was a recent graduate of the Institute of Industrial Art in Helsinki, where he studied graphic design. During the 1950s, he became known for the sculptural forms in glass that he designed for the company. In 1984, Sarpaneva designed the *Claritas* sculptural art vases, which are characterized by clear glass ovoid forms, flat on each end, with casings of black glass that have been cut, polished, and suspended inside the vase form. The black, opal, and clear glass pieces were hand-shaped using an innovative manufacturing technique that was a triumph of craftsmanship. Each vase was produced by first taking a bubble of air on the end of the blowpipe and then surrounding it with glass. The height and dimensions of the mouth-blown object varied depending on the glassblower, which meant that every piece in the series was essentially a unique work of art. The "eyes," shaped out of colored glass, are clear oval designs that were made using a steam-blowing technique—a method in which a metal spike is pushed into the surface of the molten glass and a water-soaked stick of wood is placed in the depression. The water in the wood evaporates and expands, creating bubbles of air, which become the eyes. –PM

Cesare "Joe" Colombo (1930–1971)
Gianni Colombo (1937–1993)
Acrilica Table Lamp, Model No. 281
Designed 1962
Perspex acrylic, lacquered steel
9⅛ x 9⅜ x 9½ in. (23.2 x 23.8 x 24.1 cm)
Produced by O-Luce, Milan, Italy
K2010.66

CESARE "JOE" COLOMBO AND HIS BROTHER GIANNI WERE BOTH ARTISTS, BUT JOE'S career as a painter ended in 1958 as his attention turned to architecture and design. From 1958 to 1968, the brothers had adjoining studios and worked together occasionally on such projects as this *Acrilica* table lamp. In a 1988 interview, Gianni recalled his brother's preference for the curve and his rejection of sharp corners and straight lines as he sought a "broader, freer, and more comprehensive vision of the world of forms."[7] A fluorescent bulb in the base of this lamp transmits light throughout the C-shaped curve of the acrylic shade. This simple, sweeping form reflects the Colombo brothers' austere, minimal aesthetic and won a gold medal at the 1964 Milan Triennale. –DAH

Stig Lindberg (1916–1982)
Five *Veckla* Vessels
Designed 1951
Glazed stoneware
Tallest: 14 x 6⅝ x 3¼ in.
(35.6 x 16.8 x 8.3 cm)
Produced by AB Gustavsberg, Gustavsberg, Sweden
K2013.66.1,2,4,6,7

THE INDUSTRIAL DESIGNER, CERAMICIST, AND ILLUSTRATOR STIG LINDBERG BEGAN HIS career as a ceramicist at Gustavsberg porcelain factory. In 1949, he succeeded William Kåge as artistic director, and went on to introduce several ceramic series such as the stylish *Pungo* (1953), the graphic *Domino* (1955), the *Terma* (1955), and a host of everyday ceramics. The Swedish *Veckla*—meaning "fold"—was part of the concept at Gustavsberg's Studio, an experimental studio established in 1942 by Lindberg to develop new designs and ensure the creative work at the factory. *Veckla* is typical of the designer's soft and sensual organic shapes—one that demonstrated a humanist approach to design. This particular collection of stoneware with white Carrara glaze was mass produced between 1950 and 1960. In 1957, Lindberg left his post as artistic director and became head teacher of ceramics at the University College of Arts, Crafts and Design, Konstfack, in Stockholm, until 1970. He returned to Gustavsberg in 1972 as artistic director, a position he held until 1980. He left behind a rich legacy of shapes, colors, and illustrations that are more playful, colorful, and humorous than is typical in the Scandinavian design idiom. —PM

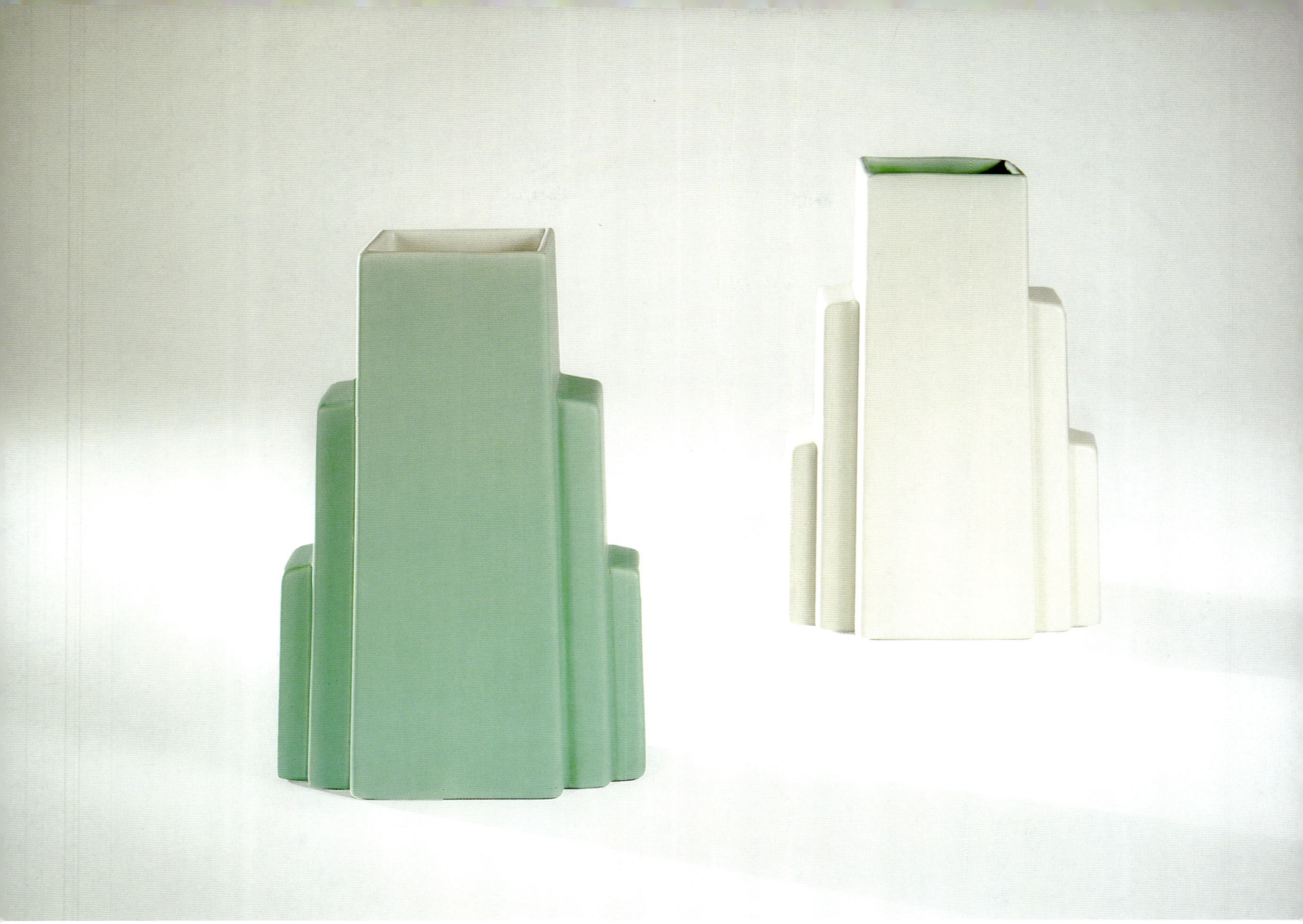

Designer Unknown

Two *Empire* Vases

Designed c. 1935

Glazed earthenware

9 x 6¼ x 3⅝ in. (22.9 x 15.9 x 9.2 cm) each

Produced by Coors Porcelain Company, Golden, Colorado

K2008.603.1-2

COORS POTTERY BEGAN IN 1915 WHEN THE STATE OF COLORADO ENACTED A statewide prohibition on alcohol, and Coors Brewing Company owner Adolph Coors recognized that he was going to need a new source of revenue. Five years earlier, Coors had helped John Herald start the Herald China and Pottery Company by providing the use of a building that he owned. The vast amounts of clay deposits found in the region were ideal for making pottery, and, when Herald decided to leave Golden in 1915, Coors's two sons, Adolph Jr. and Herman, took over the pottery company and began producing durable ceramics, including labware, dinnerware, and housewares. Herald China was renamed the Coors Porcelain Company in 1920, and continued operations even after Prohibition ended and the Coors Brewing Company reopened in 1933. This vase was most likely created during the mid-1930s at the height of the American art deco movement, when stepped skyscraper forms and motifs were popular. –PM

La Gardo Tackett (1911–1992)

Three Planters, Model Nos. TH-1, IN-1, and TH-2

Designed c. 1955

Glazed earthenware

TH-1: 8⅜ x 12⅛ x 12⅛ in. (21.3 x 30.8 x 30.8 cm)
IN-1: 8½ x 12½ x 12½ in. (21.6 x 31.8 x 31.8 cm)
TH-2: 21⅛ x 11 x 11 in. (53.7 x 27.9 x 27.9 cm)

Produced by Architectural Pottery, Los Angeles, California

K2011.206.1–2, K2011.207.2

Cover of *Architectural Pottery* catalogue, 1950.

IN 1949, WHILE TEACHING AT THE CALIFORNIA SCHOOL OF ART IN Los Angeles, La Gardo Tackett assigned graduate students the task of modernizing the design of indoor/outdoor planters. Students John Follis and Rex Goode started Architectural Pottery in 1950 to begin commercial production of the resulting geometric and biomorphic planters. According to a 2004 *Los Angeles Times* article, cofounder and business manager Rita Lawrence said she believed that "the pottery embraced the principles established by modernist architects—among them Richard Neutra and Rudolph Schindler—who sought to blur the lines between indoors and out. The pots 'provided a portable landscape and a focal point in garden plants, then carried the motif into the home and office.'"[8] Architectural Pottery's ceramics, including these large-scale cone forms, were the very picture of California mid-century modern. In 1953, Architectural Pottery added sand urns and a wide range of new designs to its catalogue, among them Tackett's inverted cone and his Trailblazer Award–winning House Glass. Over the years, the company employed a number of first-rate designers, including Paul McCobb, Raul Coronel, John Follis, Mary Kay Austin, Malcolm Leland, and Lawrence Halprin, among others. Architectural Pottery remained in business until it was destroyed by fire in 1985. –PM

James Harvey Crate (1925–2009)

Table Lamp, Model No. T-3-C

Designed 1951

Spun aluminum, enameled aluminum, stainless steel, cork

25 x 12 x 15 in. (63.5 x 30.5 x 38.1 cm)

Produced by Heifetz Manufacturing Company,
New York, New York

K2008.269

Cooper Hewitt, Smithsonian Design Museum,
promised gift of George R. Kravis II

THIS LAMP RECEIVED THIRD PRIZE (OUT OF ALMOST three thousand designs) at the 1951 Low Cost Lighting competition, sponsored by MoMA and the Heifetz Manufacturing Company, a New York lighting manufacturer founded in 1938. The competition was inspired by Marcel Breuer's comment that he could not find a single well-designed floor or table lamp for MoMA's 1949 exhibition *The House in the Museum Garden*; Breuer was among the judges. Ten of the winning designs, including this lamp, were produced by Heifetz and sold through stores that acted as co-sponsors of the exhibition and through local museums that showed the prize winners in a circulating exhibition.

The lamp had a design that was described in a MoMA press release as "a trend towards lightness of appearance, unusual flexibility in the control of light, and multiplicity of use."[9] This innovative atomic-style lamp perches lightly on cork-ball feet, and the three legs slide to adjust the angle of the light reflector, thereby controlling and customizing the degree of light reflected on the flat metal shade. Crate, a twenty-five-year-old relatively unknown designer in the styling division at General Motors, fit the profile of the typical contributor to the competition. This lamp represented a strong attempt at modern styling in lighting, a new and growing field at the time of the competition. –PM

Christian Dell (1893–1974)

Rondella Table Lamp

Designed 1928

Copper-plated brass, nickel-plated tubular steel, Bakelite

As shown: 18⅞ x 16¾ x 8½ in. (47.9 x 42.6 x 21.6 cm)

Produced by Rondella, Oberursel, Germany

K2013.133

CHRISTIAN DELL WORKED AT THE BAUHAUS METAL WORKSHOP IN WEIMAR FROM 1922 TO 1925 under the artistic director László Moholy-Nagy. He headed the metal workshop at the Frankfurt Art School from 1926 to 1933, when he was dismissed by the Nazi Party.[10] Beginning in the late 1920s, in addition to teaching and working as a distinguished silversmith, he designed lighting for mass production, including the *Rondella* table lamp.[11] The *Rondella* series of lamps was characterized by a parabolic reflector shade and a ball-and-socket joint to allow the lamp head to swivel. The circular weighted base and tubular-steel shaft were stationary, supporting a separate tubular *L*-shaped arm that was attached to the base shaft by a clamp for adjusting the height and attached to the shade by a spherical swivel joint. –DAH

left

Louis C. Kalff (1897–1976)

Desk Lamp, Model No. NB100

Designed 1957

Painted and enameled metal

13⅝ x 13⅛ x 11⅞ in. (34.6 x 33.3 x 30.2 cm)

Produced by Philips, Eindhoven, the Netherlands

K2008.267

center

Richard Kelly (1910–1977)

Table Lamp

Designed c. 1940

Aluminum, walnut, paper

15 x 15⅛ x 15⅛ in. (38.1 x 38.4 x 38.4 cm)

Produced by Kelly & Thompson, New York, New York

K2014.4

right

Max Bill (1908–1994)

IV-5-Sun Sunlamp

Designed 1951

Enameled and chromium-plated metal

As shown: 13¾ x 16 x 6 in. (34.9 x 40.6 x 15.2 cm)

Produced by Noveletric AG, Zürich, Switzerland

K2013.18

ARCHITECT LOUIS CHRISTIAAN KALFF CREATED Model No. NB100 lamp for the Dutch lighting company Philips in 1957 as a minimalist object that illuminated both the desk below it and, through the use of a circular cut-out in the aluminum hood, the room above it.

The pioneer American modernist lighting designer Richard Kelly's aluminum and walnut table lamp, with its ovoid base, is emblematic of a stripped-down, streamlined style that is typical of modernist objects of the late 1930s and early 1940s.

Swiss artist and designer Max Bill applied a clarity of design and precise proportions to objects such as the *IV-5-Sun* sunlamp, which was created by joining two cones of painted gray sheet metal with a chromed flexible metal shaft, resulting in a purely sculptural yet functional object. –PM

02
objects

HOUSEKEEPING

NOT ONLY IS HOME A HAVEN FROM THE OUTSIDE WORLD, IT IS ALSO a site where work takes place. There are many tasks that are undertaken in this arena—cleaning, vacuuming, polishing, washing—and they require a set of tools, utensils, and machines to facilitate them. Throughout most of history, these tools and machines had very basic forms, and it was sufficient that they accomplished no more than the job at hand. In the late nineteenth century, housekeeping tools started to become both more efficient and more beautiful to look at, and as a result the housewife's chores became both lighter and less arduous. New materials—brightly colored plastics among them—were harnessed to those ends, and hand-tools were transformed, through electricity, into glamorous household machines.

Henry Dreyfuss (1904–1972)
Roy H. Zinkil (1890–1968)

Neuvogue Sink

Designed 1936 (D101,446)

Vitreous porcelain, chromium-plated steel

32 x 26½ x 21 in. (81.3 x 67.3 x 53.3 cm)

Produced by Trenton Potteries Company
for Crane Co., Chicago, Illinois

K2013.51

Cooper Hewitt, Smithsonian Design Museum,
promised gift of George R. Kravis II

Neuvogue bathroom from the brochure *Crane Plumbing and Heating for the Modern Home*, Crane Co., 1940.

FOR THE *NEUVOGUE* LINE, DESIGNED BY HENRY DREYFUSS IN 1936, THE CHICAGO FIRM Crane Co. produced a harmonious and streamlined ensemble of sink, tub, and toilet that was unified by color, materials, rounded corners, and horizontal banding, as seen in the illustration from the Crane brochure. The sink provided easily cleaned shelf space around a semicircular basin. The water and waste controls were easily adjusted. Several types of bases and accessories were designed for the sink; this example has a simple vitreous porcelain pedestal and a pair of chromium-plated, strap-steel towel racks suspended from the underside of the sink. Crane presented these fixtures as part of the "Bathroom of Tomorrow." The line continued in production through the mid-1940s. The name *Neuvogue* suggests the high style of French design combined with German functionalism. Founded in 1855 by Richard Teller Crane, Crane Co. continues to be associated with the best quality in bathroom fixtures, among thousands of other products, to the present day. –DAH

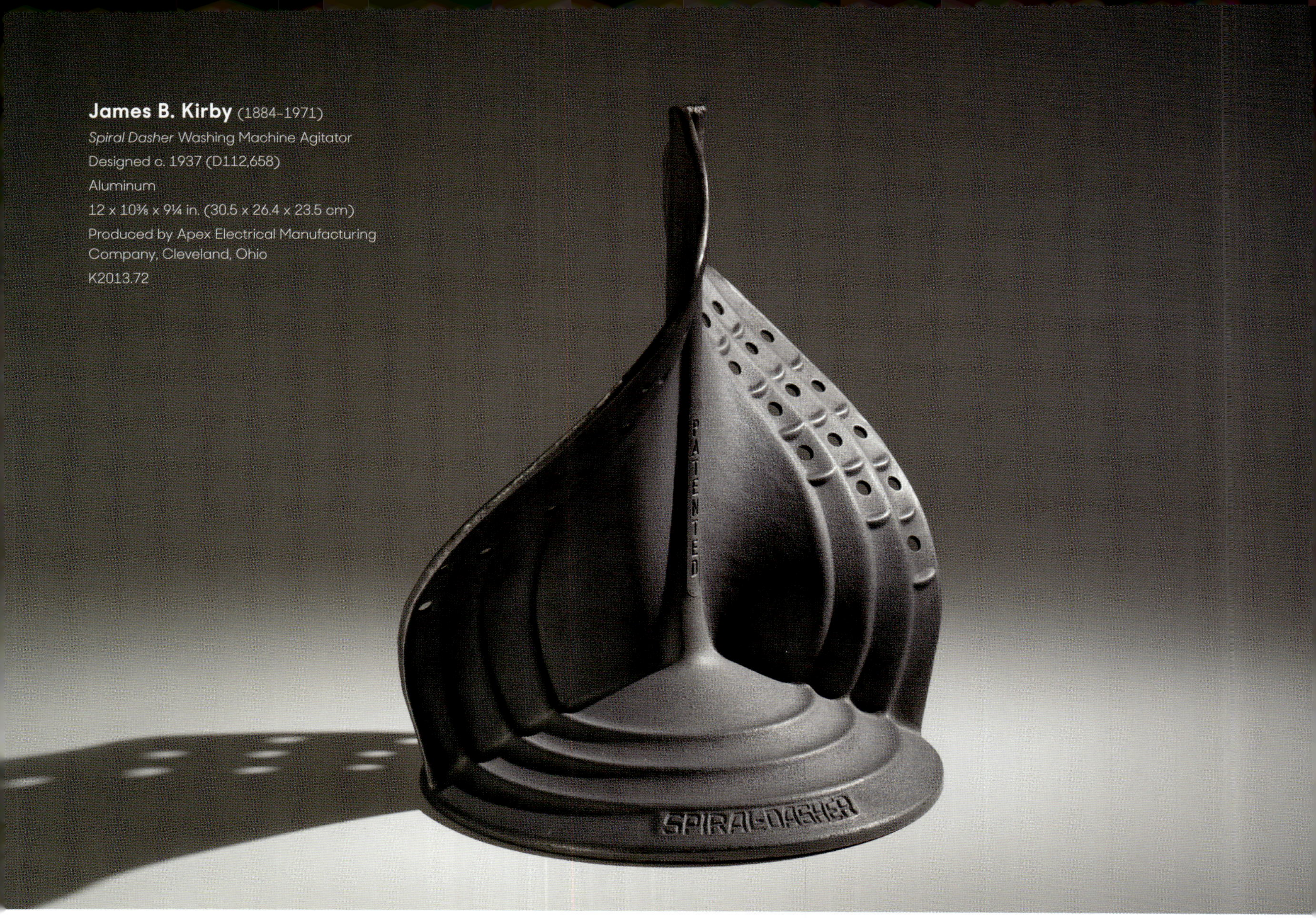

James B. Kirby (1884–1971)
Spiral Dasher Washing Machine Agitator
Designed c. 1937 (D112,658)
Aluminum
12 x 10⅜ x 9¼ in. (30.5 x 26.4 x 23.5 cm)
Produced by Apex Electrical Manufacturing Company, Cleveland, Ohio
K2013.72

THE DESIGN PATENT DRAWING FOR THIS AGITATOR PROVIDES THE NAME OF THE designer: Scottish inventor James B. Kirby, who was famous for his *Domestic Cyclone* vacuum cleaner in 1906.[1] Separated from its functional mechanism as part of the washing machine seen in the advertisement, the *Spiral Dasher* agitator has a sculptural presence, its stepped conical base supporting perforated wings. Although the design patent lists a tri-blade agitator, the Kravis agitator is a twin blade. The utility patent specifications describe its mechanical function: "corrugations which are formed on the blades of the agitator rub and scrub the clothes . . . and the perforations assist in causing the flow of the washing fluid through the pores of the fabric . . . while at the same time they cause such goods to be drawn into close contact with the blades."[2] The advertisement expresses, perhaps more eloquently, how the agitator cleaned every surface of the fabric "with 'gentle-hand' action" and "tangle-free circular motion" at sixty-two strokes per minute.[3] –DAH

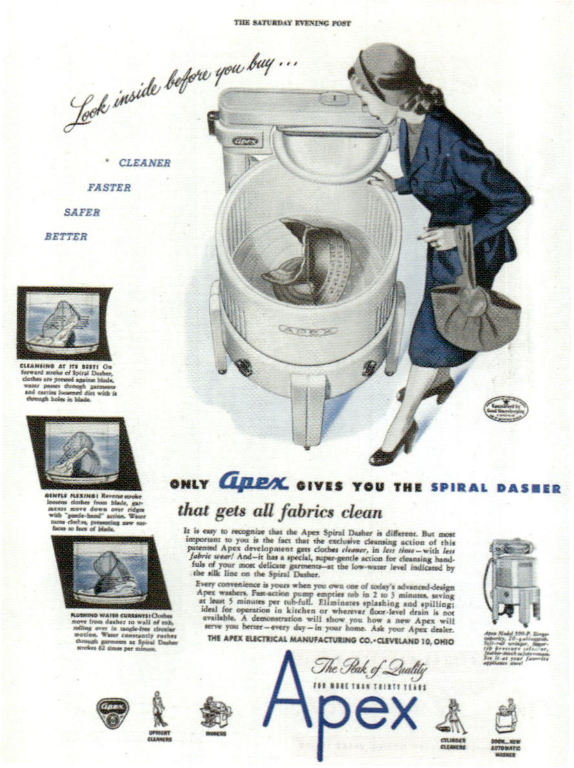

right Advertisement for the Apex *Spiral Dasher*, published in the *Saturday Evening Post*, May 10, 1947.

William Edward Gundelfinger (1900–1976)

K-M Flat Work Ironer Iron, Model No. 444

Designed c. 1939 (D116,161)

Chromium-plated steel, Bakelite

5 x 7⅜ x 7¾ in. (12.7 x 18.7 x 19.7 cm)

Produced by Knapp-Monarch Co.,
St. Louis, Missouri

K2012.205

IRONS HAVE ALWAYS BEEN TEARDROP IN SHAPE AND FUNCTIONALLY STREAMLINED, yet Knapp-Monarch promoted the innovation of this unusual flat, circular form. "Have you heard about this amazing new iron?" the accompanying brochure asks.[4] The iron sold for $5.95 and was designed specifically to make ironing flat work, such as sheets, napkins, or table cloths, "faster and smoother" through its circular, omnidirectional gliding motion. A semicircular dial on top of the iron adjusts the temperature. One of the stand-out features of the iron is the concentric rail surrounding the base. Designer William Edward Gundelfinger submitted a utility patent application in 1938—nine months before the final design was submitted to the patent office—for the "Sadiron Rest," which had a metal rail that encircled but was separate from direct contact with the heated base. The railing stayed cool, and the user could flip the iron to balance on the railing and the handle, preventing table burns when the iron was still hot but not in use.[5] The Knapp-Monarch Company was formed in 1929 through a merger of the Monarch Company of St. Louis and A. S. Knapp in Webster City, Iowa. —DAH

left

Clifford Brooks Stevens (1911–1995)
Edward P. Schreyer (1896–1981)

Petipoint Iron, Model No. W410

Designed c. 1941 (D128,267–269)

Chromium-plated steel, steel, Bakelite

5 x 10 x 5 in. (12.7 x 25.2 x 12.7 cm)

Produced by Waverly Tool Company, Sandusky, Ohio

K2011.15

center

Ralph E. Kruck (1906–1994)
William J. Russell (1894–1977)

Electric Iron, Model No. ID-72

Designed c. 1949 (D127,506; D158,295)

Chromium-plated steel, Bakelite, steel, plastic

9 x 4½ x 4½ in. (22.9 x 11.4 x 11.4 cm)

Produced by Westinghouse, Mansfield, Ohio

K2010.249

right

Frank Kuhn (1891–1975)
Laurence H. Thomas (1887–1962)

American Beauty Iron, Model No. 79AB

Designed c. 1940 (utility patent 2,235,891)

Chromium-plated steel, Bakelite, steel, plastic

4¾ x 9¼ x 4¾ in. (12.1 x 24 x 12.1 cm)

Produced by American Electrical Heater Co., Detroit, Michigan

K2010.262

THESE IRONS ARE ALL STREAMLINED, SUGGESTING THE SPEED AND EASE of the chore of clothes-pressing. Earlier irons—so called because they actually were made of iron—were heavier. Such older irons were heated on a stove and used in pairs—one being reheated while the other was in use. The modern electric pressing iron of the 1930s, its shape still recognizable today, was often made of chromium-plated steel with Bakelite handles. In their book *Art and the Machine*, the Cheneys compared a Westinghouse iron to a *Fast Commuter* train designed by Raymond Loewy. Both had "simplification, plastic order and functional expression yielding like results."[6] The *Petipoint* iron, with its horizontal speed lines and finlike side vents, is a dramatic streamlined design. The *American Beauty* is also visually compelling. In the canted handle, the use of colored plastic adds a striking contrast to the dark Bakelite, and the wedge shape under the back adds to the forward movement of the aerodynamic form. The Bakelite handle of the Westinghouse iron is integral to the body, adding to the sculptural quality while the conforming white plastic provides a stylish contrast. —DAH

Raymond Loewy's *Fast Commuter* railway car compared with a Westinghouse iron, from *Art and the Machine* by Sheldon and Martha Cheney, published by Whittlesey, New York, 1936.

Robert A. Allen (1879–1912)

Vacuum Cleaner

Designed c. 1908 (utility patent 985,765)

Aluminum, wood, steel, brass

52 x 10⅜ x 10½ in. (132x 26.4 x 26.7 cm)

Produced by Allen Vacuum Cleaner Co., Chicago, Illinois

K2012.32

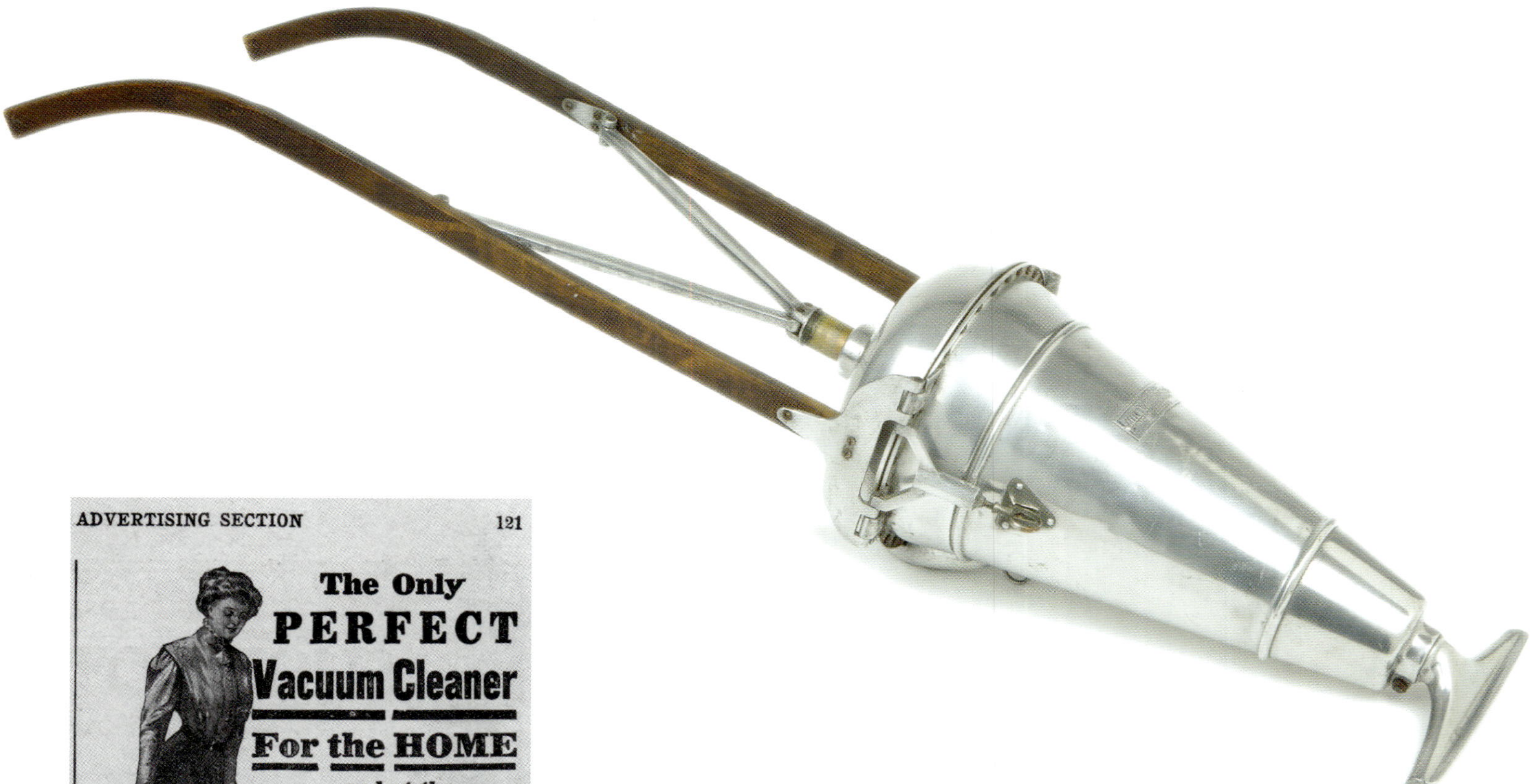

ADVERTISING SECTION 121

The Only PERFECT Vacuum Cleaner For the HOME and at the Right Price

At Last!

"So Easy—So Very Easy"

The PERFECT Vacuum Cleaner for the Home! No electricity—no pumping—takes only one person to operate—as light as a carpet sweeper and as easy to use. The easiest and most thorough known method of cleaning—that is the way of the ALLEN Vacuum Cleaner.

IN AN AGE WHEN ELECTRICITY IN SUBURBAN HOMES WAS COMMON BUT NOT YET universal, the electric vacuum cleaner was still a novel idea. This manual vacuum cleaner, referred to in its patent application as a "dust collecting apparatus"[7] but commonly called a plunger, provided the opportunity for nonelectric mechanical carpet cleaning—and a bit of aerobic exercise, as well. The user would pump the handles, which, much like a bellows, created suction when pulled apart. "Dust was sucked into the body and was trapped in a funnel-like device" with a cloth air filter to prevent the dust from being blown back out.[8] Its space-age appearance belies its early twentieth-century design. Other vacuum manufacturers produced plungers, but the version produced by Allen is noteworthy for its aesthetic appeal. —DAH

left Detail of advertisement for the Robert Allen vacuum cleaner, published in *Popular Mechanics*, November 1909.

John Crausaz (1907–1992)

Tornado Hand Vacuum

Designed c. 1951 (Swiss patent CH288698, French patent FR1047338)

Aluminum, canvas, steel, cardboard, plastic, felt

9¼ x 32 x 6¾ in. (23.5 x 81.3 x 17.1 cm)

Produced by Tornado France, Paris, France

K2016.14

THE HANDHELD TORNADO VACUUM WAS USEFUL FOR CLEANING SMALL SPACES. A period advertisement for Tornado France touted, in French, its lightness, ease of use, quietness, and sturdiness.[9] It could be used with a long or short handle that provided ease of access to more difficult areas and corners. The aerodynamic form of the *Tornado* implies futuristic innovation, with its motor encased in a zeppelin-shaped aluminum capsule with ribbing that suggests speed lines characteristic of streamlining. –DAH

far left

Lurelle Guild (1898–1986)

Vacuum Cleaner, Model No. 30

Designed c. 1937 (D106,662)

Aluminum, chromium-plated and enameled steel, steel, vinyl, plastic, rubber

9⅝ x 23¼ x 7½ in.
(24.5 x 59.1 x 19.1 cm)

Produced by Electrolux Corporation, Old Greenwich, Connecticut

K2008.282

top center

Designer Unknown

Vacuum Cleaner, Type 852

Designed 1930–50

Bakelite, vinyl, chromium-plated steel, steel

7½ x 22½ x 7½ in.
(19.1 x 57.2 x 19.1 cm)

Produced by Electrostar, Stuttgart, Germany

K2008.284

bottom center

Designer Unknown

Vacuum Cleaner, Type E 10

Designed c. 1930

Chromium-plated and enameled steel, plastic

9⅛ x 22¼ x 8¼ in.
(23.2 x 56.5 x 21 cm)

Produced by Miele, Bielefeld, Germany

K2008.285

far right

Kenneth P. Schory (1920–1991)

Royal Vacuum Cleaner

Designed c. 1949 (D158,134)

Chromium-plated and enameled steel, plastic

9¼ x 24⅛ x 7⅛ in.
(23.5 x 61.3 x 18.1 cm)

Produced by P. A. Geier Co., Cleveland, Ohio

K2012.31

BETWEEN 1937 AND 1938, LURELLE GUILD SUBMITTED A SERIES OF designs to Electrolux for a new vacuum cleaner that called for a low, horizontal case that was inspired by the design of trains. Bent steel runners beneath the body lifted it off the ground, allowing the vacuum to be dragged easily across carpeted floors. His iconic design embodies the era's fascination with speed and efficiency and incorporated new materials such as chrome. This design, with later variations, including the substitution of wheels for the gliders, remained in production for more than twenty years.

The exotic Electrostar vacuum sports a double set of speed lines at the base and is covered with vinyl imitating alligator skin, which contrasts with the dark brown Bakelite at both ends. In place of wheels, the Miele vacuum utilizes glides, like the Guild design for Electrolux. Torpedo in form, the Miele design reverses the color combination of the Electrostar with the dark brown in the middle section and chrome at each end. The *Royal* vacuum, also torpedo shaped, was referred to as the "Dirt Devil," evoked by its bright red color. The company changed its name to Royal Appliance Manufacturing Co. in 1953. For both the Miele and the Royal, the company brand is highlighted in script. Founded in 1899, Miele continues in business today. –PM

James Dyson (born 1947)
with
Alexander Stuart Knox (born 1967)
Nicholas Alexander Evans (born 1977)
DC25 Blueprint, Limited Edition Vacuum
Designed c. 2008 (D596,360)
Plastic
42¼ x 12¼ x 15¼ in. (107.3 x 31.1 x 38.7 cm)
Produced by Dyson, Malmesbury, England
K2008.143

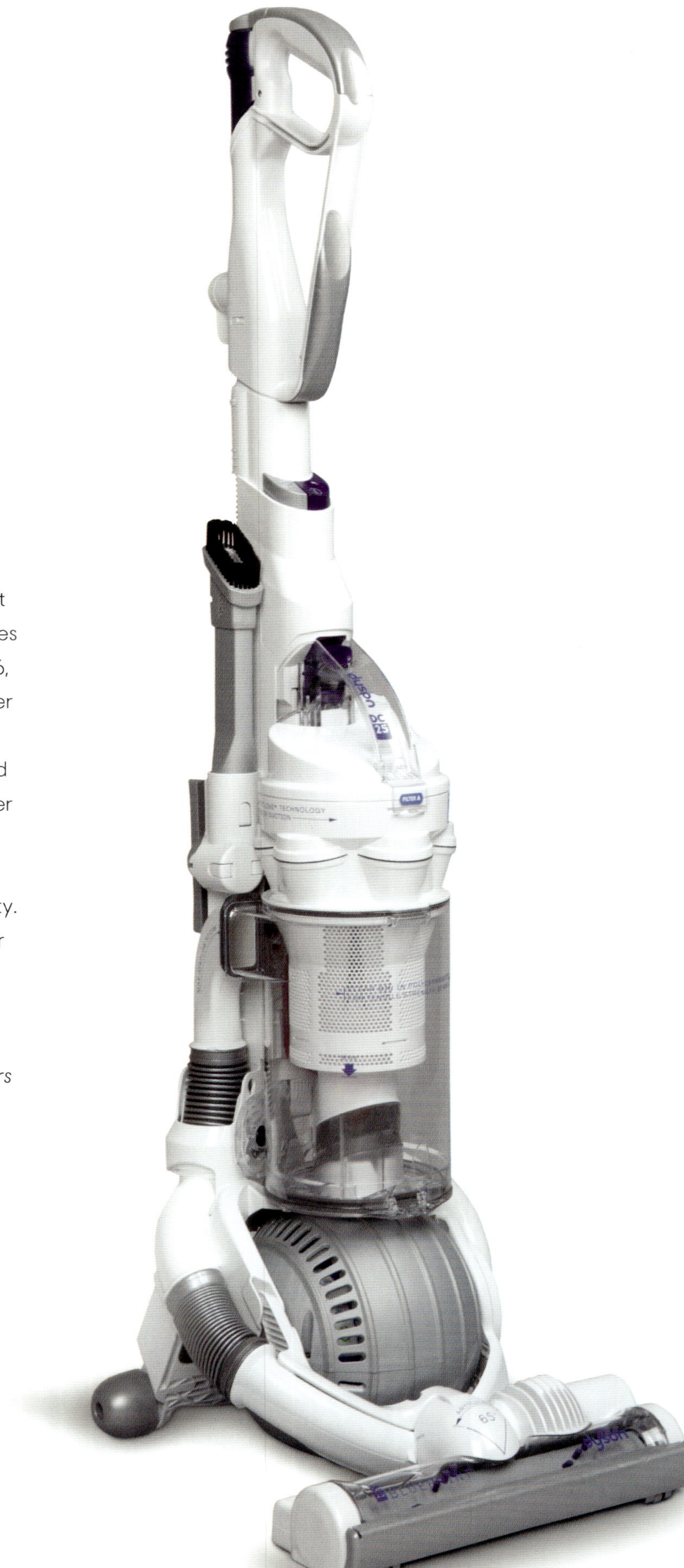

INDUSTRIAL DESIGNER AND INVENTOR JAMES DYSON WAS inspired by industrial cyclones that sucked up sawdust at a local sawmill in his design for his first vacuum cleaner. It took fifteen years and more than five thousand prototypes to devise his renowned bagless vacuum cleaner. By 1986, he launched the first Dyson vacuum, the DC01. Alexander Stuart Knox joined Dyson in 1995, working as a senior designer on most Dyson products. Nicholas Evans worked at Dyson from 2000 to 2008 and was the design manager for the DC25 project. The DC25 vacuum, pictured here, is a full-size ultra-lightweight and compact upright ball vacuum, which rides on a ball for greater maneuverability. It is lightweight, at sixteen pounds, which is convenient for those who may have difficulty maneuvering a heavier vacuum cleaner. This vacuum also steers smoothly and easily with a slight turn of the wrist, and there are no costly bags to replace. It comes in several colors: *All Floors* (yellow), *Overdrive* (blue), *Blitz It* (red), *Exclusive* (silver), *Blueprint* (white and blue), and *Animal* (purple), which was designed especially for pet owners. –PM

Pia Wallén (born 1957)

Ultrasilencer Special Edition
Vacuum Cleaner

Designed 2007

Plastic, metal

Canister: 10¼ x 15⅝ x 12⅛ in.
(26 x 39.7 x 30.8 cm)

Produced by Electrolux, Stockholm, Sweden

K2008.62

ELECTROLUX GAVE PIA WALLÉN COMPLETE FREEDOM TO CREATE HER OWN interpretation of the *Ultrasilencer*. She observed that "the special quality with this vacuum cleaner is its extraordinary silence, so I drew my inspiration from the most beautiful sound I know—that from falling snow." The vacuum cleaner body, including the insides and its packaging, are all white with bold orange detailing found around the mouth of the vacuum head and the power cord, "inspired by the sticks signaling caution when driving in snowfall on the Swedish winter highway."[10] Electrolux produced only five thousand units of the *Ultrasilencer*, the first special limited-edition product the company made in collaboration with a designer. –PM

Gino Colombini (1915–2011)
Wastebasket, Model No. 4670
Designed 1966
ABS Plastic
15 x 9¾ x 9¾ in. (38.1 x 24.8 x 24.8 cm)
Produced by Kartell, Milan, Italy
K2008.629

GINO COLOMBINI WAS AN ITALIAN ARCHITECT and industrial designer who, from 1953 to 1960, was the technical director of Kartell, an innovative plastics manufacturer of beautiful objects for the home—a distinct and colorful contrast to most typical, everyday objects produced in postwar Italy. Colombini designed many household products for the company that were revolutionary and affordable examples of everyday items. One of his most famous designs is a cylindrical plastic umbrella stand with a steel ashtray top (1965). His multipurpose round wastebasket, Model No. 4670, designed in 1966 and available in a range of colors, was stylish, minimalist, and practical. In 1968, George Beylerian of Beylerian Ltd. became the sole American agent for Kartell and sold their products at retail, as well as in his showroom, Scarabaeus. In addition to Kartell and Colombini's names on the bottom, this example is marked "by Beylerian Ltd." —PM

above Promotional photo of Kartell wastebaskets, showing the range of colors.

Ole Jensen (born 1958)
Dustpan and Brush
Designed c. 2002
Polypropylene, beech, natural bristles
Dustpan: 3½ x 11¾ x 9⅜ in. (8.9 x 29.8 x 23.8 cm)
Produced by Normann Copenhagen,
Copenhagen, Denmark
K2017.7.1ab

THE IDEA OF COMBINING A DUSTPAN AND BRUSH IS NOT NEW, BUT JENSEN'S CONTEMPORARY version provides the advantages of an extremely thin pan for brushing in crumbs in a shape that is comfortable to hold. According to the manufacturer's website, it was "inspired by a folded newspaper," which Jensen often used to catch the dust and dirt he swept up in his workshop. Jensen's method of design assigns great importance to materials: "I only trust a particular form when I have touched it with my fingers. I am able to communicate with the manufacturer as well as the end user through the particular object, and I see myself as a fertile minimalist. Everything counts. Neither more nor less! I aim at finding the uniqueness in the simple design and vice versa."[11] The simple design—a single sheet of plastic cut and joined with a simple cut-out circle near the seamed edge to hold the brush—makes manufacturing easy and inexpensive. The set comes in several colors including red and light blue. –DAH

Ole Jensen (born 1958)
Washing-Up Bowl and Brush
Designed 2002
Rubber, beech, natural bristles
Bowl: 5⅝ x 12¼ x 12¼ in.
(14.3 x 31.2 x 31.2 cm)
Produced by Normann Copenhagen,
Copenhagen, Denmark
K2007.72.4

DANISH DESIGNER OLE JENSEN HAD THE IDEA FOR THIS WASHING BOWL AND BRUSH while doing his own dishes at home. He wanted to create a flexible, durable "bowl" to better accommodate breakables, like glass and ceramics, as well as oddly shaped items that did not fit well within the rigid walls of his steel kitchen sink. Originally trained as a ceramicist, Jensen used his potter's wheel to shape the first prototype by hand, and decided that rubber would make the best material. The resulting *Washing-Up* bowl—available in red, gray, or black—is both whimsical and practical. It is easy to clean, easy to store, and kind to your dishes. The accompanying *Washing-Up* brush is made of wood and Chinese pig bristles, using traditional production techniques. In 2004, the *Washing-Up* bowl was used in the restaurant at MoMA as a wine cooler as part of the Danish Design Project. –PM

Marc Newson (born 1963)
Dish Doctor Dish Rack
Designed 1997
Polypropylene
4¼ x 18¼ x 15¾ in. (10.8 x 46.4 x 40 cm)
Produced by Magis, Motta di Livenza, Italy
K2008.313.2

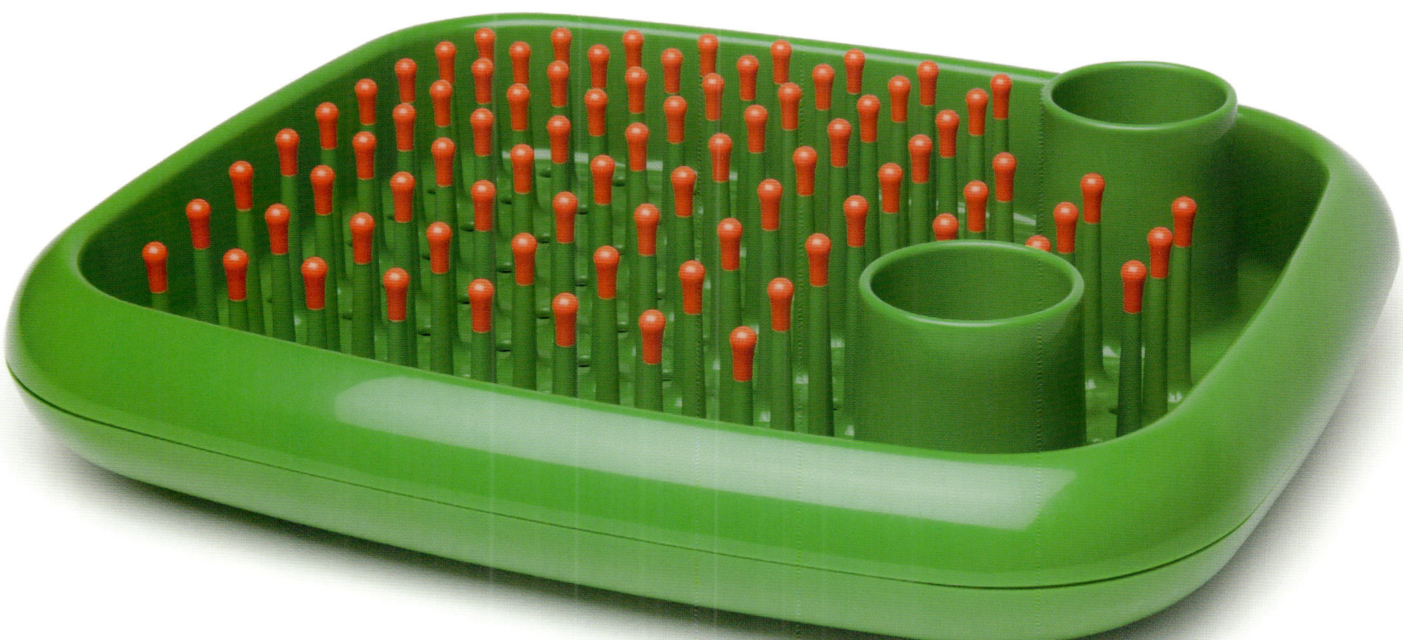

FUNCTIONAL, DURABLE, AND SPACE SAVING, MARC NEWSON'S *DISH DOCTOR* IS A DRYING rack with a deep tray and two cups for dishes and utensils. Made of injection-molded polypropylene by the Italian manufacturer Magis, the dish rack has an overall lozenge shape with rounded sides and corners. Within the enclosure are dozens of prongs with rounded tips for propping up plates and other dishes to dry, as well as two molded cups for utensils. Water that drips into the rack drains through numerous holes to the removable fitted tray below. The dish rack is available in green with orange prong tips, orange with green, and white with blue. Typical of Newson's aesthetic and styling, it transforms the dish rack into a serious object through the use of innovative plastics technology and has the designer's signature curved form, bright colorways, and quirky detail, recalling the "retro-future" space-age utopia of *The Jetsons*, the early 1960s animated cartoon about a family who lived high above the Earth's surface and zipped around in flying cars. –PM

MATERIALS AND TECHNOLOGY

by Cara McCarty

THE NEW MATERIALS AND TECHNOLOGIES OF THE MODERN WORLD FIRST BEGAN TO MAKE THEIR presence known in the midst of complex and rapid change in the nineteenth century. The most massive display of new materials and technologies appeared in a public pavilion of modular plate-glass and cast iron: Joseph Paxton's Crystal Palace, which was designed for the Great Exhibition of the Works of Industry of All Nations in London in 1851. Also around that time in London, the first design of an industrial train system that would move masses of people underground safely and rapidly throughout the city and beyond emerged. That project is now fundamental to every large urban transportation system. Perhaps less monumental in scope, and certainly in public perception, was the advent of the house as personal property. This concept arose through the invention by British bankers and builders of the residential mortgage, which gave everyday citizens access to secure, long-term loans and allowed them to engage in their largest lifetime financial investment. Not only was this the beginning of what came to be known as suburbia, but the kind of house and interior that began to evolve as a result attracted the attention of modern industrial designers and became a new platform on which to develop their ideas, skills, and talents.

Modern industrial design emerged as a consequence of a rising manufacturing culture that saw in economies of scale the rapid production of identical items for the expanding middle class market. New manufacturing technologies, materials, and optimization of work were needed to sustain mass production, as was an environment of good communication and distribution, in which advertising and packaging flourished. Nineteenth-century innovations in the iron foundry processes, particularly the cast iron industry, were critical for industrialization and such engineering feats as railways, bridges, and architectural monuments and also led to the precisely executed cast iron household wares and decorative objects that were now within reach of the growing consumer culture.[1] Coupled with these innovations and the proliferating options of goods, speed was essential for the manufacture of mass produced products. By the 1930s, fascination with the glamour and dynamism of speed itself was visibly celebrated in static objects whose rounded corners, smooth surfaces, and excessively elongated forms were described as "streamlined" or "aerodynamic."

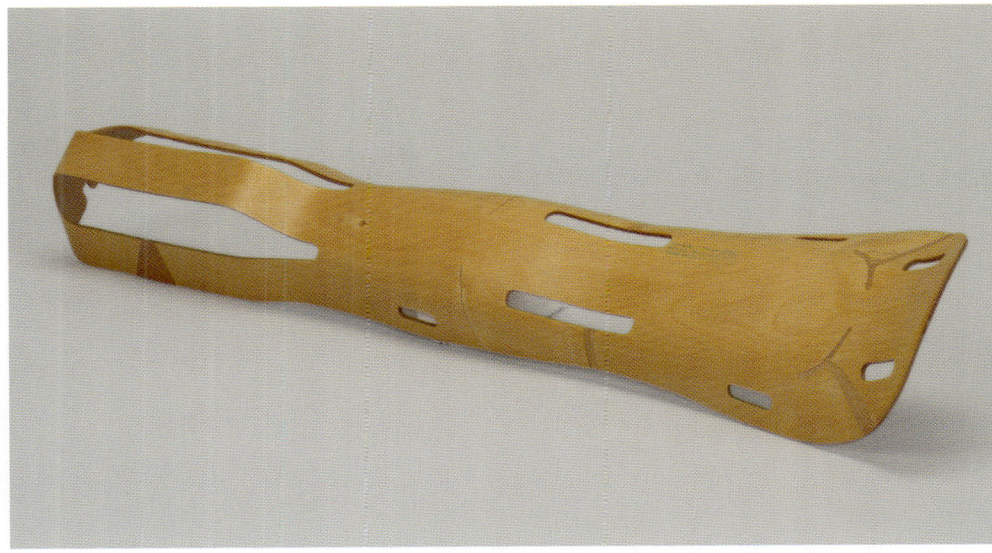

FIG 1
above left Thonet factory workers bending wood for furniture, late nineteenth century.

FIG 2
above right Leg splint designed by Charles and Ray Eames, 1942. Cooper Hewitt, Smithsonian Design Museum, gift of Barry Friedman and Patricia Pastor.

As early as the 1850s, when London was manifesting the phenomena of modernism, Michael Thonet in Germany was pioneering the new technology of bentwood furniture by steaming rods of wood and bending them around formers to reduce the number of parts and joints required, thereby dramatically increasing the speed at which they could be produced (fig. 1). At one point he was manufacturing five thousand pieces of furniture per day. Bending rods of wood was a precursor to the bent tubular steel furniture of Marcel Breuer and Ludwig Mies van der Rohe in the 1920s, and the transparent quality of their minimal, diagrammatic designs enhanced the free-flowing spaces of modern interiors. By the mid-century, through experiments manipulating sheets of plywood, Alvar Aalto and Charles and Ray Eames created several of the most iconic furniture designs of the twentieth century, emphasizing the tactile appeal of wood and the idea that there is equal value to progress in the transformation of common materials as there is in new materials. The Eames' fascination with new technology, structures, and materials was further galvanized by their wartime experiments with bent plywood (fig. 2), which they extended in the 1940s and 1950s to molded fiberglass-reinforced plastic, aluminum, and foams to achieve new designs with inviting organic contours.

Science delivered the two primary new materials for modern design: aluminum and plastics. Aluminum being strong and malleable, but lighter than steel, can be made into thin wire or rolled into foil, alloyed with other materials, and colored through anodizing. In the mid-1800s, when it first came into use, it was worth more than gold and thus restricted to luxury products such as exquisite jewelry. Its broad acceptance in the marketplace did not occur until after research by three scientists—the American Charles Martin Hall, the Frenchman Paul Héroult, and the Austrian Carl Josef Bayer—was completed in 1888, independently of one another, leading to an easier way to extract aluminum and thereby a reduction in cost. As a lustrous alternative to silver and pewter, aluminum's numerous qualities contributed to its particular appeal. By the 1920s, aluminum housewares flourished and became a standard feature of the modern home, popularized by the transformation of women's domestic lives and their emergence as important buyers. Aluminum's wide applicability led to products that became ubiquitous in the twentieth century: airplanes, automobile frames, soda cans, kitchen pans, and the fascia components of glazed skyscrapers.[2]

Perhaps no material is more readily associated with the twentieth century than plastic. In 1907, Leo Baekeland patented synthetic plastics, from which Bakelite was produced. Bakelite was the first plastic that retained its shape after being heated. This was followed by numerous material and process experiments, and by mid-century, a considerable variety of plastics was available. As is often the case, early plastics imitated other materials such as wood and tortoiseshell, and over time, some were engineered to have the more sensual look and feel of natural materials like silk, leather, and rubber.[3] Plastic was ideal for the rapid fabrication of identical products that are affordable to the masses. It can be made in high volume, molded into fluid organic shapes or crisp geometric forms, either hard or soft, opaque or transparent, and its high surface finish means that it can be used in the condition in which it leaves the mold. Molded plastics also reduce the need to assemble parts, enabling, for instance, Verner Panton's 1959 cantilevered Stacking Chair to be made as a continuous, unbroken *S*-curve. In addition to new, glossy, and oftentimes animated forms, plastics introduced an enormous range of textures and infused our interiors with a wider array of colors. As plastics grew in popularity, terms such as *nylon* and *formica* were added to the vernacular. The French philosopher Roland Barthes acclaimed its ordinariness: "Plastic has climbed down, it is a household material. It is the first magical substance which consents to be prosaic . . . for the first time, artifice aims at something common, not rare."[4]

Despite alternatives to the predominance of plastics, plastics technology has continued to evolve in large part because of its immense versatility. During the counterculture pop movement of the 1960s and 1970s, designers and manufacturers experimented with sophisticated molding techniques and new synthetic materials such as thermosetting plastics—resins, urethane foams, and unsaturated polyesters—to create bold sculptural shapes that celebrated new notions of comfort and casual living. Inspired by the polyurethane foam used in the mattress industry,[5] designers created molded furniture with varying densities, eliminating the need for an internal armature. A new take on the traditional Club Chair, the *Blow* armchair, made using polyvinyl chloride (PVC) welding, could be inflated with air like a balloon, underscoring the informality, collapsibility, and portability that characterized the new lifestyles. But the ultimate in maximum user flexibility was the Bean Bag Chair. Referred to as the *Sacco* (Sack) seating unit when it was first made in Italy in 1968, this oversized "cushion" was filled with polystyrene pellets that conformed to the body like a pillow. As a revolt against tradition, these new furniture forms, presented as hard exuberant shapes or soft, adaptable volumes that could be arranged and rearranged at will, represented a dramatic alternative to rectilinear modernism, or postwar furniture stripped down to its essentials.

The advent of the microelectronics industry in the second half of the twentieth century radically transformed not only materials and technology, but our experience of industrial design. The invention of the transistor in 1947 to control the flow of electricity, with its small size, low power consumption, and reliability, offered major advantages over bigger vacuum tubes. Such miniaturization and, soon after, the invention of the microchip, added significantly to the enlarged empire of microelectronics and reduced the mass of mechanisms that the exterior casing of most products were designed to conceal. The invention of the microchip "was a pivotal moment when discoveries in all the major sciences—physics, chemistry, mathematics, optics, and electrical engineering—converged, and combined with a culture ready to implement the technology."[6] Made of silicon crystals that were grown in a sterilized

FIG 3
Silicon wafer etched with microprocessor chips, 300 mm, 2006, produced by Intel Corporation, Santa Clara, California.

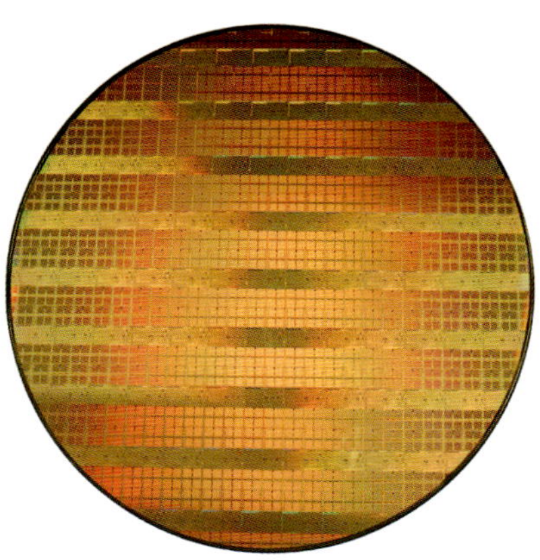

FIG 4
Heated balaclava made with a smart textile, designed c. 2016, produced by Stoll GmbH, Reutlingen, Germany.

environment to eliminate imperfections, microchips are the "brains" of our digital devices and have had a decisive impact on product design, spawning entirely new industries and products (fig. 3). Optimizing speed, efficiency, and precision, they capture the spirit of our era.

America's space program accelerated engagement with research and advancements in materials and had a transformative impact on design. Concurrent with breakthroughs in miniaturization and semiconductor electronics was the ability to manipulate the structural and chemical properties of materials on a molecular level. This created new composites and hybrid materials valued for their performance capabilities—most notably light weight, superior strength, and durability, often in extreme environments.[7] Many of the technological innovations and materials invented for aerospace found applications in everyday products, "including automobiles, airplanes, titanium medical devices, Teflon® pans, and sporting equipment, such as carbon fiber tennis racquets, skis, bicycle frames."[8]

Among the most dramatic changes in materials during the late twentieth century was in the performance capabilities "designed" into our fundamental materials—glass,

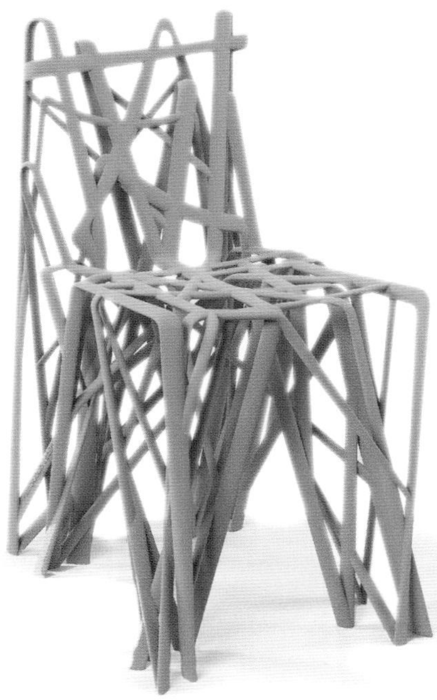

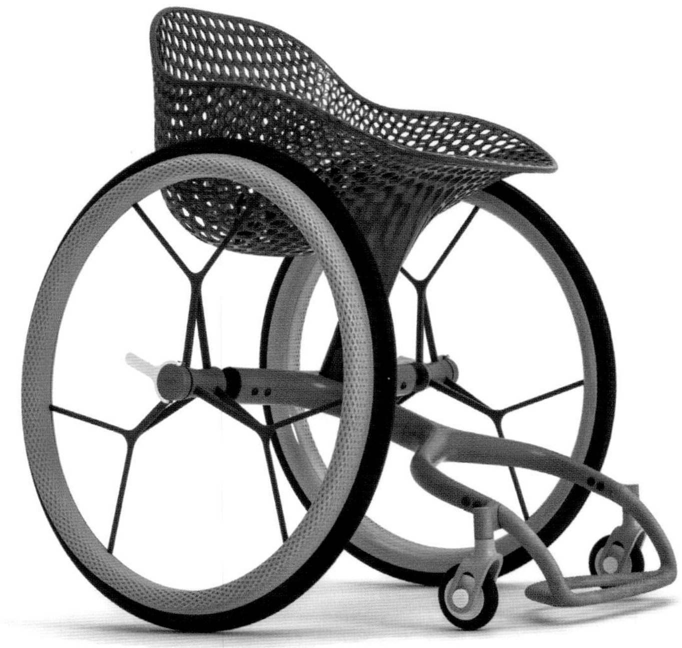

FIG 5
above left *Solid C2* chair from the *Solid* collection designed by Patrick Jouin, 2004. Cooper Hewitt, Smithsonian Design Museum.

FIG 6
above right *GO* 3D-printed wheelchair designed by Benjamin Hubert and LayerLAB, 2016.

ceramic, wood—and the new qualities they bring to the experience of design. Glass, for instance, one of the most ubiquitous materials, has been a continuous source of inspiration for designers and manufacturers, but perhaps no other traditional material is undergoing so much "boundary pushing" and transformation at its molecular structure. For centuries, the light capturing, spatial, and reflective characteristics of glass have held endless fascination. Glass has been cherished for its transparency and hygienic qualities, as well as its beauty, color, and use as a decorative object. Corning Glass Works, known for its oven-proof Pyrex® glass cookware, patented in 1915, and the mass production of incandescent light bulbs revolutionized the television industry with picture tubes that dictated the shape and size of televisions. In recent years, Gorilla® Glass, the tough, protective cover glass engineered at Corning in the 1960s, has served as the touch-sensitive window through which we interact with our handheld devices. The integration of digital technology into our daily lives is transforming the interiors of our cars and buildings with interactive glass surfaces. And, through sophisticated lamination, facades of architectural glass can now be made in extremely large panes for thermal control and increased strength, and can be structurally manipulated for dynamic behaviors that adjust and breathe like skin. Perhaps most remarkably, although glass has always hardened when cooled, ultrathin sheets of flexible glass called Willow Glass can now be rolled, shipped, then unrolled.

Miniaturization and the creation of "smart" objects, which perform multiple functions, characterizes the most advanced industrial design achievements of today, dramatically changing how we communicate, work, and socialize. Apple's iPhone, first launched in 2007, was revolutionary for being a telephone, iPod, Internet browser, and a camera all in one. With the addition of apps, or applications, today's smartphones are multi-purpose tools, capable of hundreds of functions—actions that people previously performed on many separate single-purpose objects.[9] The intelligence incorporated into many products has generated the term *smart* to describe their special performance capabilities. Smart responsive textiles—for example, a 3D-knitted balaclava made of electric-conductive yarn—can reduce the risk of chest infections in athletes by providing a consistent level of warmth in cold weather (fig. 4).

Three-dimensional printing and laser-cutting technology are today's radical alternatives to old production processes. An additive manufacturing technique developed in the 1980s, 3D printing was initially used for precision prototyping of automotive parts, small-scale architectural models, and medical devices. Today, the range of available materials and customization afforded by 3D printing provides unique alternatives to the one-size-fits-all model of assembly line mass production. As an information-driven fabrication system, it offers tremendous flexibility. It also encourages experimentation: a design can be varied merely with a change in the software, or the data information, or a material. What often start as hand-drawn gestural sketches are translated into electronic data that travels from the computer to a machine that prints the design, layer on layer, in ultrathin strata. This layering is repeated until the finished product is "built." There is no human intervention, no tooling, assembly, or intermediate steps—and no waste. With the absence of molds and a manufacturing system driven by electronic information, designers are liberated to imagine and realize complex forms previously impossible to produce by traditional methods.[10] French designer Patrick Jouin's unprecedented and elegant *Solid C2* chair was one of the first furniture designs to be printed (fig. 5). Inspired by the different patterns, shapes, bends, twists, and folds of blades of grass, the chair's complex network of crisscrossing lines could not have been molded, cast, or carved. Unlike Thonet's bentwood chairs of the 1850s and other modern furniture, with today's computation and rapid manufacturing, simplification of labor is no longer a key issue. 3D printing resurrects complexity, intricate details, and applied ornament while staying relevant to modern design.[11]

Computation and robotics have not only revolutionized and accelerated the design process, they are also defining today's designs, manufacturing, and aesthetics, generating new possibilities and unprecedented forms. The porous, Swiss cheese–like appearance of many contemporary products and architectural facades is the direct result of computer analytics that optimize materials and structures. When used in the design of new high-performance products, like prosthetic limbs or wheelchairs (fig. 6) that address individual needs, shapes, and sizes, computation is yielding new typologies that are structurally stronger, more durable, and lighter in weight. The scientific inquiry and understanding that have propelled many advancements in materials and technology have shaped much of the way we live in the twentieth and twenty-first centuries. This persistent fascination with employing technology to manipulate materials and achieve innovative creations, along with our newfound ability to control matter, will surely continue to impact the ongoing dialogue about materials, technology, form, structure, and function.

ENDNOTES

1 Elizabeth Schmuttermeier and Derek E. Ostergard, eds., *Cast Iron from Central Europe 1800–1850* (New York: The Bard Graduate Center for Studies in the Decorative Arts, 1994).

2 Sarah Nichols. *Aluminum By Design* (New York: Harry N. Abrams, 2000).

3 Rob Thompson, *Manufacturing Processes for Design Professionals* (London: Thames & Hudson, 2007), 425.

4 Penny Sparke, *The Plastics Age: From Bakelite to Beanbags and Beyond* (Woodstock, NY: Overlook Press, 1993), 111.

5 Mario Mastropietro, *An Industry for Design: The Research, Designers and Corporate Image of B&B Italia* (Milan: Edizioni Lybra Immagine, 1986), 56.

6 Cara McCarty, *Information Art: Diagramming Microchips* (New York: Museum of Modern Art, 1990), 4.

7 Cara McCarty, "NASA: Advancing Ultra-performance," in *Extreme Textiles: Designing for High Performance* (New York: Cooper-Hewitt, National Design Museum, 2005), 140.

8 Ibid., p. 141.

9 Cara McCarty and Matilda McQuaid, *Tools: Extending Our Reach* (New York: Cooper Hewitt, Smithsonian Design Museum, 2014), 128/82.

10 Cara McCarty, *Patrick Jouin: Currents 101* (St. Louis: Saint Louis Art Museum, 2007), 1.

11 Ibid., p. 1.

GILBERT

03
objects

CONTROLLING CLIMATE

WITH THE ADVENT OF NEW TECHNOLOGIES FROM THE LATE nineteenth century onward, the ability to control the temperature of our environments has become ever more sophisticated—from wood and coal-driven fires and stoves to electricity and gas—and new objects, machines, devices, and processes were developed to facilitate it. Free-standing electric heaters and fans adopted a myriad of forms, often becoming sculptural objects in their own right. Central heating and cooling systems, complete with radiators and thermostats, followed soon thereafter. Included here is a selection of the many fans in the Kravis collection, as well as humidifiers, heaters, and thermostats, all elegant enough to be on full view in any home.

Henry Dreyfuss (1904–1972)

Honeywell Round Thermostat, Model No. T86A

Designed c. 1942 (D136,848–52)

Plastic, enameled metal

3¼ x 3¼ x 1½ in. (8.3 x 8.3 x 3.8 cm)

Produced by Minneapolis-Honeywell, Minneapolis, Minnesota (Honeywell after 1963)

K2016.30.1

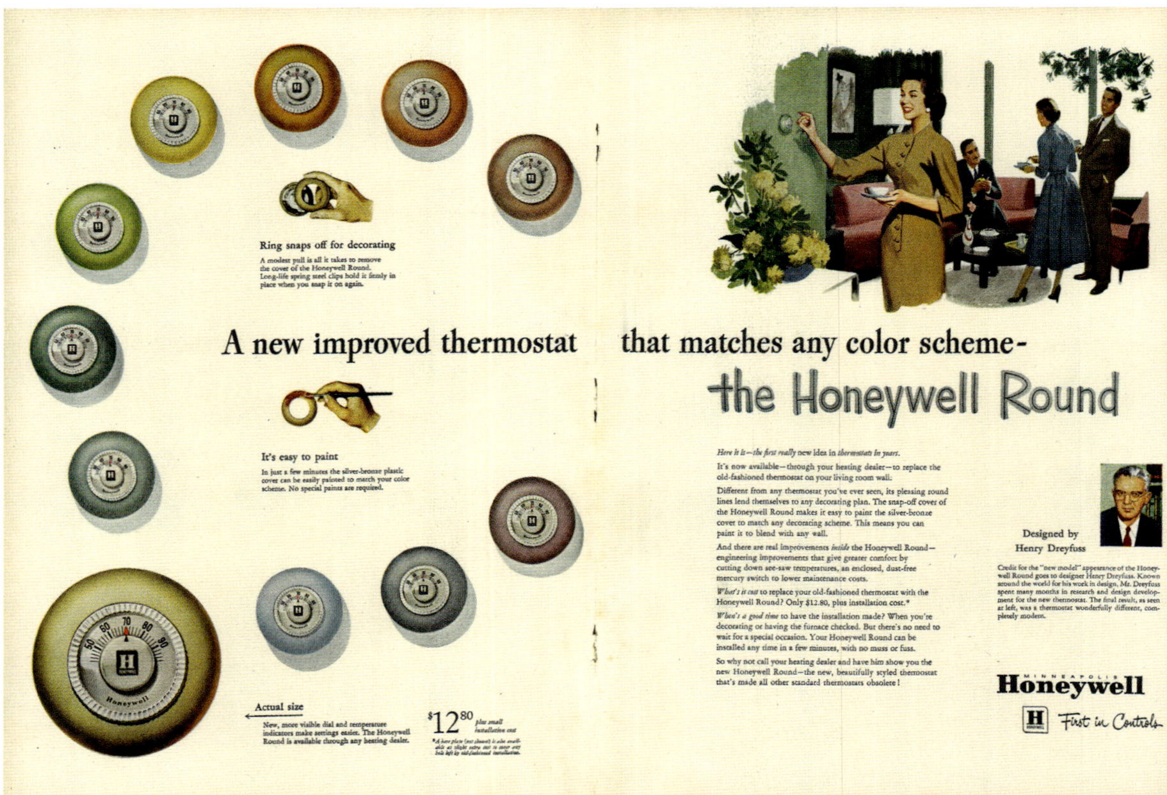

Advertisement for *Honeywell Round* thermostat, published in *LIFE* magazine, November 9, 1953.

RECTANGULAR THERMOSTATS COULD BE DIFFICULT TO HANG SQUARELY ON THE WALL, and with the *Honeywell Round* T-86, Henry Dreyfuss solved this design problem. The idea of the *Round* was conceived more than a decade earlier: Dreyfuss submitted design patent applications for multiple variations of the domed thermostat form in June 1942,[1] but production was delayed until 1953. Another iconic Dreyfuss design of just a few years earlier, the 1939 *Big Ben* alarm clock, also had a domed front. A colorful 1953 advertisement illustrates the thermostat's snap-off ring, which "makes it easy to paint the silver-bronze cover to match any decorating scheme," illustrating the possibilities in a circular display mimicking the shape of the thermostat. It also mentions the *Round*'s internal innovation of "an enclosed, dust-free mercury switch to lower maintenance costs."[2] –DAH

60
70
80
90
H
HONEYWELL
60
90
COMFORT RANGE
Honeywell

Christian Barman (1898–1980)

H.M.V. Room Heater

Designed c. 1934

Chromium-plated steel

11⅝ x 17⅞ x 12⅝ in. (29.5 x 45.4 x 32.1 cm)

Produced by His Master's Voice, Hayes, Middlesex, England

K2008.294

Advertisement for the *H.M.V.* room heater, c. 1938.

THIS SPACE HEATER IS AN EXAMPLE OF 1930S STREAMLINING, here achieved by stacking four shining, chromed plates into a rounded and slightly conical shape. It is a convection heater, which draws cool air in by means of a fan, heats it, and returns it to the room.[3] A 1936 British utility patent (GB48472A) credits Norman Barnes and Alfred Whitaker for the invention of this heater's internal mechanism, which is perhaps the "entirely new heating principle" mentioned in the ad seen here, which introduced the heater in 1938. The exterior design is by the modernist designer Christian Barman, who was admired for his range of industrial designs. He was an architect and product designer, as well as editor of *The Architectural Review*. He later became a publicity officer for London Transport, helping to formulate their design policy from 1935 to 1941.[4] The manufacturer was the phonographic equipment company His Master's Voice, later known as H.M.V.,[5] as seen in the advertisement, which states that this product "warms the room in half the time, and therefore at half the cost. It is safer, being constructed to circulate heat without itself becoming hot enough to scorch," making it suitable for use in a home with children or pets. –DAH

Michael W. McArdle (1874–1935)

Sunbeam Electric Heater

Designed c. 1928 (D77,423)

Chromium-plated steel, enameled cast iron, enameled steel, painted wood

8¾ x 5½ x 5 in. (22.2 x 14 x 12.7 cm)

Produced by Chicago Flexible Shaft Co., Chicago, Illinois

K2008.574

THE GRILLE ON THE FRONT OF THE *SUNBEAM* HEATER IS A stylized sun with radiating rays and an *S* for Sunbeam at the center. The circular heater is suspended in a shaped frame. Organized in 1897, the Chicago Flexible Shaft Company was a major manufacturer of electrical appliances such as toasters, irons, and mixers. Although the name did not officially change to the more popularly known Sunbeam until 1946, both company names were used concurrently for a period, as indicated by the marks on the bottom of this heater: "Sunbeam/HEATER AND DRYER/CHICAGO FLEXIBLE SHAFT CO/CHICAGO USA." The Chicago Flexible Shaft Company produced its first Sunbeam branded household appliance in 1921. Among its famous designers were Ivar Jepson and Robert Davol Budlong. The designer of this heater, Michael W. McArdle, grew up in Baileys Harbor, Wisconsin, the son of Irish immigrant parents. He moved to Chicago in 1903 and worked for the Chicago Flexible Shaft Company, eventually becoming CEO and president. He was an educator, author, lawyer, and philanthropist, as well as an inventor of Sunbeam products, with patents filed for a juice extractor in 1932 and a lawn sprinkler in 1933.[6] —DAH

Ralph Cameron Ormiston (1921–2000)
John Hardie Smellie (1915–2001)

Sofono Spacemaster/Challenge Heater,
Model No. PRC402ST

Designed 1959 (British patent GB878577)

Enameled and chromium-plated steel, melamine

27 x 25⅜ x 7¾ in. (68.6 x 64.5 x 19.7 cm)

Produced by Sofono Electrical Division of
Federated Foundries Ltd., Falkirk, Scotland

K2010.26

THE SUCCESSFUL LAUNCH OF SPUTNIK I, THE FIRST MAN-MADE SATELLITE, IN 1957 IGNITED the space age and inspired many designs, including the *Sofono Spacemaster*. The form, reminiscent of a flying saucer, consists of two large steel disks joined by a frame with central chromium-plated grilles on each side. Chromium-plated legs—tapered and canted—support the lightweight, portable structure. Scottish-born designer Ralph Cameron Ormiston attended the Glasgow School of Art. His studies were interrupted by his service in the Royal Air Force during World War II, but he returned to the Glasgow school to study industrial design. After graduating in 1949, he worked at various jobs in England for several years before taking a job at Grange-Camelon Iron Company in Falkirk, part of the Federated Foundries Group that would later produce Ormiston's *Sofono* heating designs.[7] –DAH

Dieter Rams (born 1932)
Electric Space Heater, Model No. H2, Type 4510
Designed 1959
Plastic, painted metal
3¾ x 10½ x 5⅜ in. (9.4 x 26.7 x 13.7 cm)
Produced by Braun AG, Frankfurt, Germany
K2010.236

GERMAN ARCHITECT AND INDUSTRIAL DESIGNER DIETER RAMS IS THE SON OF AN ELECTRICAL engineer, but it was from his grandfather that he learned woodworking and the ideal of simplicity in design. He began to work for Braun as an architect in 1955, and, by 1961, he had become the chief design officer at the company, where his functionalist and "less but better" approach to design, characterized by the active avoidance of superfluous detail, was applied to all Braun products. Due to its shape and, at the time, remarkably small size, Braun's first heater, the model *H1/11* designed by Rams, was known as "the brick." This H2 is nearly identical to the original, save for updated markings on the control dial. Consumers were at first unconvinced that such a small device could produce the same 2,000 watts of heat as its much larger counterparts on the market. The heater's high performance and low noise were made possible by a "cross-flow blower," and the flow of air could be directed by adjusting the base. Its distinctive colors and pattern of ventilation openings combine to achieve the minimalist aesthetic that makes it immediately recognizable as a Braun product. The resulting modernist rectangular box recalls architecture in miniature—complete with minimalist facade. –PM

Reinhold Weiss (born 1934)
Jürgen Greubel (born 1938)

Two Table Fans, Model No. HL 70, Type 4550

Designed c. 1969 (D217,460)

Plastic, acrylic

Vertical position: 6 x 3½ x 2¾ in.
(15.2 x 8.9 x 7 cm) each, fan only

Produced by Braun AG, Frankfurt, Germany

K2008.214 and K2008.397

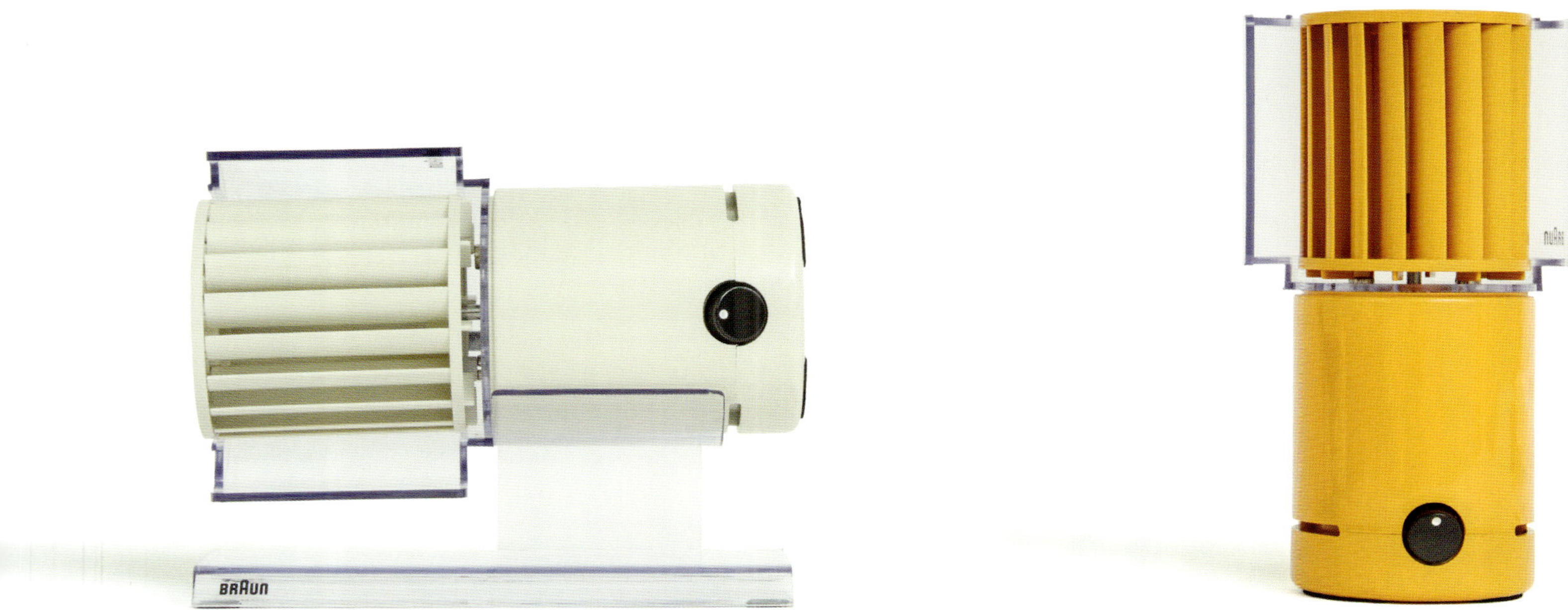

DIETER RAMS WAS NOT THE ONLY INDUSTRIAL DESIGNER WHO CONTRIBUTED TO THE DESIGN legacy of Braun. Reinhold Weiss and Jürgen Greubel also played an important role in creating everyday household appliances for the consumer market in the postwar period. Rams credited the many freelancers who trained at the Ulm School of Design for inspiring the company's modernist design philosophy in the 1950s, prior to the establishment of Braun's in-house design department in the 1960s. Reinhold Weiss, an Ulm graduate, was an influential member of the new internal design team. Weiss's 1961 HL 1 Multiwind table fan, on which this HL 70 fan is based, reflects the pure functionalism and simplicity of the company's philosophical design at the time. Weiss and Greubel's HL 70 fan, with its innovative blower, replaced the gray exterior of the HL 1 with colors such as white, brown, and yellow, and simplified the controls. –PM

AEG

opposite

Peter Behrens (1868–1940)

Table Fan

Designed 1912

Enameled cast iron and brass

15¾ x 13⅜ x 9½ in. (40 x 34 x 24.1 cm)

Produced by Allgemeine Elektricitäts-Gesellschaft (AEG), Berlin, Germany

K2008.287

PETER BEHRENS'S FAN IS A FUNCTIONAL OBJECT with a cast-iron base and four brass fan blades. The one element of whimsy is the wire armature surrounding the blades. In another AEG fan design, Behrens used straight ribs; here he employs curved, playful forms that recall his earlier jugendstil work. The center medallion with the AEG logo is a harbinger of the corporate branding that would become so ubiquitous in Western consumer products later in the twentieth century. –RCM

right

Leslie J. Roberts (dates unknown)

Kingsway 10 Table Fan

Designed c. 1950

Enameled metal, rubber

12½ x 9 x 7¾ in. (31.8 x 22.9 x 19.7 cm)

Produced by Electric Fans & Controls Ltd. for General Electric Company Ltd., High Wycombe, England

K2010.109

AMERICAN STREAMLINING ARRIVED LATE IN ENGLAND but flourished after World War II with a number of aerodynamic designs for industrial products. The *Kingsway 10* fan was exhibited in the 1951 Festival of Britain and published in *Design in the Festival* of the same year.[8] The fan displays characteristics of the streamlined style: the motor is sheathed in a cowling that hides the functional works and looks very much like an airplane propeller. The fin at the rear with sweeping speed lines has no function other than to add to the sense of aerodynamic speed. The head can be rotated to direct the flow of air. No protective grille was required for the rubber blades. –DAH

left

Max Weber (1888–1955)

Roto-Beam Table Fan

Designed c. 1937 (utility patent 2,123,448)

Steel, aluminum, chromium-plated steel

12¾ x 13½ x 8 in. (32.4 x 34.3 x 20.3 cm)

Produced by Roto-Beam Company, Chicago, Illinois

K2010.160

center

Designer Unknown

Zephyr Airkooler Fan

Designed c. 1946

Aluminum, brass, metal

12¼ x 10½ x 10¼ in. (31 x 26.7 x 26 cm)

Produced by Wright Industries, Inc., Lisbon, Ohio

K2012.221

right

Robert Heller (1899–1973)

Airflow Table Fan

Designed c. 1937 (D104,259)

Steel, aluminum, chromium-plated steel, enameled aluminum, plastic, rubber

13⅜ x 11 x 7½ in. (34 x 27.9 x 19.1 cm)

Produced by A.C. Gilbert Company, New Haven, Connecticut

K2010.158

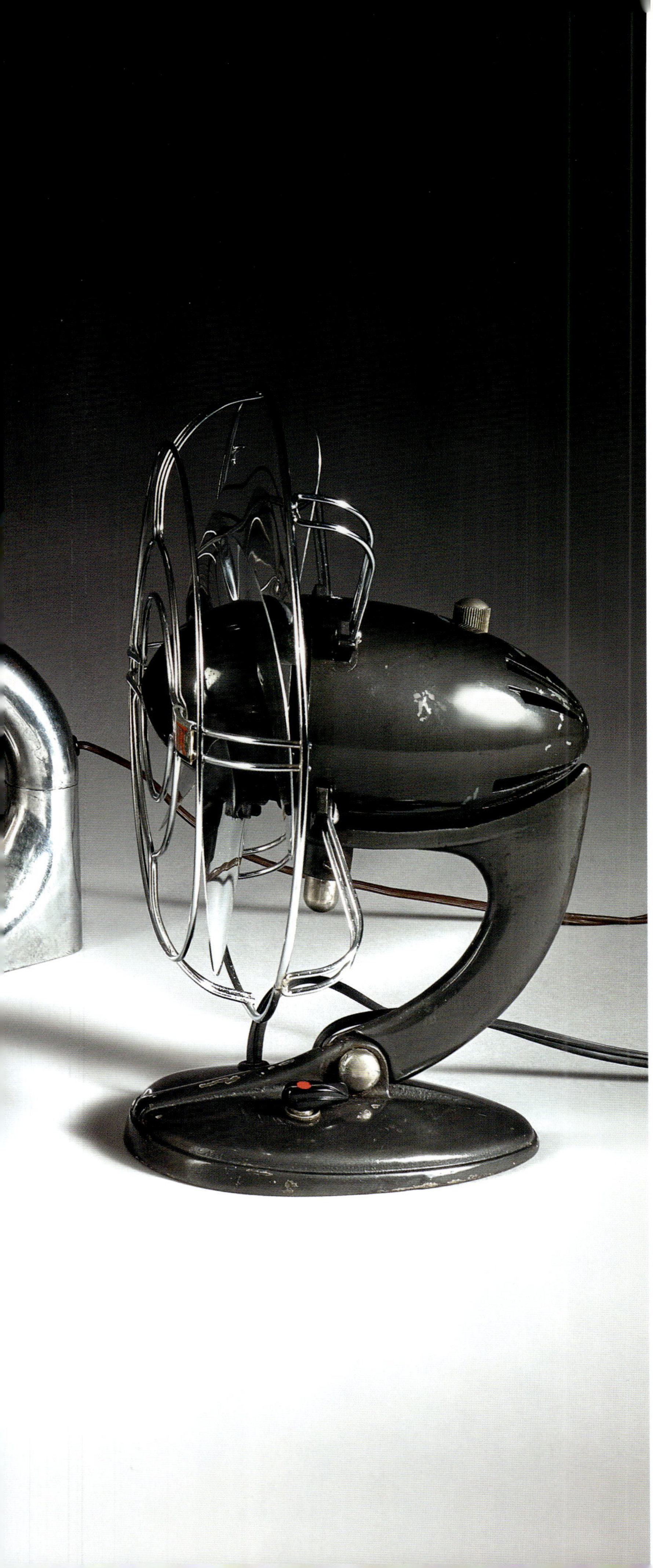

THESE THREE FANS REPRESENT DESIGN ADVANCEMENTS IN the 1930s as the form evolved from stepped pedestal bases and cylindrical motor housings to dynamic streamlined designs inspired by the new airplanes being produced during the same decade, such as the Douglas DC-3. Airplane propellers and fan blades are functionally the same in that they are intended to move air, except with different goals. The shaft of Robert Heller's aerodynamic *Airflow* sweeps in a dramatic cantilever from the teardrop form of the base to the ellipsoid motor housing.[9] According to a contemporary article, "The redesign of this mechanism has contributed to the beauty and simplicity of the streamlined form of the fan. . . . The 10 in. blade size, finished in telephone black with metal parts chromium plated, will retail for $9.95."[10] The *Zephyr Airkooler* has the Heller orientation but is more solid in its thick shaft and base, which is made of a single section of metal, rising cobra-like to support the fan blades and guard. Max Weber's fan is less streamlined and more geometric. A utility patent filed by Weber claimed its invention "operates with less vibration and noise" than other fans because "the wire formed guard carries the electric motor" and "serves not only yieldingly to support the motor but also to absorb any vibration resulting from drive or operation of the motor."[11] –DAH

Designer Unknown

Libelle (Dragonfly) Table Fan

Designed c. 1953

Enameled and chromium-plated metal, Bakelite

10⅜ x 8½ x 7 in. (26.4 x 21.6 x 17.8 cm)

Produced by Schoeller Elektrotechnische, GmbH, Frankfurt, Germany

K2008.16.1

THE *LIBELLE* FAN IS A DRAMATIC EXAMPLE OF A postwar German aerodynamic design that was influenced by American streamlining of the 1930s and 1940s. Its sculptural teardrop base of wine-red enameled metal has three deep ridges on each side, which serve as speed lines above a fitted chrome base. The tri-sectioned guard protecting the Bakelite fan blades is supported by an arced adjustable shaft. The *Libelle* fan had two speeds and could be mounted on the wall. The electronics company Schoeller was founded in 1911. –DAH

Ezio Pirali (born 1921)
Zerowatt Table Fan, Model No. VE505
Designed 1953
Aluminum, chromium-plated steel, rubber
5¾ x 10⅛ x 6⅝ in. (14.6 x 25.7 x 16.8 cm)
Produced by Fabbriche Elettrotecniche Riunite, S.p.A., Milan, Italy
K2008.215

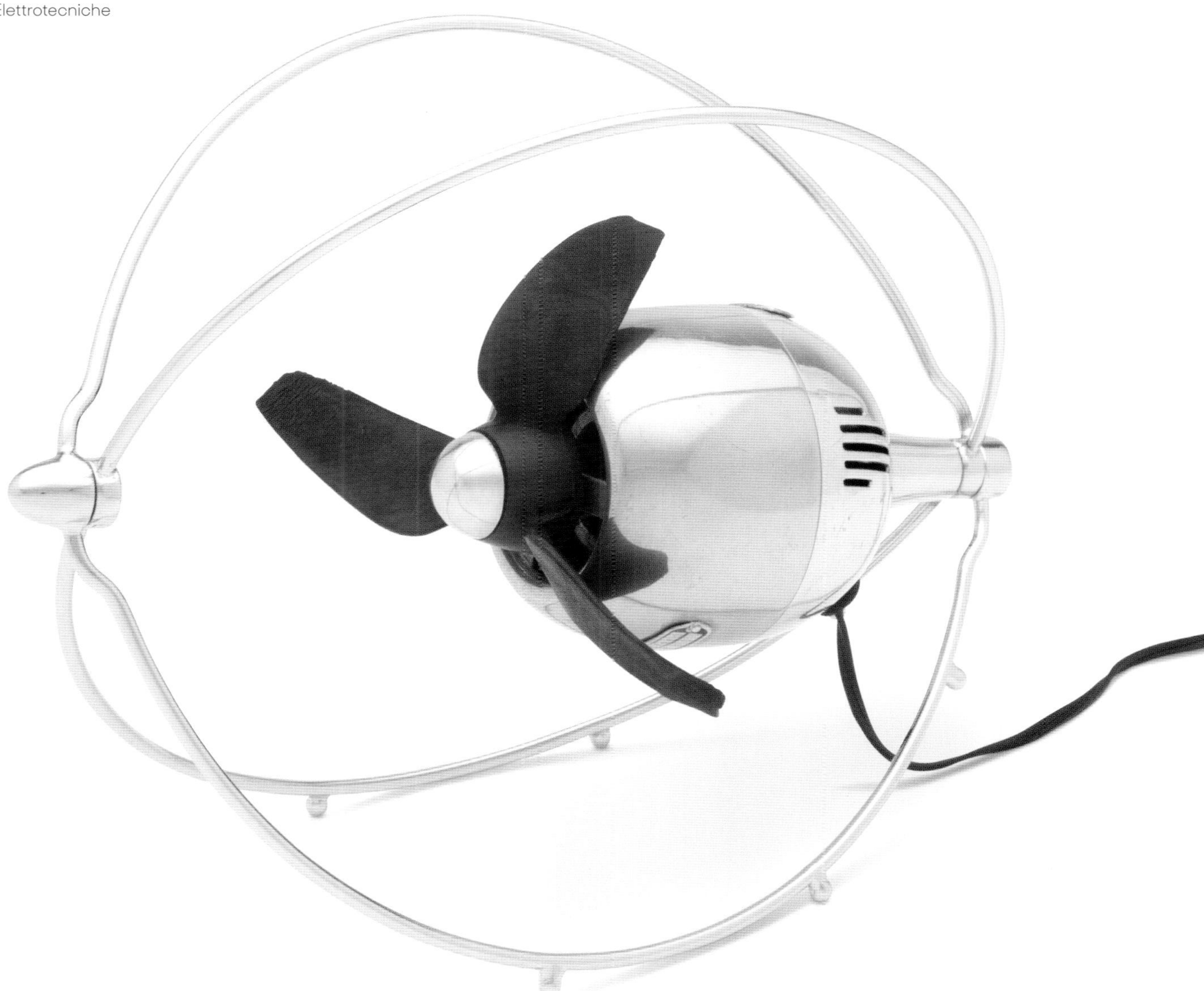

SHAPED LIKE AN AIRPLANE PROPELLER, EZIO PIRALI'S FAN FOLLOWS THE PRINCIPLES OF aerodynamic design and provides a sense of speed. Placing the motor in the center of the design with two oval rods both supporting and forming a protective frame around it was a unique and daring idea. In 1954, the *Zerowatt* fan won the Compasso d'Oro and a silver medal at the Milan Triennale. It was produced by Fabbriche Elettrotecniche Riunite (FER), where Pirali was managing director and CEO, as well as the designer of the company's distinctive products.[12] *Zerowatt* had been a tradename for FER since 1949, and the company name was officially changed to Zerowatt, S.p.A, in 1954 –DAH

Joy B. Schmitt (1894–1985)

Electric Steam Vaporizer, Model No. 148
Designed c. 1943 (utility patent 2,363,365)
Bakelite
7½ x 4⅛ x 4⅛ in. (19.1 x 10.5 x 10.5 cm)
Produced by the DeVilbiss Co., Toledo, Ohio
K2010.55

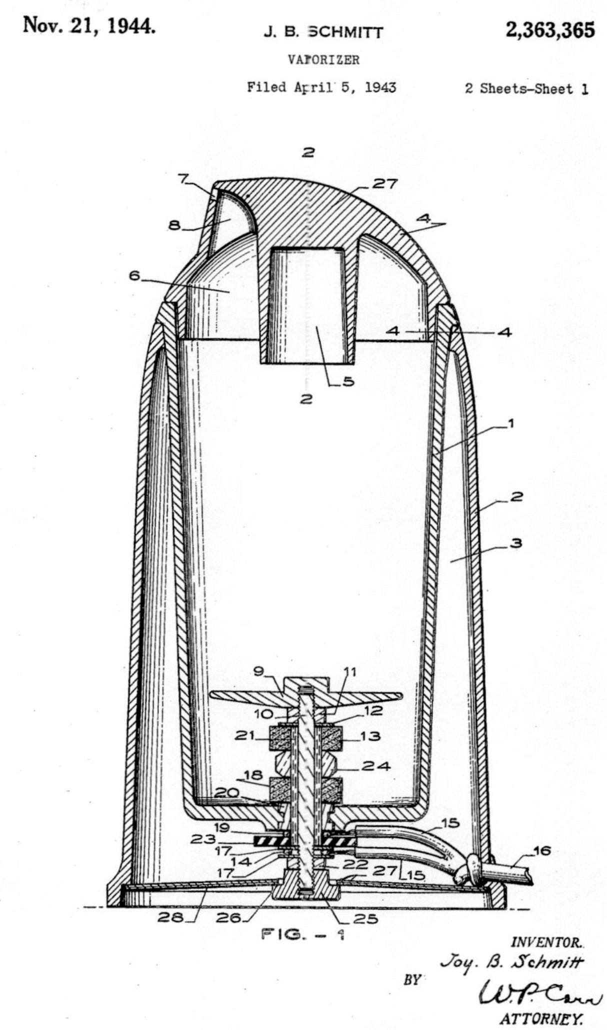

Detail of U.S. utility patent 2,363,365 for the Schmitt vaporizer.

DEVILBISS COMPANY'S VAPORIZER ACCOMMODATES MEDICINAL MIXTURES INSIDE ITS LID AND, when the water is heated, it vaporizes the medicine and shuts off automatically when all the water has evaporated. According to the accompanying instruction brochure, "Inhalation of warm medicated vapor to the inflamed respiratory membranes is a treatment recommended for coughs, bronchitis, colds, asthma, deep-seated sore throat, whooping cough and laryngitis."[13] The black plastic bullet form of the vaporizer exemplifies streamlining with its cylindrical form that tapers inward as it rises to a domed cover.[14] According to the Durez plastics company's monthly newsletter, in the section titled "Notes on Finishing Phenonlic Plastics," molded products like this vaporizer required buffing and "cutting"—the removal of excess plastic "when the mold parting line is actually on an important surface."[15] The utility patents marked on the vaporizer list Joy B. Schmitt as the inventor of the mechanism and assignor to DeVilbiss; one of the two patents, seen above, shows the form and interior workings of this vaporizer in the application drawing patents.[16] –DAH

Matti Walker (born 1966)

Fred Humidifier, Model No. F-000

Designed 2008

ABS Plastic, metal

11 x 14¾ x 14¾ in. (27.9 x 37.5 x 37.5 cm)

Produced by Stadler Form Aktiengesellschaft, Zug, Switzerland

K2010.335.1

STADLER FORM, FOUNDED BY MARTIN STADLER IN 1998, WORKED WITH SWISS designer Matti Walker to create its first household appliance, the *Fred* humidifier. Stadler's products are given the names of people and deities; other humidifiers in the collection include *Eva*, *Hera*, and *Anton*. *Fred* is a germ-free, warm-mist humidifier with a hygrostat with external sensors to measure ambient humidity. The device is nearly silent, a great advantage for those using it to combat the effects of excessively dry air while sleeping. Its styling is unique and playful, recalling a retro-futuristic organic shape, one that echoes the space-age progress of the 1950s yet still looks forward. –PM

Naoto Fukasawa (born 1956)
Plus Minus Zero Humidifier
Designed 2003
Polycarbonate plastic
6⅛ x 12 x 12 in. (15.6 x 30.5 x 30.5 cm)
Produced by Plus Minus Zero Co., Ltd, Tokyo, Japan
K2017.6.1

THIS HUMIDIFIER WAS DESIGNED FOR THE JAPANESE COMPANY PLUS MINUS ZERO, founded in 2003, which established its mission as the act of breathing new life into everyday objects such as doormats, calculators, and umbrellas. As a typology of design, most domestic humidifiers tend to be generally bland and lack contemporary styling. Fukasawa sought to make his humidifier beautiful enough to be a welcome addition to any room, rather than just an appliance to be hidden away when not in use. It is the simplest rounded form with the steam vent nearly hidden at the center and the seam between the molded top and bottom similarly almost invisible. On sight, it can be difficult to guess what this shining monochrome bubble might be. The humidifier was made in six colors, including leaf green, sky blue, and white. –PM

04
objects

MEASURING

ASSIGNING A NUMBER TO SOMETHING AS A MEASURE OF ITS WEIGHT, size, or quantity allows us to make comparisons and thereby exert a level of control over our environment. Measuring is hugely important in the world of science, but it is equally important in domestic environments, for instance when baking a cake or determining the time to catch a train. The Kravis collection includes many types of measuring objects, among them items destined for a laboratory—glass flasks, for instance—and others destined for a domestic or office setting, such as clocks or scales. As long as it allowed a user to tell the time, a wall clock could, as George Nelson so ably demonstrated, embody a vast array of forms, while a mundane scale for measuring food, as Marco Zanuso showed us, could be made from red plastic to brighten any kitchen.

Isamu Noguchi (1904–1988)
Measured Time Clock and Kitchen Timer
Designed c. 1932
Plastic, metal, glass, painted metal
6⅛ x 5⅜ x 3⅛ in. (15.6 x 13.7 x 7.9 cm)
Produced by Stevenson Manufacturing Company, La Porte, Indiana
K2016.18

THE JAPANESE-AMERICAN ARTIST ISAMU NOGUCHI CONSIDERED HIMSELF TO BE NOT only a sculptor, but also an engineer, inventor, and visionary designer. During the 1930s, he was encouraged by the educator Dr. Edward Rumely to start a business called Time Design, devoted to producing "mechanized production of household things" for the American consumer. It combined the visionary ideas of Noguchi's friend Buckminster Fuller and the business savvy and pragmatism of Rumely.

Measured Time was one of the first projects that Noguchi developed for his venture, and it was the first of his designs to go into mass production. The semicircular dial above the clock face is a 120-minute timer with a Bakelite setting knob and an activation switch to the left. The aim was to help ease a homemaker's schedule, as well as to ensure that meals were not overcooked. Five hundred timers were produced in the first run at a price of $9.50; it was projected that at least fifteen thousand units would be produced. They were available in black or white Bakelite—an efficient, inexpensive, and lightweight but durable plastic—with either a wind-up or electric movement. A utility patent for the movement, applied for on December 1, 1930, shows a design for a rectangular casing. Noguchi's design was more curvaceous and streamlined, reflecting a more contemporary style of the period. –PM

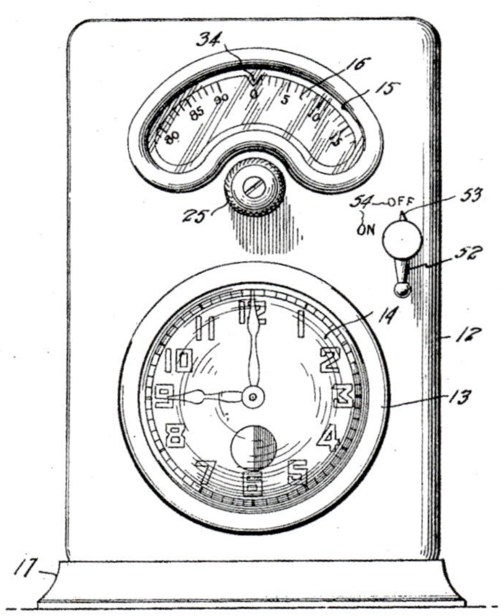

Detail of U.S. utility patent 1,961,417 for an "electric interval timer," granted to Chester R. Gains, June 5, 1934.

Dieter Rams (born 1932)

Personenwaage Bathroom Scale, Model No. HW 1, Type 4960

Designed 1968 (German patent DE6800958U)

Enameled metal, plastic, chromium-plated steel

12 x 10⅝ x 2¾ in. (30.5 x 27 x 7 cm)

Produced by Braun AG, Frankfurt, Germany

K2010.89

BETWEEN 1955 AND 1995, UNDER THE DIRECTION OF DIETER RAMS, BRAUN MANUFACTURED OR sold more than 1,200 products. Rams and his team—or teams in which Rams was a member—designed more than 500 of those products. Simplicity is a key characteristic that refers to the relationship between the product's function and form, and this scale is an example of the classic minimalism for which Braun is renowned. Only the essentials matter. In the design for this bathroom scale, for example, Rams incorporated five movable colored markers across the front to "remember" the weights of different people. –DM

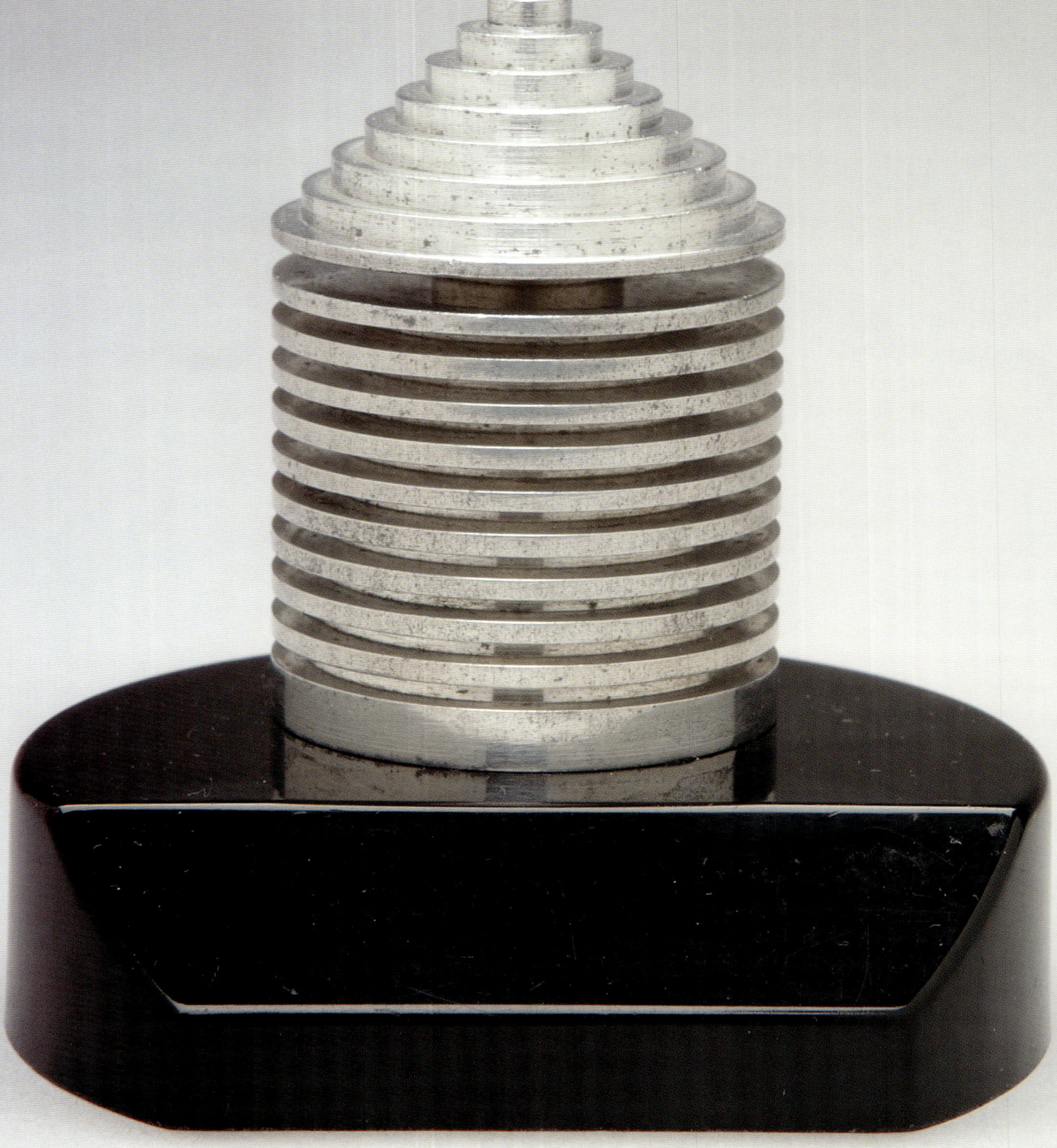

Arnold Melvin Thompson (1920–1995)
Melvin Delbert Thompson (1895–1955)

Postal Scale

Designed c. 1948 (D155,944)

Chromium-plated steel, Bakelite

Closed: 3 x 3 x 2⅝ in. (7.5 x 7.5 x 6.7 cm)

Produced by Thompson Tool and Manufacturing Company, Franklin Park, Illinois

K2010.315

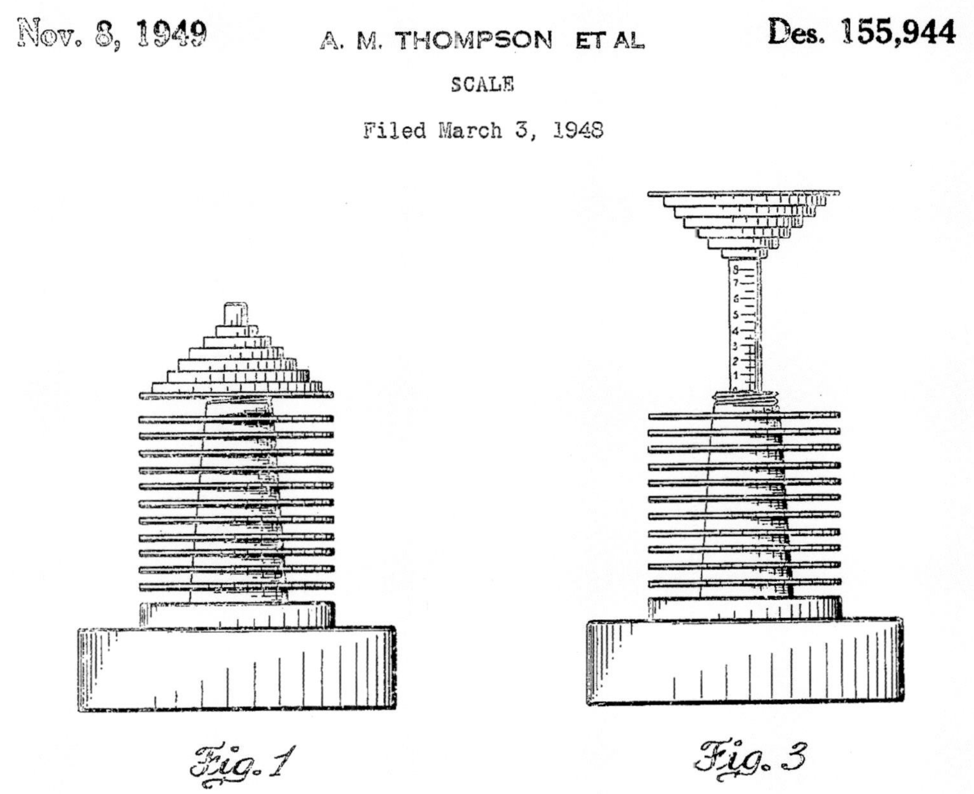

Detail of U.S. design patent D155,944 for the Thompson postal scale, showing the top flipped and mounted for use.

THIS STREAMLINED POSTAL SCALE GIVES THE IMPRESSION THAT THE MAIL BEING weighed will be delivered with great speed. Although the teardrop form associated with this aerodynamic style is not utilized here, the rows of multiple concentric rings suggest the rapid movement of streamlining. The slanted section on the Bakelite base was often reserved for a company name and given away as a gift or for promotion. An elaborate floral label is attached to the bottom of the scale with instructions for its use: "To weigh, unscrew top and invert." The application drawing for the patent granted to the father-and-son design team shows the scale both closed and open for use and gives a sense of its architectonic quality.[1] –DAH

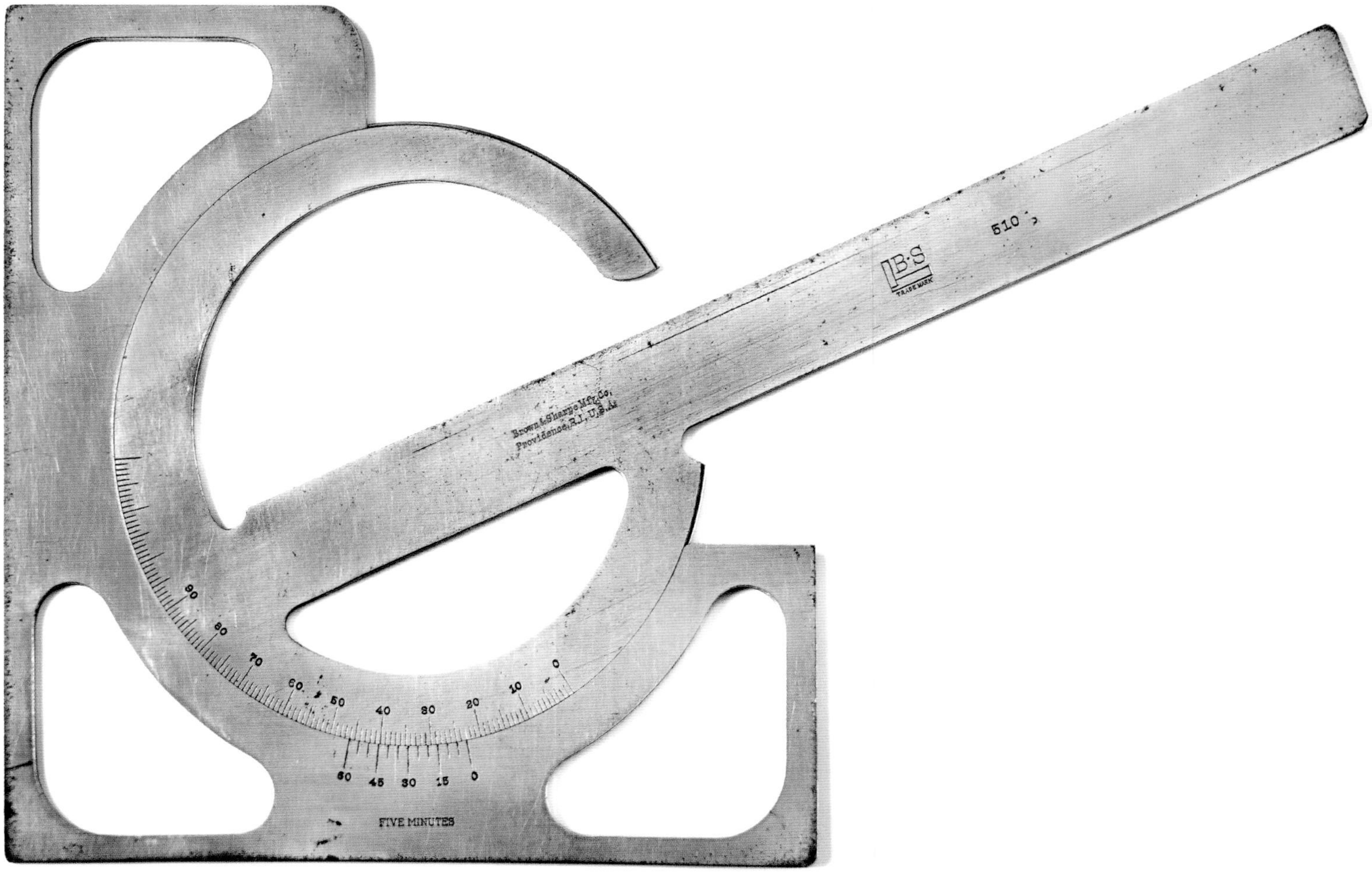

Alton J. Shaw (1858–1895)
Protractor, Model No. 510
Designed c. 1887 (utility patent 367,673)
Stainless steel
6 x 10¼ in. (15.2 x 26 cm)
Produced by Brown & Sharpe Mfg. Co., Providence, Rhode Island
K2016.32

THIS DRAFTSMAN'S PROTRACTOR, AN INSTRUMENT FOR MEASURING AND MARKING angles, combines a perforated square with an inner circle that can be adjusted within the square. An identical protractor, produced by Eugene Dietzgen Co., Chicago, was exhibited in MoMA's *Machine Art* exhibition in 1934.[2] This is actually an example of nineteenth-century modernism, but such simple geometric shapes are timeless, as was reflected in the content of the exhibition, which placed emphasis on aesthetic value rather than designer status. Russel Wright wrote in the 1934 article "Unintentional Modern" about "things which unconsciously and unintentionally express all that designers consciously strive for . . . which, without meaning to, embody the essentials of contemporary design. Consider the lowly paper clip for example. Offhand I cannot think of a better demonstration of functionalism minus adornment."[3] Although this protractor was crafted of stainless steel, today's protractors are often made of transparent plastic. This broad, flat instrument, which can be damaged if stored or transported unprotected, is accompanied by its original, velvet-lined carrying case. –DAH

Designer Unknown

4-Liter and 10-Liter Chemical Flasks

Designed 1930–40

Pyrex glass

4-Liter: 19¾ x 5⅞ x 5⅞ in. (50.2 x 15 x 15 cm)
10-Liter: 19¾ x 8 x 8 in. (50.2 x 20.3 x 20.3 cm)

Produced by Corning Glass Works,
Corning, New York

K2012.27.1–2

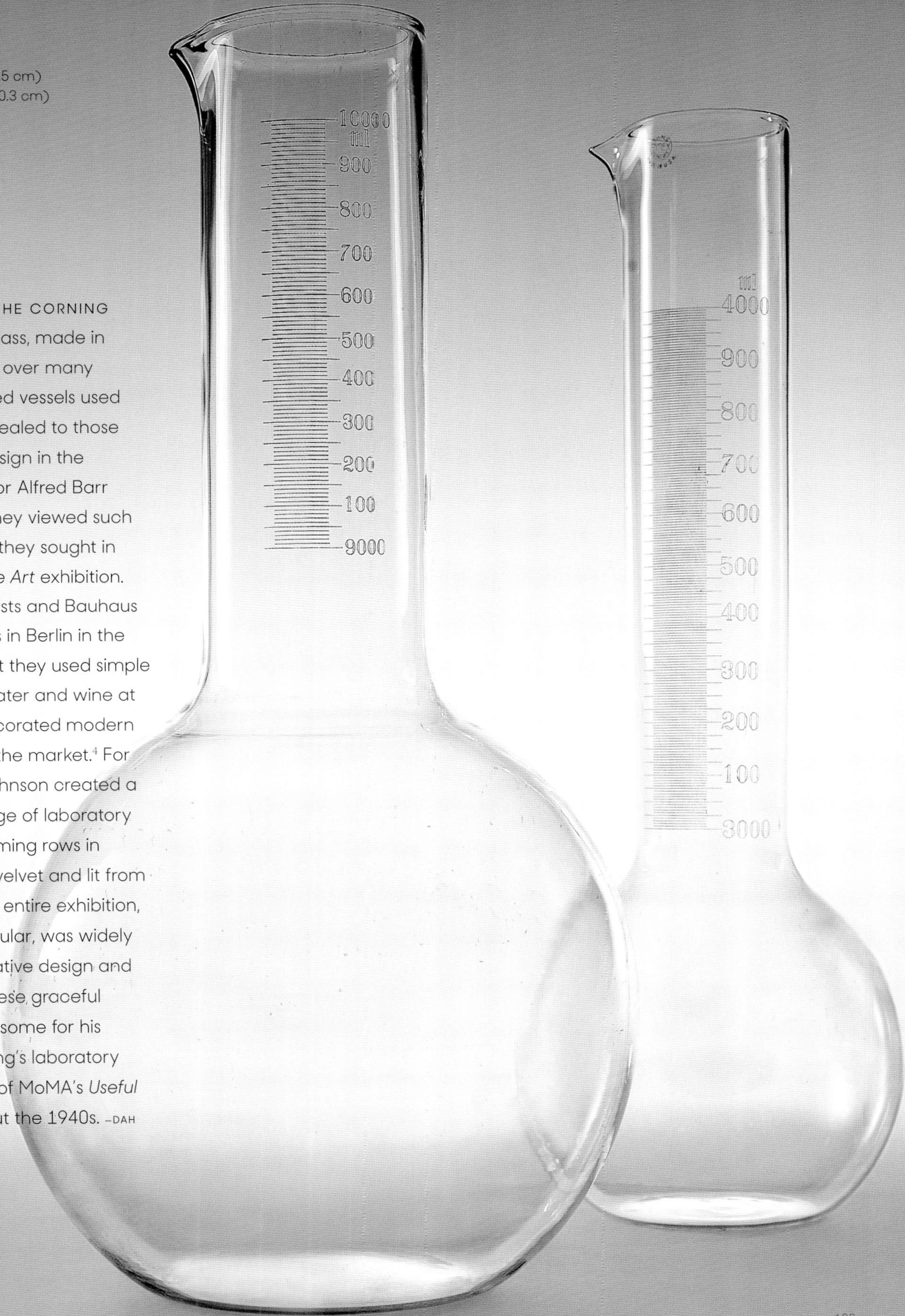

ONE OF THE SPECIALTIES OF THE CORNING Glass Works was laboratory glass, made in forms that changed very little over many years. These simple, unadorned vessels used for scientific experiments appealed to those who championed modern design in the 1930s, including MoMA director Alfred Barr and curator Philip Johnson. They viewed such industrial designs as the ideal they sought in objects for their 1934 *Machine Art* exhibition. Johnson visited prominent artists and Bauhaus masters Anni and Josef Albers in Berlin in the summer of 1933 and saw that they used simple unmarked boiling flasks for water and wine at their dining table, as no undecorated modern decanters were available on the market.[4] For the *Machine Art* exhibition, Johnson created a "Jewel Room" to exhibit a range of laboratory glass forms, displayed in gleaming rows in ascending size on rich black velvet and lit from above.[5] The installation of the entire exhibition, and the Jewel Room in particular, was widely praised by critics for its innovative design and striking beauty. Inspired by these graceful forms, Alfred Barr purchased some for his own personal use, and Corning's laboratory glass was included in several of MoMA's *Useful Objects* exhibitions throughout the 1940s. –DAH

Heinrich Jacob (dates unknown)
Barometer
Designed c. 1932 (D88,298)
Chromium-plated metal, glass,
painted metal
6⅜ x 4 x 4 in. (16.2 x 10.2 x 10.2 cm)
Produced by Zeiss Ikon, Dresden, Germany
K2010.316

THIS SIMPLE DESIGN IS COMPOSED OF A RECTANGULAR shape—a double-glazed gauge with a numbered dial on the interior of the clear glass—on a stepped cylindrical base. The silver needle shows the current atmospheric pressure, and a tiny knob on the back of the vertical support allows adjustment of the black needle to mark the pressure at any given time for later comparison. An American patent was taken out by Heinrich Jacob from Friedenu, Germany, for Zeiss Ikon, presumably to protect this design for sales in the United States. Zeiss Ikon, founded in Dresden as a subsidiary of Carl Zeiss in 1926, was a leading manufacturer of cameras.[6] The business later expanded to include a wider range of products, such as this elegant barometer. The sculpture *Study in Pure Form* of 1924 by the America artist John Storrs also consists of basic geometric shapes in a vertical composition, reminding us of the close connections between design and the fine arts. –DAH

Robert Newton Falge (1892–1981)

Light Output Meter

Designed c. 1935 (utility patent 2,111,585)

Enameled, painted, and chromium-plated steel; metal; Bakelite; glass

12⅜ x 12⅝ x 9¼ in. (31.4 x 32.1 x 23.7 cm)

Produced by Guide Lamp Corporation, Division of General Motors, Anderson, Indiana

K2010.184

THIS METER, WITH ITS STREAMLINED DESIGN AND ANTHROPOMORPHIC FEATURES, provided a way of measuring the light output of any headlamp, and specifically for the light emitted from car headlights. It was originally made for use at car service stations. The test was done by holding the open end of the output meter against the lens of a headlight and reading the value on the light unit meter. Light rays would project through a diffusing lens on the back side of the meter and then converge on a photo-sensitive cell within the housing. The reading was then compared to the output rating for the specific type of headlamp being used. It came with two easily adjustable rheostats to compensate for the different car battery voltages and bulb candle powers. The meter was marketed as a "silent salesman," because the customer could see for themselves whether or not their lights were up to par, which in turn allowed a car service owner to sell replacement items such as lenses, reflectors, and bulbs. –PM

Dr. Scholl's
Automatic
SHOE SIZER

John L. Cobb (1885–1958)
Harry C. Ogle (1882–1949)

Dr. Scholl's Automatic Shoe Sizer
Foot Measure
Designed c. 1927 (utility patent 1,792,892)
Aluminum, stainless steel, enameled metal, glass
2⅝ x 7 x 15⅛ in. (6.7 x 17.8 x 38.4 cm)
Produced by Reliance Die and Stamping Co. for the Scholl Manufacturing Company, Inc., Chicago, Illinois
K2010.191

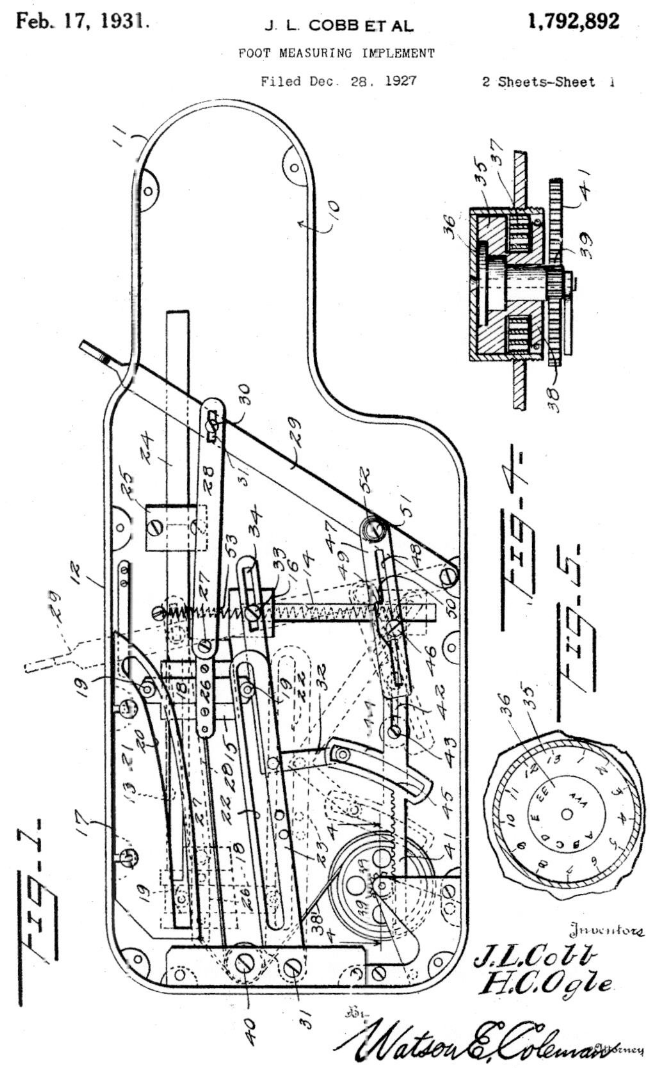

Detail of U.S. utility patent 1,792,892 for a "foot measuring implement."

MEASURING INSTRUMENTS ARE DESIGNED TO PRODUCE INFORMATION THAT ENHANCES the efficient progression of daily activities. During the first part of the twentieth century, manufacturers came up with new devices to measure a person's shoe size. This allowed shoemakers to move away from costly custom-made shoes, and allowed customers to pick their shoes from a pre-established range of sizes. In 1906, podiatrist William Mathias Scholl started the Scholl Manufacturing Company in Chicago. The company created more than one thousand products over the course of a century, and Dr. Scholl's, the internationally known trade name, is still a go-to brand for footwear today. According to the patent application, Cobb and Ogle, both of Baker, Oregon, invented an improved device to more accurately measure and register the width and length of a foot. As competition increased, manufacturers would develop different measuring models for fitting different types of footwear for both adults and children. –PM

Gilbert Rohde (1894–1944)

above left

Table Clock, Model No. 6356

Designed c. 1937

Chromium-plated steel, enameled bronze, aluminum, glass

6⅜ x 11⅜ x 5 in. (16.2 x 28.9 x 12.7 cm)

Produced by Herman Miller Clock Company, Zeeland, Michigan

K2012.10

above right

Table Clock, Model No. 4706

Designed 1933–34

Chromium-plated and enameled metal, glass

6¼ x 6 x 2⅞ in. (15.9 x 15.2 x 7.3 cm)

Produced by Herman Miller Clock Company, Zeeland, Michigan

K2008.617

GILBERT ROHDE'S LINE OF DISTINCTIVE CLOCKS WAS LAUNCHED AT CHICAGO'S 1933 *Century of Progress* exposition, where seven of them were displayed in rooms in the *Design for Living* house.[7] These two clocks were designed in the 1930s, and Model No. 4706 is similar to one that was exhibited. Elegant and stylish, the clocks used stylized or abstract numerals. The simple, modern designs of Rohde's clocks, based on geometric forms, were appropriate for MoMA's 1934 *Machine Art* exhibition, which included an even more austere example in the show—a flat black circle with chromed spheres for numbers. By contrast, Model No. 6356, which was shown in the Herman Miller Clock Company's 1937 sales catalogue, is especially dramatic, with the clock itself suspended over a curved bronze support and aluminum base. A 1937 edition of the Aluminum Company of America's *Aluminum News-Letter* praised it thus: "In support of the argument that a piece of 'moderne' household equipment or furniture doesn't have to have the appearance of being spawned in a hardware store, may we offer the Herman Miller Clock Company's new creation, designed by Gilbert Rohde."[8] Rohde's work for the Herman Miller Furniture Company began in 1932. The company had been producing traditional period-style furniture, but with the Depression forcing many manufacturers out of business, Herman Miller accepted Rohde's proposal for a line of modern furniture. The clocks were to accompany this new line. Herman Miller's bet on modern design paid off; the company remains in business to this day and is regarded as one of the leading furniture companies of the twentieth century. –DAH

Gilbert Rohde (1894–1944)

Table Clock with Thermometer/Hygrometer, Model No. 6381

Designed c. 1932

Nickel-plated steel, Macassar ebony, lacquered wood, enameled aluminum, glass

5½ x 10¼ x 2⅜ in. (14 x 26 x 6 cm)

Produced by Herman Miller Clock Company, Zeeland, Michigan

K2012.52

IN 1929, GILBERT ROHDE BECAME AN INDUSTRIAL DESIGNER AFTER A CAREER in advertising and display design. He is best known for his innovative furniture designs for such renowned companies as Herman Miller, Heywood-Wakefield, and Thonet. In the early 1930s, the Herman Miller Clock Company produced a series of Rohde's sleek, modern table clocks. This clock with thermometer and humidity gauge is made of materials that typify the kind of furnishings that complemented modernist interiors of the 1930s. –PM

Peter Behrens (1868–1940)
Synchron Electric Clock
Designed c. 1910
Enameled steel, glass, brass
10½ x 10½ x 4 in. (26.7 x 26.7 x 10.2 cm)
Produced by Allgemeine Elektricitäts-Gesellschaft (AEG), Berlin, Germany
K2010.73

PETER BEHRENS, GERMAN architect, designer, and teacher, was one of the first modern industrial designers. From 1907 to 1914, he worked for AEG in Berlin, designing factory and office buildings as well as graphics, electrical appliances, and clocks, such as this *Synchron*. He designed several clock variations for AEG, but all had the same simple functional appearance with clearly indicated large black numbers and prominent, angular hour and minute hands.[9] The inner circle of red numerals on the *Synchron* makes it a twenty-four-hour clock, as is commonly used in Europe. –DAH

Max Bill (1908–1994)

Wall Clock

Designed 1954

Glazed porcelain, aluminum, glass, enameled metal

7 x 7 x 2¼ in. (17.8 x 17.8 x 5.7 cm)

Produced by Gebrüder Junghans AG, Schramberg, Germany

K2011.132

THIS CLOCK WAS DESIGNED in 1954 by the artist, designer, educator, and Ulm school founding director Max Bill. The Ulm School of Design, founded in 1953, collaborated with manufacturers such as Gebrüder Junghans, for which Bill designed a collection of watches and clocks, to produce its designs. For Bill, and for Ulm designers broadly, the goal was to apply "good design"—a combination of utility and a simple, clean aesthetic—to everyday objects such as kitchen appliances, audio-visual equipment, tools, and shelving, as seen in his famous 1949 exhibition *Gute Form* with the Deutscher Werkbund. –PM

George Nelson & Associates
(active 1947–83)

right

Charles Pollock (1930–2013)

Popsicle Wall Clock, Model No. 2257

Designed 1957

Walnut, enameled aluminum, brass, map pins

14 x 14 x 3¼ in. (35.6 x 35.6 x 8.3 cm)

Produced by Howard Miller Clock Company, Zeeland, Michigan

K2012.47

below

Asterisk Wall Clock

Designed c. 1953

Painted zinc-plated steel

10 x 10 x 3 in. (25.4 x 25.4 x 7.6 cm)

Produced by Howard Miller Clock Company, Zeeland, Michigan

K2008.161

opposite

Spider Web Wall Clock, Model No. 2214

Designed 1954

Birch, cord, plastic, painted steel

19⅛ x 19⅛ x 2¾ in. (48.6 x 48.6 x 7 cm)

Produced by Howard Miller Clock Company, Zeeland, Michigan

K2010.68

AS WORLD WAR II CAME TO AN END, D. J. DE Pree, president of Herman Miller, hired George Nelson, then a writer at *Fortune* magazine, as the company's design director after seeing photographs of his Storage Wall design in the January 22, 1945, issue of *LIFE* magazine. It impressed De Pree so much that he commissioned Nelson to design a new furniture line for Herman Miller. Soon after, Nelson signed up the Howard Miller Clock Company account (Howard was Herman's brother), and he gave Irving Harper—who, along with Ernest Farmer, was Nelson's only employee at the time—the responsibility of overseeing the design of products for this account. Harper offered several proposed designs to Howard Miller each year. The first clock—Clock 4755, also known as the *Ball* clock—was reminiscent of Alexander Calder's sculptures of the 1930s and 1940s, with its contrast in materials, color, and movement. It was to become an icon of atomic-age design.

Harper designed iterations of the *Ball* clock, many of them with whimsical shapes and names like *Popsicle*, *Spider Web*, *Pretzel*, and *Asterisk*. There was even an eye-shaped clock. At the launch of one of the new collections in 1949, the *New York Times* described one of the clocks as having "a handsome futuristic appearance that should make them suitable even in tomorrow's world."[10] The clocks were radically modern in their day because they have no traditional clock face with numbers. The clear position of the hands in relation to the spokes made numbers unnecessary in the idiom of this design. They were available with a cord and plug or as a "Chronopak," which was mounted in a standard outlet box. The clock faces were like pieces of functional abstract sculptural art, and they became part of the architecture of an interior rather than merely a hanging object. While Harper worked on most of the clock designs, other Nelson employees, such as Lucia DeRespinas and Charles Pollock, contributed designs. Altogether, 150 objects were designed by members of the Nelson office over four decades. –PM

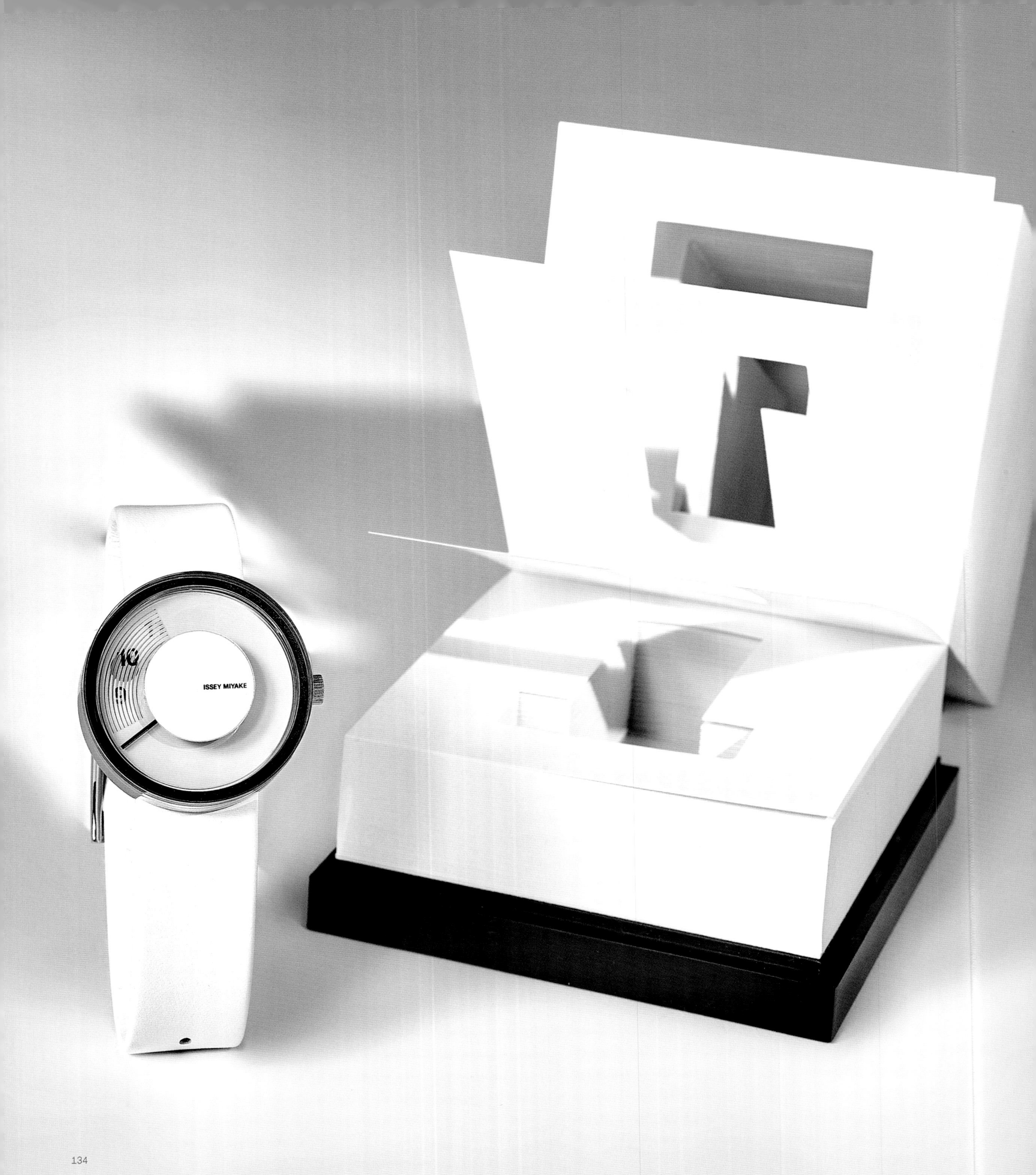
ISSEY MIYAKE

Yves Béhar (born 1967)
for Issey Miyake Inc. (founded 1971)
Vue Watch with original box
Designed c. 2010 (D672,256S)
Plastic, leather, metal, hard-coated mineral glass, stainless steel, paper
Watch: 9½ x 1⅝ x ½ in. (24.1 x 4.1 x 1.3 cm)
Produced by Seiko, Tokyo, Japan
K2011.156

THIS WATCH IS A COLLABORATION BETWEEN JAPANESE FASHION DESIGNER ISSEY MIYAKE, Seiko, and the San Francisco-based designer Yves Béhar, who saw the collaboration as an opportunity to work with a clothing designer whom he admired. In an interview with VernissageTV in 2010, Béhar talked about the concept: "The challenge for me was how can I do something that is worthy of such a master in the world of creativity. The feel of his clothes and of his brand is just as much spiritual as it is physical, so I was trying to match that with a watch." For Béhar, time is precious: "Time is what people are most struggling to find more of; when you ask them, they probably want more time than money." Béhar decided to design a watch that expressed a romantic side to time, rather than focusing on a purely functionalist solution. This is an analog watch, but Béhar wanted the user to experience time without the traditional two hands. While the minutes are shown with a conventional minute hand, the hour is indicated by a number on the dial that appears and disappears through the slit circle on a rotating disc hovering over the face. As the rotating disc turns a full circle in twelve hours, the center of the slit circle reveals the current hour. The end result is a watch that makes time seem to come and go, suggesting a sense of both the past and future in the present. The packaging—also innovative—is composed of a pad of recycled paper with a cut-out for the watch, which can also be used as a notepad. –PM

SANAA
Kazuyo Sejima (born 1956)
Ryue Nishizawa (born 1966)

Three *Neko* Watches
Designed 2011
Polyurethane, acrylic, chromium-plated steel
2⅞ x 1⅜ x 2¾ in. (7.3 x 3.5 x 7 cm) each
Produced for Alessi, Milan, Italy
K2011.127.1–3

THE ARCHITECTURAL FIRM SANAA (KAZUYO SEJIMA AND RYUE NISHIZAWA ASSOCIATED) specializes in urban planning, landscape and interior design, and industrial design. Many well-known architects and designers have created watches for Alessi, including Andrea Branzi, Karim Rashid, Achille Castiglioni, and Marcel Wanders, and the 2011 *Neko* watch is SANAA's addition. What is innovative about the cat-like *Neko* watch—available in pink, green, brown, yellow, and black—is its watchband, which is made of a flexible polyurethane strap that wraps around the wrist without a fastening. –PM

Alfred Hofkunst (1942–2004)

Three *Swatchetable* Watches,
from the One More Time series:
Bonjuhr, *Verduhra*, and *Guhrke*

Designed 1989

ABS plastic, silicone, leather

Gurhke: 9⅞ x 2⅛ x ½ in. (25.1 x 5.6 x 1.3 cm)

Produced by Swatch AG, Biel, Austria

K2014.14.1–3

THE SWATCH WATCH WAS CONCEIVED AS A TRENDY PLASTIC WATCH IN THE EARLY 1980S, AND was developed as a response to Asian-made digital watches that were competing against traditional European-made mechanical watches. The low-priced Swiss watch became popular first in Europe and was cheaper to produce than a conventional watch, because the company was able to reduce the number of parts from the usual ninety-one or more to only fifty-one components. The company collaborated with many artists—painters, sculptors, musicians, filmmakers—to create watch designs, beginning with Kiki Picasso and Keith Haring. The watches became known as "the world's smallest canvas" and the company collaborated with other artists, including Jean-Michel Folon, Mimmo Paladino, Not Vital, Akira Kurosawa, and Pedro Almodóvar. The watches shown here, by Swiss draughtsman and artist Alfred Hofkunst, resemble a strip of bacon with egg, a chili, and a cucumber, respectively, and their names are a play on the German word *uhr*, meaning hour. The irregular shapes and colors made these watches a truly original accessory and a popular collectible. The production run was limited to 9,999 pieces. –PM

Marco Zanuso (1916–2001)
Richard Sapper (1932–2015)
Terraillon 2000 Kitchen Scale
Designed 1969
Plastic
4¾ x 6½ x 4¼ in. (12.1 x 16.5 x 10.8 cm)
Produced by Terraillon, Paris, France
K2008.315

MARCO ZANUSO OF ITALY AND RICHARD SAPPER OF GERMANY DESIGNED the *Terraillon 2000* scales. The weights and copper used in traditional scales are replaced here with a curved elegance, making this everyday object something that might have been displayed on a kitchen counter rather than hidden away in a cabinet. In these scales, Zanuso and Sapper unified for the first time the previously separate elements of display and food receptacle. The conforming top is reversible; the user can choose the appropriate orientation for the quantity and type—solid, liquid, granular, etc.—of the material to be weighed. A magnifier set over the weight gauge allows the user to read it from above rather than bending down to work-surface level to see it properly. This scale represents an example of designing for use and affordability, and millions were sold internationally. —PM

Gino Valle (1923–2003)

Cifra 3 Clock

Designed 1965 (German patent DE1938589U)

Plastic

3⅞ x 7 x 3¾ in. (9.8 x 17.8 x 9.5 cm)

Produced by Solari di Udine, Udine, Italy

K2008.491

ALTHOUGH GINO VALLE PRIMARILY CONSIDERED HIMSELF AN ARCHITECT, HE WAS also a successful designer of appliances. He designed the *Cifra 3* (Digit 3) flip clock for Solari di Udine in 1965. Like many early digital clocks, the *Cifra 3* displays the time using the same split-flap mechanism found in old-style departure/arrival boards in airports and train stations. The clear plastic front allows space for the flipping number flaps. The printed numbers are in Helvetica, a sans serif font championed by Valle's sister and brother-in-law, Lella and Massimo Vignelli, with the hour display numbers in a bolder type than the minutes, which presents the time data in an information hierarchy. It is also a 24-hour clock, as is standard in most of Europe. Later models added a second smaller cylinder as the base support. This bestselling clock was produced from 1966 to 1989, and reissued again in 2015. –PM

DESIGN FOR THE SOCIAL GOOD

Pat Kirkham and Tom Tredway

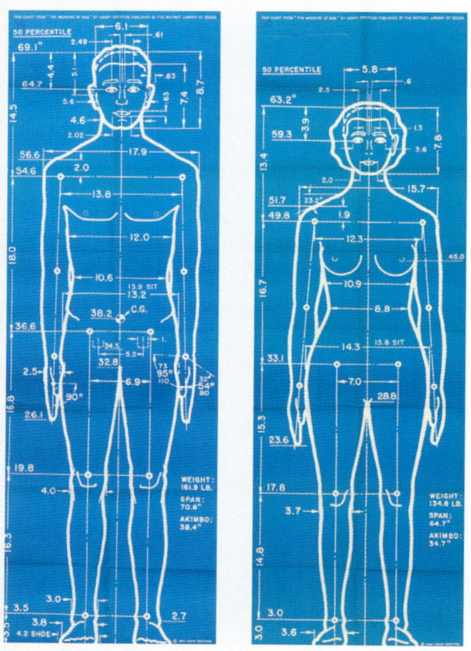

FIG 1
The Measure of Man (Male and Female) posters designed by Alvin Tilley and Henry Dreyfuss Associates, published by Whitney Library of Design, 1969 (first published 1959). Cooper Hewitt, Smithsonian Design Museum.

THIS SHORT SURVEY OF SOCIALLY RESPONSIBLE DESIGN CONCENTRATES ON THE PERIOD from the end of World War II to the present day, when anxieties about the sustainability of the planet are greater than ever and issues of "green," "accessible," and "inclusive" design are a regular feature in discussions among industrial designers and manufacturers.[1] In the postwar period, a small but growing number of consumers and industrial designers criticized particular goods on social and ecological grounds, and beginning in the 1950s, increasingly decried the sheer number of goods being produced. First published in 1957 in the United States, journalist and social critic Vance Packard's bestselling *The Hidden Persuaders* raised questions about the morality of advertisers (and, by implication, the manufacturers who hired them) who used techniques prompted by consumer motivational research to manipulate expectations and desire in a booming economy.[2] His *The Waste Makers*, published three years later, criticized planned obsolescence.[3] The increasing awareness about the deleterious effects of industrial capitalism led to calls to end toxic manufacturing processes, expand user-centered design practices, and increase social sustainability.

Designers interested in user interactions looked to ergonomics, the scientific study of human performance, which had been used during World War II when military personnel, psychologists, engineers, and industrial designers improved the efficiency of all manner of military hardware. Notable non-military applications in the 1950s benefitted from Henry Dreyfuss's book *Designing for People*, which presented ergonomic data in relation to the archetypal users Joe and Josephine (fig. 1).[4] The life-size posters seen in fig. 1 are a later version of Joe and Josephine, first published in 1959. Such data culled from the close study of human form and motion helped Dreyfuss and others design for productivity, comfort, and greater ease of use, and covered all manner of objects, from tractors and telephones to vacuum cleaners and cameras.

Significant changes vis-à-vis socially responsible design occurred during the late 1950s and 1960s, a period of considerable affluence and materialism in North America and Western Europe, when people began calling for greater attention to be paid to the less fortunate and to people with special physical and cognitive needs. Previous improvements for this group had come mostly from family members, caregivers, and people involved in medical or medical-related work, whereas by the 1960s designers took a more active role, both adapting existing products and designing new objects.

Public concern about the health and environmental repercussions of industrialisation continued to grow. In 1969, architect, designer, polymath, and technological utopian R. Buckminster Fuller published *Operating Manual for Spaceship Earth* (fig. 2), which presented the planet as an "integrally-designed machine which . . . must be comprehended and serviced in total."[5] Together with his call for designers to be mindful of the earth's finite resources, this spawned a generation of "design outlaws." Fuller viewed production and consumption as part of a larger, interrelated ecological system, and his interest in alternative social and physical structures for modern living influenced the counterculture movements of the 1960s and 1970s. This growing critical attention to the ecological and social effects of modern lifestyles is best evidenced in the *Whole Earth Catalog* (published in the United States 1968–72 and intermittently until 1998), which advocated conserving the environment; self-sustaining and simple living; non-material-intensive building types; solar energy; organic food; and socially responsible design (fig. 3).

Fuller and the counterculture had an impact on a number of Scandinavian designers who were critical of contemporary Western design and consumption and who, in turn, influenced the Austro-American designer and educator Victor Papanek. Papanek saw design as a cross-disciplinary tool that should be responsive to all human needs, not just the desires of the privileged, and he promoted useful, low cost, non-elite, and ecologically sound products. His 1971 *Design for the Real World: Human Ecology and Social Change* (with an introduction by Fuller) (fig. 4) was first published in Sweden and attacked contemporary advertising practices as fraudulent, and industrial design's focus on styling and rapid turnover as morally irresponsible.[6] He also criticized industrial designers for colluding in the creation of unnecessary objects for the affluent.

FIG 2
below left Cover of R. Buckminster Fuller's *Operating Manual for Spaceship Earth*, published by Southern Illinois University Press, Carbondale, 1970.

FIG 3
below center Cover of the first issue of *Whole Earth Catalog*, 1968.

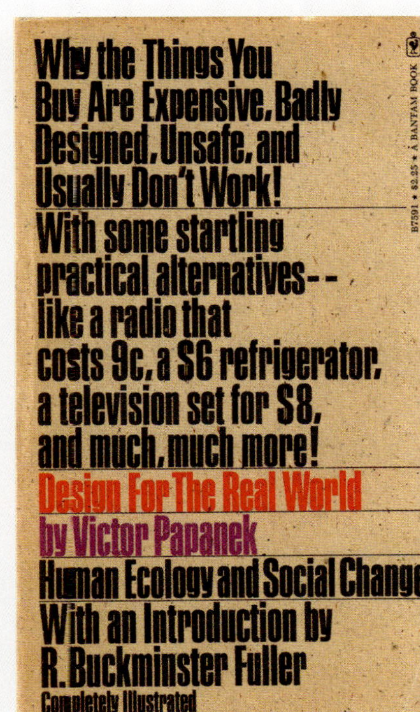

FIG 4
below right Cover of Victor Papanek's *Design for the Real World: Human Ecology and Social Change*, published by Bantam Books, New York, 1973.

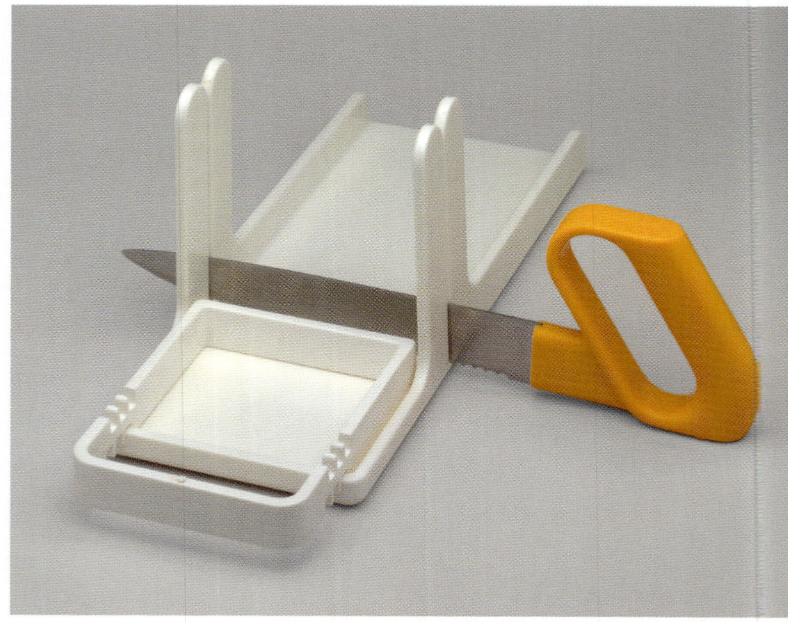

FIG 5
above left Tin-can radio designed by Victor Papanek and George Seeger, North Carolina State College, 1962–67.

FIG 6
above right Kitchen knife and cutting board designed by Ergonomi Design Gruppen, Maria Benktzon, and Sven-Eric Juhlin, 1973, produced by AB Gustavsberg, Sweden. The Museum of Modern Art, gift of RFSU Rehab.

A bestseller, *Design for the Real World* provided examples of products for the "developing world," such as a cooling unit for perishable foods that ran without electricity. Papanek and his student George Seeger designed a radio receiver (1962–67) made of reused tin juice cans that used accessible energy sources like wax and dung, and enabled isolated communities to gain access to modern means of mass communication (fig. 5). These and other examples, such as a portable exercise and play environment designed specifically for children with cerebral palsy in 1968, inspired a new generation of designers seeking to engage in participatory design processes involving potential users. Papanek also called for abolishing design and intellectual property rights so that more people might benefit from designs and ideas, a viewpoint found in contemporary open source design projects.

Universal design, inclusive design, and design for all approaches seek to create products and environments that are useable by as many people as possible, regardless of age or ability. Originating primarily in the United Kingdom and Scandinavia in the 1960s and 1970s, these ideas were taken up internationally by the 1990s. Swedish designers Maria Benktzon and Sven-Eric Juhlin of design firm Ergonomi Design (now Veryday) helped establish these more user-centered, research-based methods in the early 1970s. Their designs resulted from extensive ergonomic studies of users with a range of abilities in their own homes, and their attention to both the formal and functional qualities of products, including a knife and cutting board for Gustavsberg in 1973 (fig. 6), helped bring goods that were appropriate for special needs users with limited physical abilities into the mainstream marketplace. Released in the United States in 1990, OXO Good Grips kitchen tools (fig. 7) were developed by Smart Design in collaboration with Samuel Farber, who lobbied for better products after he observed the frustrations experienced with everyday kitchen tools by his wife, an arthritis sufferer, and others with limited strength and manual dexterity. The improved grips and ease of use ensured that the items sold well, and their successful distribution through big-box mass-market retailers demonstrated that this approach to design appealed to a variety of consumers and users.

FIG 7
above left *Good Grips* peeler designed by Tucker Viemeister, c. 1990, produced by OXO International and Smart Design. Cooper Hewitt, Smithsonian Design Museum, gift of OXO.

FIG 8
above right SL48 portable solar lantern designed by Hedda Beese and Charles Ash, 1985, produced by BP Solar International. The Museum of Modern Art, Emilio Ambasz Fund.

The idea of design as a tool for social good increased across the board in the 1980s and 1990s. The SL48 solar lantern (1985), a collaborative project by Hedda Beese and Charles Ash of Moggridge Associates and the United Kingdom's BP Solar International, was developed in response to requests from humanitarian field workers in Africa and provided a durable, reliable light source that utilized photovoltaic cells to generate power stored by nickel cadmium batteries, all housed in an impact- and water-resistant ABS/polycarbonate shell (fig. 8). Architect-designer Shigeru Ban has used paper tubes to build inexpensive, largely recyclable disaster relief housing, first in Kobe, Japan, following an earthquake in 1995, and more recently in the Philippines, New Zealand, and Haiti. Open Source Ecology, established in the United States in 2003 by Marcin Jakubowski, is developing the Global Village Construction Set, a group of open source blueprints for the fifty machines deemed most essential to contemporary life with the aim of providing low cost tools that allow for the development of decentralized, closed loop production around the world (fig. 9).

The green movement, which began in North America and Europe in the 1960s and early 1970s in response to pesticide use, counterculture ideas, the 1973 Organization of the Petroleum Exporting Countries (OPEC) oil embargo, and the 1979 "oil crisis" (as it was known in the West) led to louder calls for an ecologically sustainable and socially responsible society. Today, new approaches in industrial design that address these issues are arising out of critical analyses of where and how materials are obtained and processed, how products are disposed of, and manufacturing's impacts on human, plant, and animal populations, as well as on scarce natural resources. Analysis of the ecological and social costs of industrial production goes by many names, including cradle-to-grave analysis, life-cycle assessment, and eco-balance analysis.

Many designers now advocate for a complete reconceptualization of industrial systems and processes. They seek a transition from linear systems of production—materials are extracted and processed into products which are then distributed, consumed, and disposed of—to closed loop systems that resemble the cycles of biological nutrients in nature, wherein chemical components are endlessly reused. The biomimetic Cradle to Cradle framework developed in the early 1990s by architect-designer William McDonough and chemist Michael Braungart of McDonough Braungart Design Chemistry is among the most influential of these approaches.

Pressures from environmentally aware consumers also encouraged certain manufacturers to change their policies and products. Herman Miller, an internationally renowned furniture making company based in the United States and long committed to ethical values, established an environmental quality action team in 1989. Two years later, the firm stopped using rosewood veneers on the iconic Eames Lounge Chair and Ottoman (in production since 1956) because the Brazilian trees supplying them were under threat from the slash-and-burn methods of clearing tropical rainforests that were endangering the local ecosystem. The company reintroduced rosewood veneers in 2006 only when a sustainable alternative harvested from certified forests became available. Today, the company's bestselling Aeron Office Chair, first designed in 1994 by Bill Stumpf and Don Chadwick then remastered by Chadwick in 2016, boasts 39 percent recycled materials and is up to 91 percent recyclable. The German design firm Studio 7.5 used the Herman Miller Design for Environment product assessment tool, which was created in collaboration with McDonough Braungart Design Chemistry, when developing the Mirra Chair in 2003 (fig. 10). A Cradle to Cradle product, it made extensive use of recycled materials and could be easily and quickly disassembled at the end of its life using standard household tools, allowing for efficient reuse, recycling, and minimal waste.

Eco-labeling schemes forced manufacturers to offer more environmentally friendly products in order to compete in environmentally aware consumer markets. The New

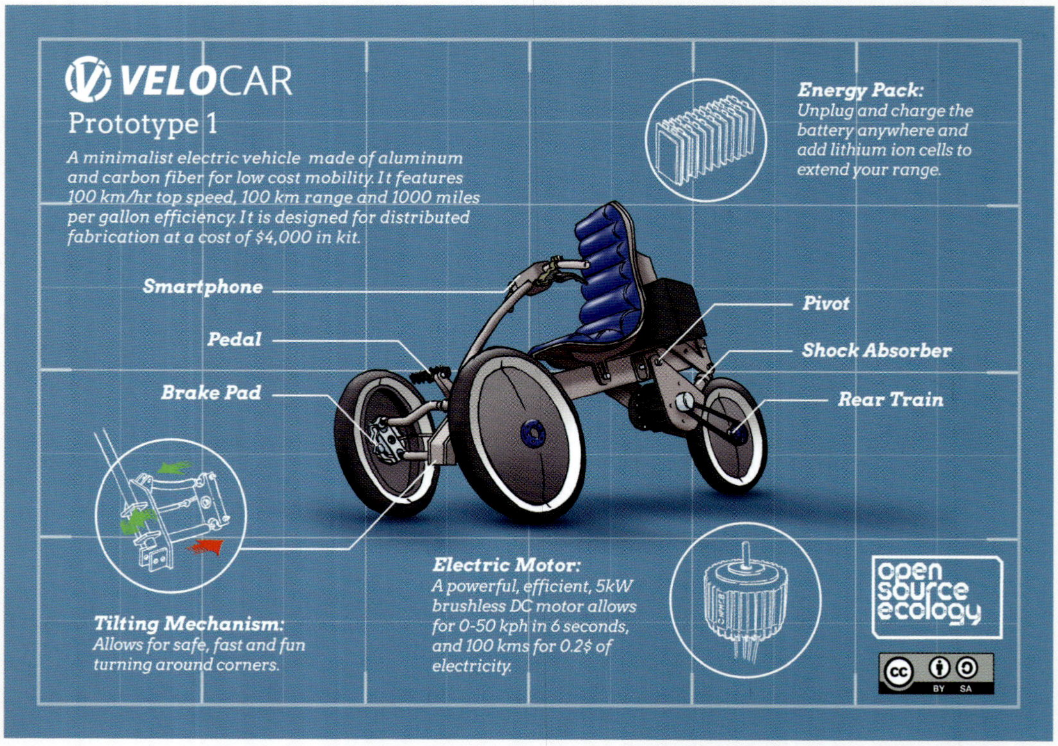

FIG 9
VeloCar Prototype 1 designed by Marcin Jakubowski, 2014, Open Source Ecology, Maysville, Missouri.

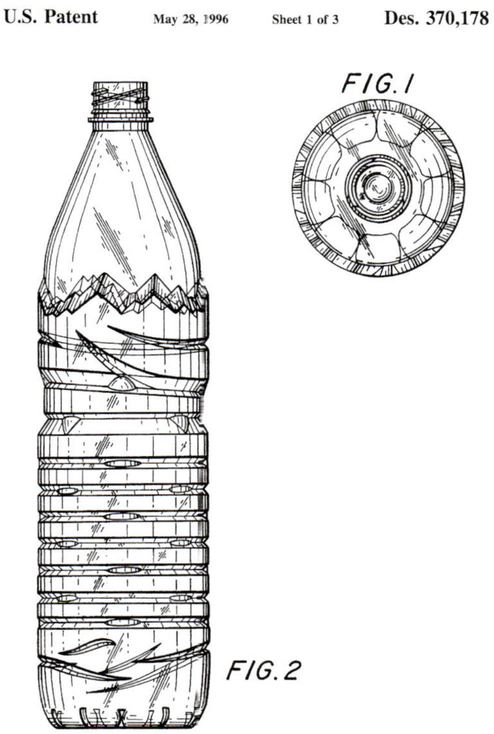

Wave washing machine, developed by Hoover and released in 1993, set a high standard for energy and water efficiency and low detergent consumption without sacrificing high wash performance. It also helped expand the US company's market share in eco-conscious European markets. The Greenfreeze refrigeration system, developed in part by Greenpeace, replaced ozone-depleting chlorofluorocarbons with propane and butane refrigerants, and major manufacturers such as Siemens and Bosch introduced Greenfreeze products in 1993. Italian appliance producer Zanussi redesigned its washing and dishwashing machines in order to save water, energy, and detergent, while incorporating more recyclable materials and modular, easy-to-disassemble components into its products.

Since the 1970s, eco-friendly consumers have campaigned against unnecessary and excessive packaging in plastic (derived from oil) and card (derived from timber) that add to transportation, recycling, and disposal costs. Alternatives available since the 1990s include biodegradable, plant-based plastics, such as Mater-Bi and Biopol. Bioplastics are increasingly used for packaging, although the percentage remains small compared to petroleum-based

FIG 10
above left Recyclable parts of Herman Miller's *Mirra* chair designed by Studio 7.5, 2003.

FIG 11
above right Detail of U.S. design patent D370,178 for a "collapsible bottle," produced as a recyclable PET Evian water bottle, 1995.

CLEVER
LITTLE BAG
WELL IT'S SMARTER THAN AN OLD FASHIONED SHOEBOX BECAUSE IT USES 65% LESS PAPER. EVEN BETTER, IT MEANS YOU DON'T NEED AN EXTRA CARRIER BAG AND YOU CAN USE IT OVER AND OVER AGAIN. CLEVER HUH? FOLLOW THE PUMA ECO-TABLE. REUSE THIS BAG.
www.puma.com/cleverlittlebag
THE PUMA ECO-TABLE
CLEVER
LITTLE BAG

FIG 12
opposite *Clever Little Bag* designed by Yves Béhar and fuseproject for Puma, 2010.

plastics, and their creation, as with biofuels like ethanol, increases the global rate of conversion of grasslands and woodlands into agricultural land, thereby complicating their positive ecological impacts. Improvements in petroleum-based plastic packaging focus on increasing recyclability, reducing amounts of material, and removing chemical components that are hazardous to health. In 1995, the French multinational Danone redesigned its Evian plastic water bottle as a collapsible version in PET (polyethylene terephthalate), a clearer, less rigid, and more impact resistant, chemically inert, UV-stable, and recyclable plastic than the PVC (polyvinyl chloride) previously used (fig. 11). The result was a lighter, recyclable bottle made of 37 percent less material. The Clever Little Bag (2010) by pro-sustainability designer Yves Béhar (fig. 12) replaced Puma's conventional laminated cardboard shoeboxes, which were not recyclable and were resource- and energy-intensive to make. The "bag," a cardboard cradle in a recycled plastic bag, eliminated the need for laminated board, reduced the cardboard use by 65 percent, and reduced the water, energy, and fuel involved in manufacturing and shipping it by about 60 percent.

Other contemporary approaches include an emphasis on closed loop systems like Cradle to Cradle, biomimetic approaches championed by Janine Benyus, materials certification by organizations such as the Forest Stewardship Council, and modularity in design components, as seen in Google's Project Ara cell phone. Grassroots sustainable design developments continue, as do attempts to increase the psychological connections between users and products in order to reduce waste through emotionally durable design strategies, including the use of narrative, intentional changes in surfaces over time, and emotional attachments between users and objects. The impacts of such efforts, however, still pale in comparison to the scope of the social and ecological challenges facing humanity in the twenty-first century. As William Morris recognized in the 1880s, design alone may not be equipped to bring about the broad shifts necessary to create sustainable systems of production and consumption that truly address human needs and sustain the planet.

ENDNOTES

1 Special thanks to Christian A. Larsen, Sarah A. Lichtman, and Catherine L. Whalen, whose contributions to "Europe and North America 1945–2000," in *History of Design: Decorative Arts and Material Culture, 1400–2000*, ed. Pat Kirkham and Susan Weber (New Haven: Yale University Press, 2013) informed this chapter.

2 Vance Packard, *The Hidden Persuaders* (New York: David McKay, 1957).

3 Vance Packard, *The Waste Makers* (New York: David McKay, 1966).

4 Henry Dreyfuss, *Designing for People* (New York: Simon and Schuster, 1955).

5 R. Buckminster Fuller, *Operating Manual for Spaceship Earth* (Carbondale: Southern Illinois University Press, 1969), 52.

6 Victor Papanek, *Design for the Real World: Human Ecology and Social Change* (New York: Pantheon Books, 1971).

05
objects

PREPARING & STORING FOOD

OBJECTS LINKED TO THE PREPARATION OF FOOD ARE AMONG THE most ubiquitous in our everyday lives, and the proliferation of them is a measure of the way in which, over time, preparing food has become ever more sophisticated. Whether in a domestic or a commercial kitchen, making food is undertaken by almost everyone, and the range of objects linked to that activity is enormous. The Kravis collection contains a wide range of these tools, machines, and objects, from hand whisks, potato peelers, and Tupperware containers to electrically-powered machines, such as blenders, toasters, and refrigerators. While some are meant to adorn kitchen counters, and are designed accordingly, others are more humble, anonymous tools intended for the cook's eyes only.

Designer Unknown

Balanced Recipes Cookbook by
Mary Ellis Ames (1887–1968)

Published 1933

Aluminum, paper, cardboard, steel

9⅛ x 6½ x 1½ in. (23.2 x 16.5 x 3.8 cm)

Produced for Pillsbury Flour Mills
Company, Minneapolis, Minnesota

K2008.207.2

THIS HINGED BINDING FOR A RECIPE BOOK IS among the many inventive uses of aluminum from the 1930s. Its streamlined design focuses on the title within a black circle with five bold speed lines to the left, providing a dynamic sense of movement. Inside is a multi-ringed binder with tabbed sections for different types of recipes, from soups to breads to cookies. Manufacturers promoted the use of aluminum for modern kitchenware through aggressive advertising. With declining domestic help, consumers responded to the advertising claim that the bright new material would provide cleanliness, efficiency, and health.[1] The aluminum-bound cookbook was part of this promotion, shared by two related industries: baking and aluminum cookware manufacturers. *Balanced Recipes* sold for $1.25 when it was produced in 1933, the first of three editions.[2] The author, Mary Ellis Ames, was the director of Pillsbury's Cooking Service and ran its Cookery Club. Members could send in vouchers from Pillsbury products in exchange for recipes and tutorials, which they would collect in the aluminum binding.[3] –DAH

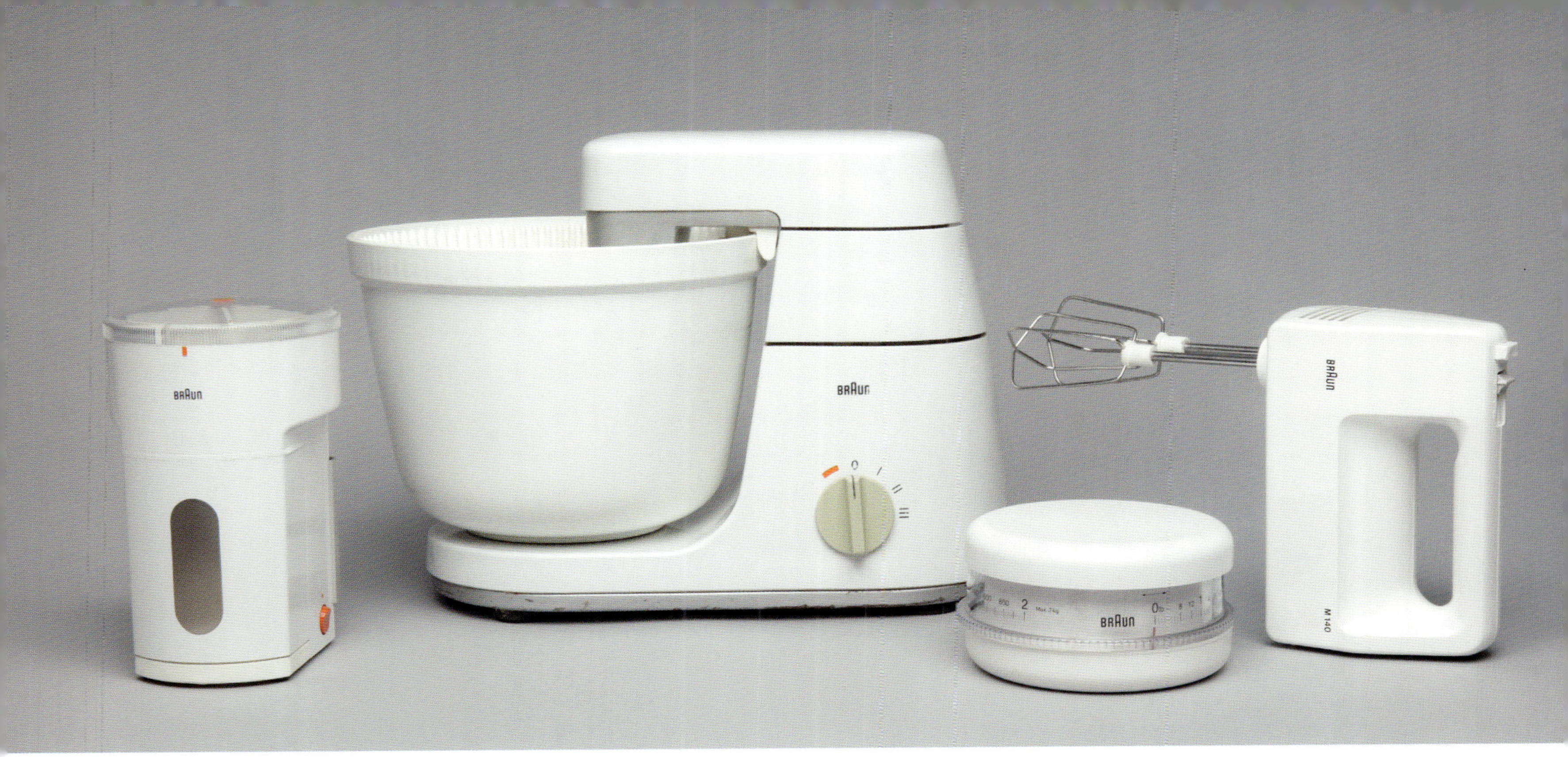

left to right

Dieter Rams (born 1932)
Coffee Grinder, Type KMM 2
Designed 1969 (D218,257S)
Plastic, stainless steel
7¼ x 4¾ x 4¾ in. (18.6 x 12.1 x 12.1 cm)
Produced by Braun AG, Frankfurt, Germany
K2008.398

Gerd Alfred Müller (1932–1991)
Robert Oberheim (born 1938)
Stand Mixer, Type KM 32
Designed 1964
Plastic, metal, glass
10¼ x 14 x 9½ in. (26 x 35.6 x 24.1 cm)
Produced by Braun AG, Frankfurt, Germany
K2010.110

Hartwig Kahlcke (born 1942)
Kitchen Scale, Model No. 4243
Designed 1984
Plastic
2⅞ x 5⅝ x 5⅝ in. (7.5 x 14.3 x 14.3 cm)
Produced by Braun AG, Frankfurt, Germany
K2008.399

Reinhold Weiss (born 1934)
Hand Mixer, Type M 140
Designed c. 1966 (D207,859)
Plastic, stainless steel
6¾ x 5 x 11⅜ in. (17.1 x 12.7 x 28.9 cm)
Produced by Braun AG, Frankfurt, Germany
K2010.81

SOME OF THE MOST ICONIC EXAMPLES OF POSTWAR GERMAN INDUSTRIAL DESIGN were designed for Braun by a stable of designers including Reinhold Weiss, Gerd Alfred Müller, Robert Oberheim, Hartwig Kahlcke, and Dieter Rams. Weiss was the first of two graduates of the important Ulm School of Design, and upon graduation in 1959 to 1967, he worked as a product designer for Braun. In 1962, he became the deputy director of the company's design department, and was responsible for the company's household appliances and personal care products. In 1967, he moved to Chicago and started his own firm, Reinhold Weiss Design, and ultimately settled in Evanston, Illinois, where he worked with companies such as the hi-fi devices manufacturer New Acoustic Dimension (NAD) and Proton.

When Dieter Rams joined Braun in 1955 (he became chief design officer in 1961, a position he retained until 1995), he and his colleagues produced electronic gadgets that were distinguished by their austere aesthetic and user friendliness. They were so popular that Braun became a household name in the 1950s. Rams would later develop his ten principles of design in answer to the question "What is good design?" His response was that it must be innovative, useful, beautiful, understandable, unobtrusive, honest, long-lasting, thorough down to the last detail, environmentally friendly, and designed with a "less but better" mindset. Among Braun's mainstay product lines was kitchen equipment; many of these forms continue to be produced today. White was a preferred color scheme for the company's minimalist designs. –PM

Henry Dreyfuss (1904–1972)

Five Kitchen Utensils

Designed c. 1935 (D116,555)

Chromium-plated steel, painted wood, wire mesh

Large spoon: 11¾ x 2⅝ x 1¾ in.
(29.8 x 6.7 x 4.5 cm)

Produced by the Washburn Company,
Worcester, Massachusetts

K2012.166.8,9,11,13,15

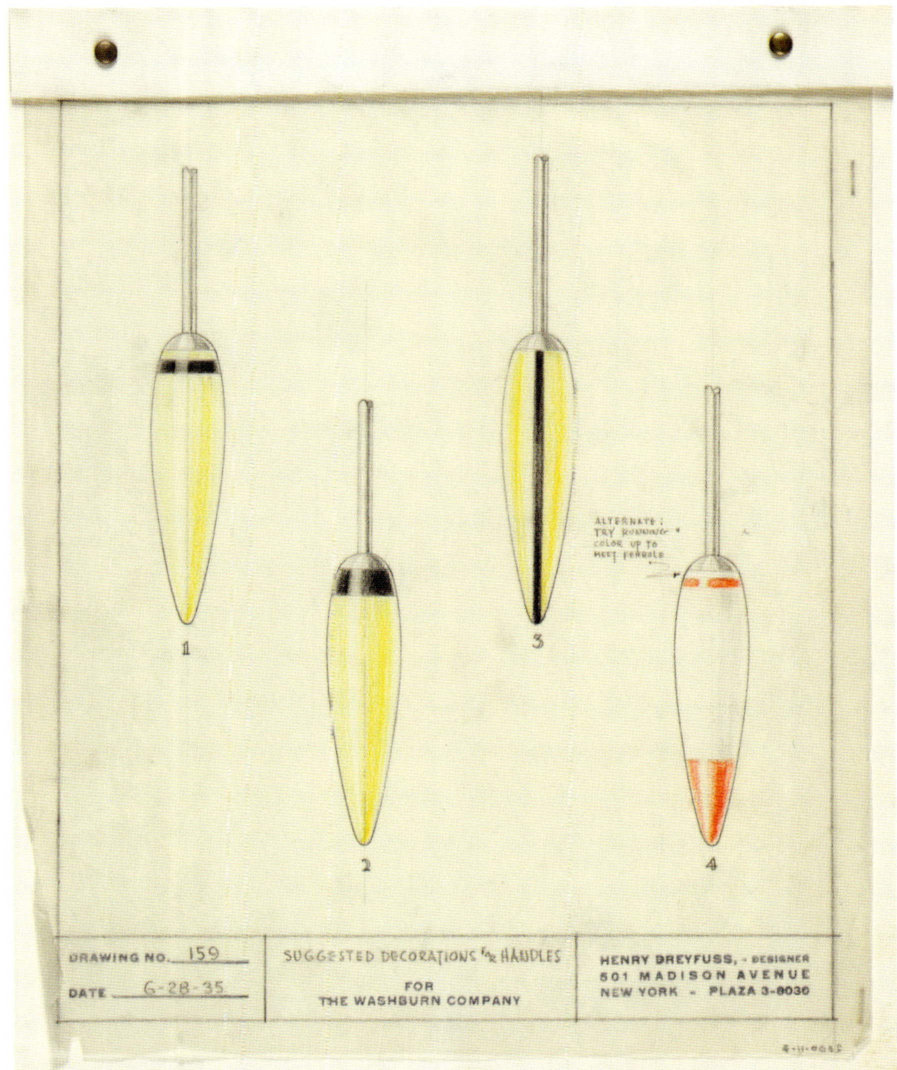

Henry Dreyfuss design drawing *Suggested Decorations for Implement Handles* for the Washburn Company, 1935. Cooper Hewitt, Smithsonian Design Museum.

THIS UPDATED DESIGN OF THE WASHBURN COMPANY'S KITCHEN UTENSILS BY HENRY DREYFUSS was quite an improvement over the original: the decoratively grooved handles of the earlier set were replaced by simple teardrop-shaped handles in green, canary yellow, or red.[4] They reflected the ergonomic principles Dreyfuss advocated, put forth in his 1955 book *Designing for People*. The fussy perforated designs of the earlier utensils were replaced with an integrated, geometric configuration. The Washburn Company's president was impressed with the Dreyfuss design firm's sensible approach, and the commission was for an entire line of inexpensive kitchen implements.[5] The new forms and colors made the line modern. –DAH

Donald E. Grove (1907–1973)
Jackson D. Comstock (1890–1973)

Two *Hollywood Liquefier* Blenders
Designed c. 1941 (D132,762 and utility patent 2,328,526)

opposite left
Model No. 41
Bakelite, glass, plastic, steel, rubber, aluminum
14¼ x 8½ x 5¾ in. (36.2 x 21.6 x 14.6 cm)
Produced by Hollywood Liquefier Corp., Glendale, California
K2012.189

opposite right
Model No. 51
Bakelite, plastic, steel, rubber
14¼ x 8½ x 6⅛ in. (36.2 x 21.6 x 15.6 cm)
Produced by Hollywood Liquefier Corp., South Pasadena, California
K2008.230

Folded recipe brochure accompanying the *Hollywood Liquefier*, Model No. 41.

LIKE A SHIP'S PROW BREAKING THROUGH WAVES, THESE BLENDERS TYPIFY AMERICAN streamlining with their teardrop forms and sense of forward movement—reminding us that streamlining began with modes of transportation, including ocean-going ships. The shaped lids have rudder-like finials, reinforcing the nautical theme. Functionalists at mid-century objected to the use of aerodynamic forms for objects not intended to move, but today the wit and humor of such designs is appreciated. The patent drawing for these blenders claimed that the invention allowed one to extract juice quickly, which helps preserve the vitamins that start to break down when vegetable juices are allowed to stand rather than consumed immediately. The illustrated leaflet for Model No. 41, shown here, demonstrates how the liquefier could be used to provide "greater strength, vitality and energy from fresh fruit and vegetables." The two blenders are from the same original design but were produced at different periods and at different production locations, hence the numerous small differences, including color, model number, and jug materials, one plastic and one glass. –DAH

Attributed to

Raymond L. Barton (1885–1960)

Top-O-Stove Potato Baker with original box

Designed c. 1935 (utility patent 2,096,726)

Aluminum, metal

3⅞ x 7⅛ x 4⅛ in. (9.8 x 18.1 x 10.5 cm)

Produced by Na-Mac Product Corporation, Los Angeles, California

K2010.13.1

IF YOUR GOAL IS TO BAKE A SINGLE POTATO, this gadget is for you. Skewered in this ovoid capsule, the potato could be cooked faster in the aluminum enclosure over a gas or electric stove burner. The accompanying directions promised better flavor, retention of vitamins, shorter baking time, and ease of cooking on top of the stove rather than heating the whole oven, thus saving energy. Such appeals to better health, speed, and convenience were often selling points for modern design in the 1930s. The drawing on the packaging shows the cooker opening horizontally rather than vertically and with two narrow supports rather than a row of glides at the base. Although Na-Mac specifically marketed this gizmo as a "kitchen accessory," the patent listed one of the invention's objectives as for use over campfire or other open flame.[6] While no design patent has been found, a 1935 utility patent drawing filed by Raymond L. Barton depicts the potato baker in a design similar enough to suggest he was the designer of the form. This small object has a commanding presence due to its aerodynamic shape, which embodies key characteristics of American streamlined design. With its curved form, reminiscent of a Zeppelin airship, and conforming handle, it looks ready for take-off. –DAH

William P. Cameron

(dates unknown)

Clover Leaf Portable Electric Range, Model No. D

Designed c. 1946 (utility patent 2,446,063)

Aluminum, porcelain, rubber, plastic, steel

Closed: 5⅞ x 8 x 11¾ in. (14.9 x 20.3 x 29.8 cm)

Produced by Camco Products, Inc., New York, New York

K2008.293

PERFECT FOR SMALL APARTMENTS, THIS PORTABLE electric range made it possible to heat up to three vessels simultaneously on circular ceramic plates with electrical coils. The three plates open out from a single fulcrum for use, and close to a compact form for storage.[7] Whether open or closed, the structure looks architectonic, even futuristic, and characterizes the aerodynamic ideals of streamlining. "Electric range" usually refers to an electric stove installed in a kitchen, but this portable range could perform the same function anywhere with access to an electrical outlet. –DAH

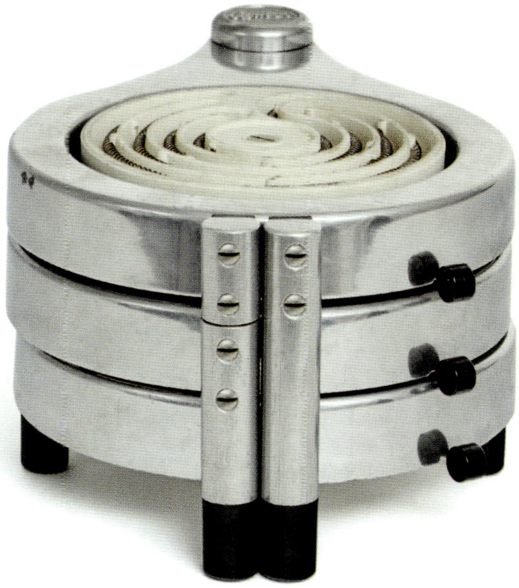

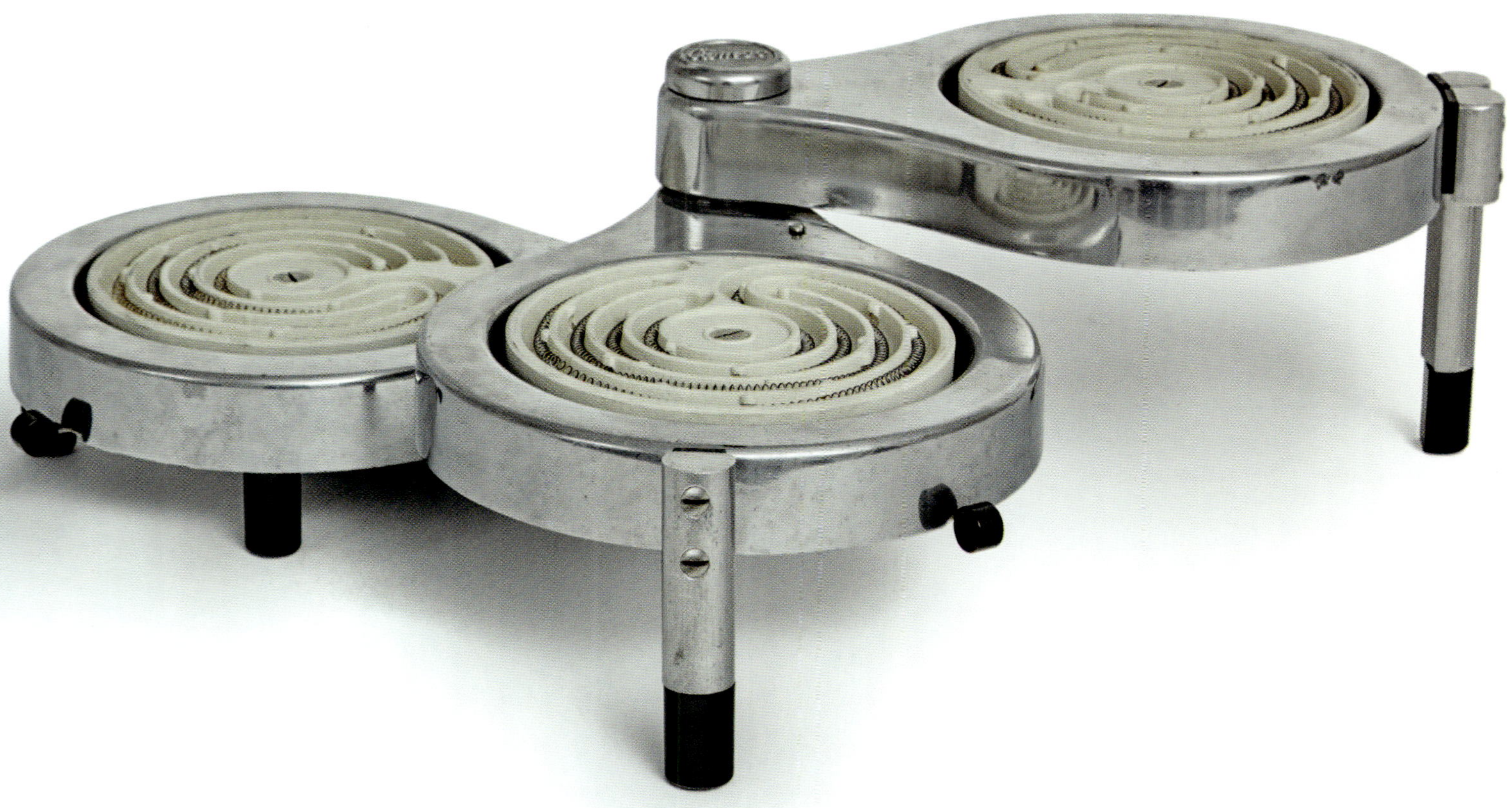

Designer Unknown

Elenova Coffee Grinder

Designed c. 1950

Chromium-plated steel, plastic, Bakelite, rubber

6½ x 9½ x 4¼ in. (16.5 x 24.1 x 10.8 cm)

Produced by KYM (Kissing & Möllmann), Islerlohn, Germany

K2010.49

IN THE NINETEENTH CENTURY, GRINDING ROASTED COFFEE BEANS FOR BREWING GREW in popularity, and a variety of machines were developed for this purpose. The forms tended to be awkward, with containers for the whole coffee beans leading down into the grinding mechanism, which was operated by a crank on the top or side, and a container below to catch the grounds. Kissing & Möllmann, established in 1832, was a leading manufacturer of these products. The shape of the coffee grinder evolved to this unified, streamlined form, which replaced the cumbersome hand crank with electrical power and a simple on/off switch near the base. A hinged plastic lid on the top opens to a reservoir for the whole beans. A separate plastic cup with a punched metal lid and fin handle fits neatly beneath the grinding mechanism, completing the ovoid shape of the machine's front. Its production in the early 1950s reflects the effect of American streamlining on European design after the war; it resembles a train engine or automobile. A pair of long grooves, one on each side, serves as handles. –DAH

left

Designer Unknown

Speedie Toaster

Designed 1950s

Chromium-plated metal, plastic

6½ x 7⅞ x 4⅞ in. (16.7 x 20 x 12.4 cm)

Produced by Speedie Electrical Industries, Cheltenham, Victoria, Australia

K2008.163

right

Designer Unknown

UFO Toaster, Model No. 581

Designed 1952

Chromium-plated and enameled metal, Bakelite

7½ x 8⅛ x 4½ in. (19.1 x 20.6 x 14.4 cm)

Produced by Maybaum, Germany

K2008.612

BOTH OF THESE MID-CENTURY TOASTERS REFLECT THE NEW AESTHETIC WHEN THE STREAMLINED designs of the 1930s and 1940s gave way to more geometric or bulbous forms. Vertical perforations on the doors of both toasters serve as vents: seven straight lines grace the bulbous body of the German *UFO*, while five straight lines of different lengths adorn the Australian *Speedie*, as well as four more lines wrapped around each curved corner from the sides to the top. The yellow enameled metal base and Bakelite handles and switch of the *UFO* toaster contrast with its chromium-plated body. Both machines have toasting compartments on each side accessed by spring-loaded hinged doors that snap shut when released. The bread was toasted by placing it against the vertical grille that guarded the central heating element, which is made up of a series of thin wires. Two slices of bread could be toasted simultaneously, but unlike American-invented automatic pop-up toasters, these required the user to manually flip each slice to toast the second side and turn off the toaster when done. –DAH

Alfonso Iannelli (1888–1965)
Toaster, Model No. T-9
Designed c. 1939
Chromium-plated steel, Bakelite
7½ x 10 x 6¾ in. (19.1 x 25.4 x 17.1 cm)
Produced by Sunbeam Corporation,
Chicago, Illinois
K2008.165

ALFONSO IANNELLI IMMIGRATED TO THE UNITED STATES FROM ITALY IN 1898. HE WAS trained as a sculptor, and it was his figural sculpture for Frank Lloyd Wright's Midway Gardens in 1913–14 that brought him to Chicago and made him famous. Beginning in the 1920s, Iannelli designed products for Sunbeam and other firms. Although his sculpture and graphic designs were cubist, his industrial designs were streamlined. In the late 1930s, most toasters were rectangular, but Iannelli's semicircular form provided a sense of aerodynamic movement.[8] Automatic pop-up toasters were not new, but the innovative option to have the toast pop up when done or be kept warm inside the toaster, along with the curved exterior, gave this design a sense of modernity. The toaster was featured in the monthly newsletter of Durez, the company that made the Bakelite base and lever covers, with commentary that, though somewhat hyperbolic, revealed the importance of aesthetics in industrial design at the time: "The majority of product designers long ago came to the conclusion that wherever a combination of self-insulation against moderately high heats and a fine permanent finish is required, molded Durez parts are the ideal solution."[9] –DAH

right Advertisement for *Sunbeam Automatic* toaster, published in the *Saturday Evening Post*, September 13, 1947.

Jasper Morrison (born 1959)
Toaster
Designed c. 2002 (D497,279)
Polypropylene, stainless steel
7⅝ x 15⅛ x 4½ in. (19.5. 38.6 x 11.4 cm)
Produced by Rowenta, France
K2008.290

BRITISH DESIGNER JASPER MORRISON ATTENDED THE SCHOOL OF ART AND DESIGN AT KINGSTON College and the Royal College of Art. Working with well-known brands such as Alessi, Vitra, and Muji, he has designed everything from cutlery to phones, bus shelters to prefab houses. Morrison's philosophy of design centers around what he calls "objectality": "Nowadays, even a power drill needs to send out a message to prospective customers. While a product is born for industry, it lives the rest of its life with the person who buys it, and it's the emotional response it elicits, or 'objectality,' that is the key to its success."[10] For the design of this stripped-down and functional yet elegant toaster for the French household appliances manufacturer Rowenta, Morrison studied other toaster appliances on the market. His minimalist aesthetic transforms this everyday object into one of timeless beauty, with its slim lines, a quiet mechanism, and a gentle beeping sound when the toast is done. The motorized operation of the toaster quietly lowers and raises the bread at the touch of a button. A brushed steel warming tray top efficiently and rapidly warms, and keeps warm, all types of rolls and breads, and a photo-sensor browning device carefully measures the browning process, no matter what type of bread. Morrison also designed a kettle and coffeepot as part of the same set for Rowenta. —PM

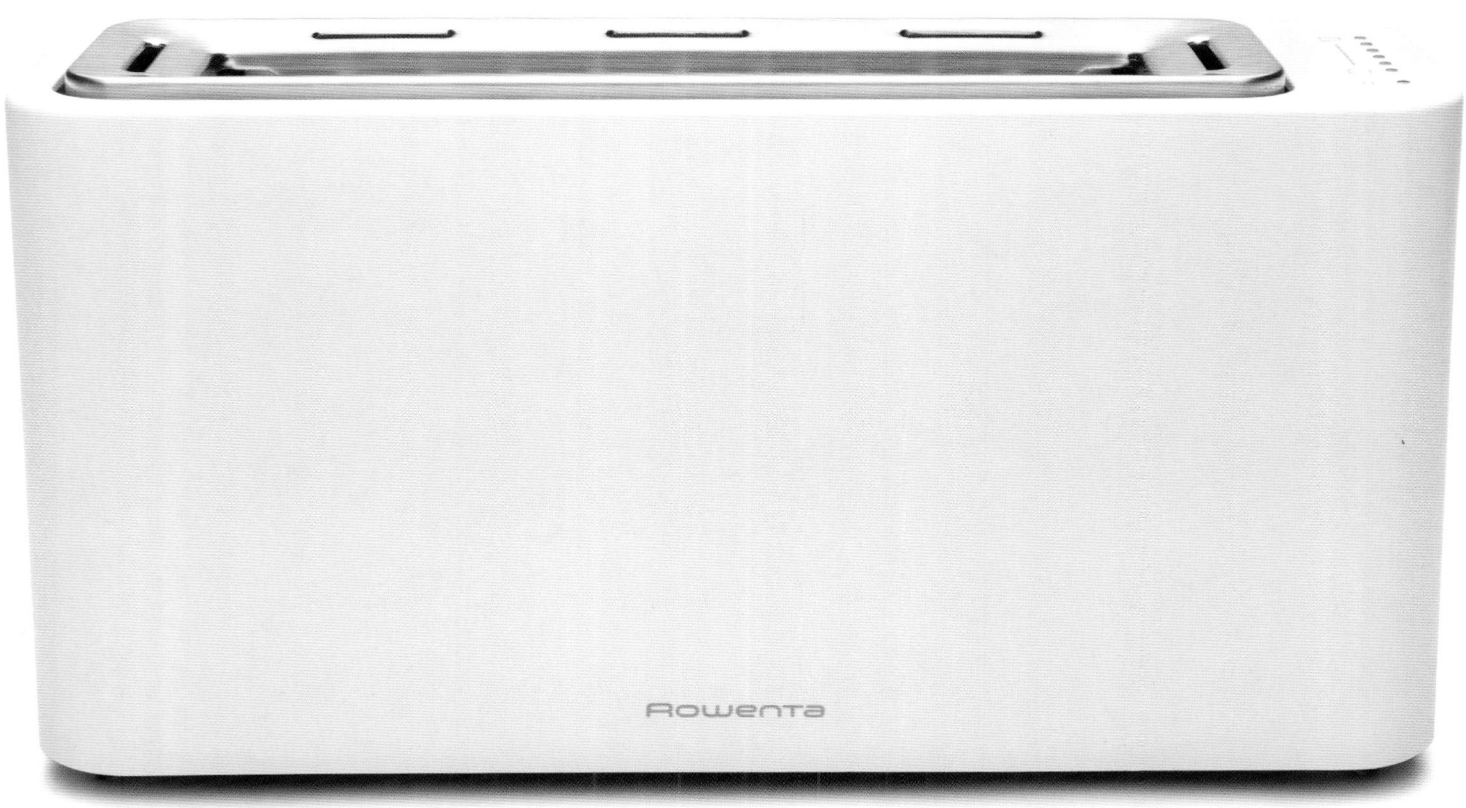

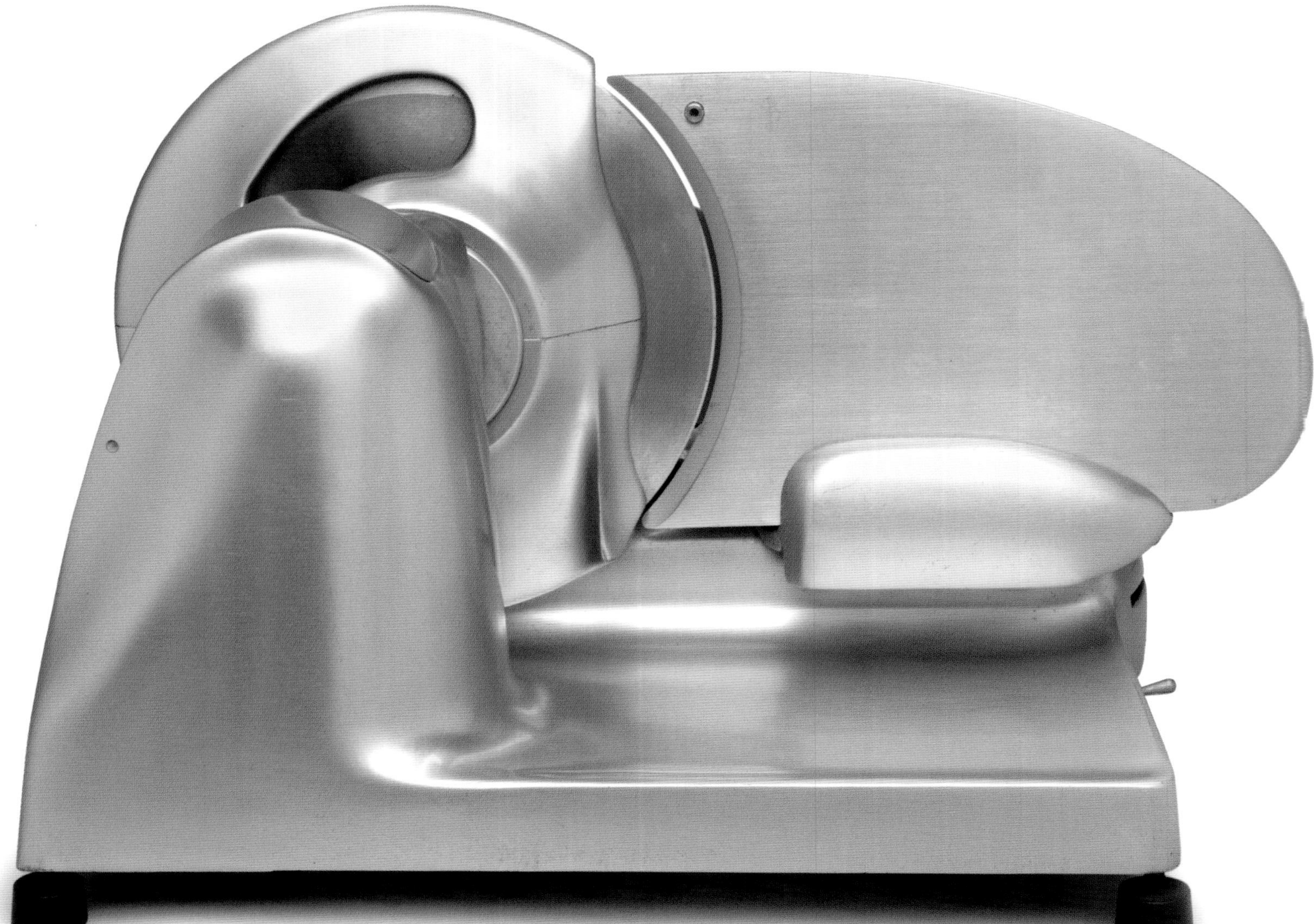

Egmont H. Arens (1888–1966)
Theodore C. Brookhart (1898–1942)

Streamliner Meat Slicer, Model No. 410
Designed c. 1942 (D135,739)
Aluminum, steel, rubber
13⅛ x 20 x 17¼ in. (33.3 x 50.8 x 43.8 cm)
Produced by Hobart Manufacturing Company, Troy, Ohio
K2012.155

Boxing by Alexander Archipenko, illustrated in *Art and the Machine* by Sheldon and Martha Cheney, 1936.

THIS MEAT SLICER WAS INTENDED FOR COMMERCIAL USE IN SHOPS AND RESTAURANTS rather than in domestic kitchens. Rounded cast aluminum forms sheath the motor, blade, and other moving parts. Streamlining gives this static object an aerodynamic form. The multiple smoothly curved parts transform a functional object into sculpture—comparable to Alexander Archipenko's *Boxing* from 1935. The meat slicer's curved, continuous surfaces were easy to clean, eliminating crevices where food might become lodged. Its production continued from 1944 to 1985, an indication of its usefulness.[11] Egmont H. Arens, a leading American industrial designer, headed the industrial styling division of the advertising firm Calkins & Holden in New York from 1929 to 1935 before opening his own design firm. In 1932, with Roy Sheldon, he coauthored *Consumer Engineering: A New Technique for Prosperity*, which promoted economic recovery during the Depression by appealing to consumer psychology through improved engineering and streamlined design. The slicer was designed in collaboration with Hobart employee Theodore C. Brookhart, who was responsible for the mechanical details. —DAH

Timo Sarpaneva (1926–2006)

V-40 Covered Cooking Pot

Designed 1959

Enameled cast iron, teak

6½ x 8 x 7¾ in. (16.5 x 20.5 x 19.7 cm)
(with handle in carrying position)

Produced by W. Rosenlew and Co.,
Pori, Finland

K2014.17

Cooper Hewitt, Smithsonian Design Museum,
promised gift of George R. Kravis II

THIS CAST-IRON CASSEROLE FOR W. ROSENLEW and Co. has become an iconic work of organic design. Timo Sarpaneva's grandfather was a blacksmith and the inspiration for this creative reinterpretation of tradition. The versatility and beauty of enameled cast iron makes the pot ideal for use in the oven, on the stovetop, and at the table. It comes with a removable teak handle both for lifting the lid (as shown here) and for carrying the pot. The pot won a prize at the Milan Triennale in 1960 and an International Design Award from the American Institute of Interior Designers in 1963. It was even featured on a 1998 Finnish postage stamp. It continued in production until the late 1980s. –PM

left

Stefano Giovannoni (born 1954)
Mami Kettle, Model No. SG65
Designed 2003
Stainless steel, polyamide
8⅛ x 8½ x 7½ in. (20.6 x 21.7 x 19.2 cm)
Produced by Alessi, Milan, Italy
K2010.101

right

Philippe Starck (born 1949)
Hot Bertaa Water Kettle
Designed 1987
Enameled aluminum, plastic
9¾ x 12½ x 6½ in. (24.8 x 31.8 x 16.5 cm)
Produced by Alessi, Milan, Italy
K2016.29.13

DESIGNED BY PHILIPPE STARCK IN 1987, THE *HOT BERTAA* KETTLE REPRESENTED a postmodern sculptural design meant for display in any kitchen. Though it remains an icon of design of the late 1980s and early 1990s, manufacturer Alessi ceased production of the kettle after seven years and called it "our most beautiful fiasco." Complaints about the kettle included inconveniences like no way of viewing the water level, and an awkwardly placed too-short handle, which introduced the very real danger of burns while lifting and pouring from a boiling-hot kettle gushing steam.

The softness and roundness of the *Mami* kettle, by the architect turned product designer Stefano Giovannoni, embodies a "maternal code" that is characterized by a search for classicism and harmony. In a 2014 interview with *designboom*, Giovannoni offered this explanation: "In a sophisticated society we need objects which move further and further away from basic functional needs, objects which correspond to our identity. The objects we own have to communicate messages about who we are."[12] –PM

left to right

Peter Behrens (1868–1940)
Electric Kettle
Designed 1909
Nickel-plated brass, rattan, wood
8⅝ x 8⅜ x 6¼ in. (21.9 x 21.3 x 15.9 cm)
Produced by Allgemeine Elektricitäts-Gesellschaft (AEG), Berlin, Germany
K2010.23.1

Designer Unknown
Therma Electric Kettle (Model 11148 F)
Designed c. 1900–1910
Brass, rattan
8¾ x 9 x 6½ in. (22.2 x 22.9 x 16.5 cm)
Produced by Therma AG, Schwanden, Switzerland
K2008.459.1

Peter Behrens (1868–1940)
Electric Kettle
Designed 1909
Brass, rattan, wood
8⅜ x 7¾ x 5½ in. (21.3 x 19.7 x 14 cm)
Produced by Allgemeine Elektricitäts-Gesellschaft (AEG), Berlin, Germany
K2009.15.1

ALTHOUGH HE IS PERHAPS BEST REMEMBERED FOR HIS PROTO-MODERNIST work from the first decade of the twentieth century, Peter Behrens, like other designers of his generation, worked in many styles over the course of his long career, from jugendstil to expressionism to neoclassicism. Behrens came out of the German arts and crafts movement in which designers worked in multiple media, designing every element from the largest to the smallest in scale. With his appointment as artistic consultant for Allgemeine Elektricitäts-Gesellschaft (AEG) in 1907, he adapted a similar design approach for industrial production. He created one of the first "corporate identities for industry" for AEG between 1907 and 1914, designing everything from marketing materials to buildings, housing, showrooms, furniture, and products. Behrens also embraced the concept of component standardization, creating pieces with interchangeable features, which allowed for a wider variety of products for consumers.

Behrens designed three standard models for his electric kettles for AEG: cylindrical, half-oval, and octagonal. There were also three options each to choose from in terms of size, finish, and surface. The octagonal kettle is like a miniature building in its beautiful and complex play of geometric forms. The rattan handle adds a subtle but elegant profile to the composition. The sinuous *Therma* kettle was not part of the industrial AEG electric kettle series, but was designed as part of a tea service, which had flowing and fluid forms. The dramatic asymmetrical curved handle gives the design great presence. –RCM

Christopher Dresser (1834–1904)

Water Kettle

Designed c. 1885

Copper, brass, ebony

9¼ x 8⅝ x 5⅞ in. (23.5 x 21.9 x 14.9 cm)

Produced by Benham & Froud Ltd.,
London, England

K2012.179.1

THIS SPHERICAL WATER KETTLE IS THE EMBODIMENT of Christopher Dresser's style, which helped foster the ideal that spare forms are the most appropriate shapes for industrial designs. Dresser is remembered as a "godfather" of modernism for his simple and unadorned geometric forms. He was one of the most influential English designers in the second half of the nineteenth century due to his designs, publications, and design concepts. He was enormously prolific in multiple media, designing everything from furniture, metalwork, ceramics, and glass to wallpaper and textiles. One of his goals was to make well-designed products accessible to the average consumer, which would become a central tenet of twentieth-century industrial design. Dresser cultivated a keen interest in botany, and he "believed that the abstraction of natural forms provided the basis for ornamentation, a view that ran counter to the prevalent Victorian obsession with historical eclecticism." It was a compelling idea that would continue to have a pervasive influence on designers at the turn of the century, from Louis Sullivan to Frank Lloyd Wright. –RCM

opposite left to right

Attributed to

Paul V. Gardner

(c. 1908–1994)

Pyrex Flameware Double Boiler, Model No. 6763

Designed 1940–45

Pyrex glass, wood, metal

9 x 10¾ x 6¼ in. (22.9 x 27.3 x 15.9 cm)

Produced by Corning Glass Works, Corning, New York

K2011.162

Designer Unknown

Pyrex Flameware Saucepan with Detachable Handle, Model No. 832B

Designed 1936; handle designed c. 1943

Pyrex glass, steel

3⅛ x 8⅜ x7⅞ in. (7.9 x 21.3 x 20 cm)

Produced by Corning Glass Works, Corning, New York

K2010.118.2

Attributed to

Paul V. Gardner

(c. 1908–1994)

Pyrex Flameware Double Boiler, Model No. 6773-L

Designed c. 1939 (D118,445, saucepan)

Pyrex glass, stainless steel

9 x 10¾ x 6 in. (22.9 x 27.3 x 15.2 cm)

Produced by Corning Glass Works, Corning, New York

K2011.163

PYREX, INVENTED IN 1915 AND PRODUCED BY CORNING GLASS WORKS, IS A TEMPERED glass used for cooking and in laboratories. This sturdy glass has continued in popularity to the present day (though it has been reformulated), with many of the bakeware forms still produced. Its appeal to a modern aesthetic is seen in its simple, unadorned geometric forms. Such objects for everyday use were championed by MoMA, which included minimal, utilitarian objects in their design shows of the 1930s and 1940s.

Corning introduced its *Flameware* for stovetop use in 1936. The *Flameware* saucepan has a patented detachable handle, designed to fit each of three pieces in the set—two saucepans and a frying pan—interchangeably. The standard-issue double boiler's *Flameware* label proclaims the virtues of the Pyrex cookware for modern life—timesaving, easy cleanup, better taste, and transparent material—all appealing to the efficient servantless households of the ever-growing middle class in the twentieth century. Of particular interest in this grouping is the variation seen in the replacement of the double boiler's steel banding with wood during World War II, when use of metals was restricted for military purposes. Design and utility patents document the inventions of this cookware, some by well-known Smithsonian curator and glass expert Paul V. Gardner.[13] —DAH

above Paper insert from Pyrex *Flameware* double boiler.

Wilhelm Wagenfeld (1900–1990)

Kubus Food Storage Containers

Designed 1938

Glass

8¼ x 10¾ x 7 in. (21 x 27.3 x 17.8 cm) overall

Produced by Vereinigte Lausitzer Glaswerke,
Weisswasser, Germany

K2012.85

IN 1935, WILHELM WAGENFELD WAS APPOINTED ART DIRECTOR OF A LARGE glassware manufacturer, Lausitzer Glaswerke, where in 1938, he designed *Kubus*, an inexpensive set of modular glass storage containers for kitchen use. They were made of a newly developed heat-resistant pressed glass, and consisted of ten separate, stackable, hygienic storage containers: seven boxes and three jars, all with air vents and interchangeable lids. These storage containers were beautiful enough to be used as serving dishes, and encouraged leftovers. When not in use, they could be stacked efficiently together to form neat geometric shapes. This mass produced modular system, manufactured until 1968, exemplifies the form-follows-function ideals of the Bauhaus, where Wagenfeld studied and taught. –PM

Earl S. Tupper (1907–1983)

Tupperware Millionaire-line Sugar Bowl and Creamer

Designed c. 1946 (D154,348; D156,855; D158,155)

Polyethylene

Sugar bowl: 4½ x 4¾ x 3⅜ in. (11.4 x 12.1 x 8.6 cm)
Creamer: 4⅝ x 4¼ x 3 in. (11.8 x 10.8 x 7.6 cm)

Produced by Tupper Corporation, Farnumsville, Massachusetts

K2016.31.1–2

THE ARTISTIC QUALITY OF THE SIMPLE AND UNADORNED TUPPERWARE CONTAINERS appealed to modernists. In "Fine Art for 39¢," *House Beautiful* editor Elizabeth Gordon said of the plastic containers, "Above all else, the bowls have a profile as good as a piece of sculpture."[14] MoMA first acquired examples of Tupperware for their permanent collection in the 1940s. Several pieces, including bowls, tumblers, and utensils, were shown in MoMA's 1955 *American Art of the Twentieth Century* exhibition, presented in Paris as part of the "Salute to France" festival.[15] Earl Tupper, the designer and inventor, experimented in the early 1940s with the new soft, flexible plastic polyethylene. Among the early designs were this creamer and sugar, with handles that are integral to the body. The combination of two of his inventions made the later airtight Tupperware possible: a method for molding polyethylene for pliability and the patented Tupper Seal, a "type of finger-operable seal for containers capable of repeated use . . . to effectuate a fluid- and air-tight seal."[16] He founded Tupperware Corporation in 1942, but his commercial success came after the 1950 Tupper Seal invention and a unique direct-marketing strategy: housewives were employed to sell directly to their friends and neighbors at Tupperware parties held in their homes.[17] –DAH

left Demonstration of the tight Tupper Seal at a Tupperware home party, 1950s.

opposite left

Attributed to
Ralph E. Kruck (1906–1994)

General Refrigerator Butter Dish (green)

Designed c. 1939

Glazed earthenware

3⅞ x 7⅝ x 4½ in. (9.8 x 18.7 x 11.4 cm)

Produced by the Hall China Company, East Liverpool, Ohio, made exclusively for Westinghouse Electric and Manufacturing Company, Meadville, Pennsylvania

K2008.112.13

opposite center

Designer Unknown

Hercules Refrigerator Leftover Dish (orange)

Designed c. 1940

Glazed earthenware

4 x 7 x 4⅜ in. (10.2 x 17.8 x 11.1 cm)

Produced by the Hall China Company, East Liverpool, Ohio, made exclusively for Westinghouse Electric and Manufacturing Company, Meadville, Pennsylvania

K2008.113.1

opposite right

Ralph E. Kruck (1906–1994)

General Refrigerator Pitcher (light blue)

Designed c. 1939 (D115, 833)

Glazed earthenware

8½ x 7¼ x 4¾ in. (21.6 x 18.4 x 12 cm)

Produced by the Hall China Company, East Liverpool, Ohio, made exclusively for Westinghouse Electric and Manufacturing Company, Meadville, Pennsylvania

K2008.128

below left

Attributed to
Ralph E. Kruck (1906–1994)

General Refrigerator Leftover Dish (yellow)

Designed c. 1939

Glazed earthenware

3¾ x 6¾ x 4½ in. (9.5 x 17.1 x 11.4 cm)

Produced by the Hall China Company, East Liverpool, Ohio, made exclusively for Westinghouse Electric and Manufacturing Company, Meadville, Pennsylvania

K2008.112.4

right

Joseph Palin Thorley (1892–1987)

Peasant Ware Refrigerator Pitcher (dark blue)

Designed c. 1940

Glazed earthenware

9¼ x 7¾ x 3⅞ in. (23.5 x 19.7 x 9.8 cm)

Produced by the Hall China Company, East Liverpool, Ohio

K2008.262

WHAT BETTER WAY TO PROMOTE A REFRIGERATOR THAN BY means of these colorful storage containers, which were offered free with the purchase of an appliance? The Hall China Company provided exclusive custom designs for Westinghouse, General Electric, Solar, and other manufacturers. Three basic items—water pitchers, leftover dishes, and butter dishes—were offered in bright colors in compact forms for efficient storage, and the thick earthenware helped to reduce chips and breakage. The curved forms, often with speed lines, epitomize American streamlined design.

The design masterpiece of refrigerator ware is the *Peasant Ware* pitcher by Joseph Palin Thorley, shown here in cobalt blue, an object included in the collection of the Metropolitan Museum of Art since 1981. The hollow handle, which also holds water, is integral to the body of the vessel, balancing the vestigial spout and a conforming fitted lid. Grooves provide speed lines above the stepped base and from the lid down the handle, providing a sense of movement. British born and trained, Thorley had worked as an apprentice at Wedgwood before immigrating to the United States in 1927.[18] His skills in ceramic design covered a range of styles, from revivals of earlier forms to modern lines for Hall. His traditional designs might have been used in more formal rooms of the house, but no one questioned using his refrigerator ware in the kitchen, where modern design first entered the household. The green butter dish and yellow leftover dish are attributed to Ralph Kruck because he patented the design for this light blue pitcher in the same series for Westinghouse.[19] –DAH

Attributed to

Raymond Loewy (1893–1986)

Refrigerator

Designed c. 1935

Painted steel, aluminum, rubber, chromium-plated steel, plastic

Closed: 43¾ x 24½ x 23¼ in. (111.1 x 62.2 x 59.1 cm)

Produced by Frigidaire, Division of General Motors Corporation, Dayton, Ohio

K2013.55

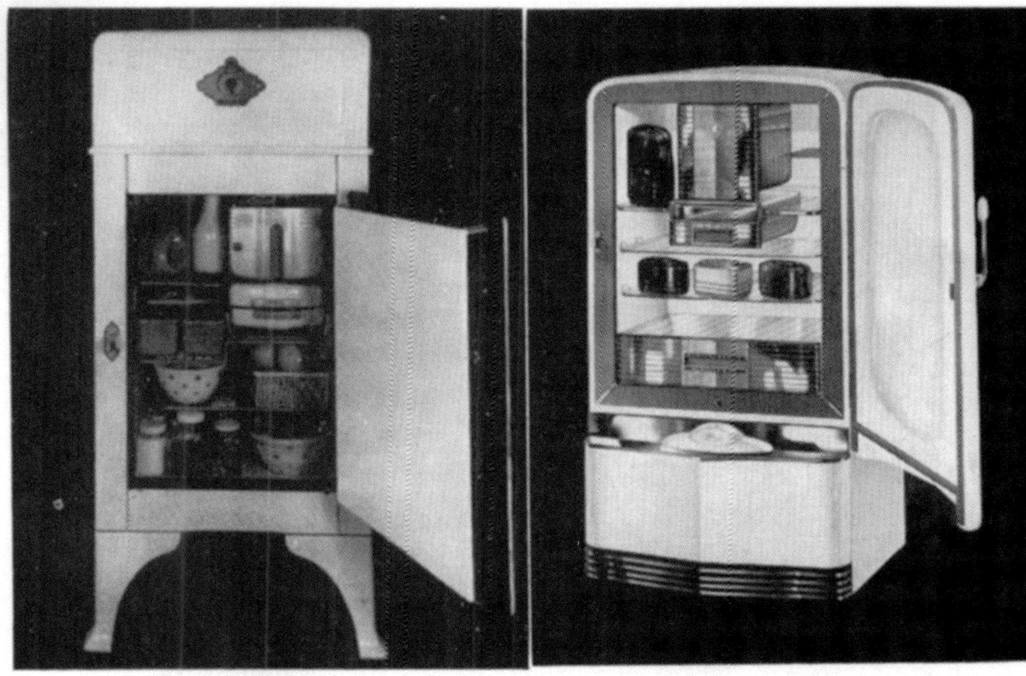

Old style refrigerator . . . and the newer Kruck model, embodying utility, handsome design and ample room.

An updated Westinghouse refrigerator (right) showing Hall China's refrigerator-ware for Westinghouse inside (pages 172–173), from "Ralph Kruck," by Dorothy Grafly, published in *Design* magazine, February 1948.

RAYMOND LOEWY'S MOST FAMOUS DESIGN IS THE *COLDSPOT* REFRIGERATOR FOR SEARS: HE DESIGNED four versions between 1935 and 1938. Although he is known to have designed a line of appliances for Frigidaire in 1955, this apartment-size Frigidaire dates back to about 1935 and is attributed to Loewy on the basis of a design element similar to his *Coldspot*: a cut-out section at the base. Loewy himself described the improvements in his *Coldspot* refrigerator design: earlier refrigerators were decorated with moldings, which he eliminated, and he introduced perforated aluminum shelving to replace the steel wire, which rusted.[20] In the Frigidaire design shown here, the aluminum freezer door is adorned with the company's logo below the temperature-control knob. It combines "Frigidaire" with "GM," for parent company General Motors. Originally white, a coat of red paint was applied to the exterior of this particular refrigerator at a later date. It was originally owned by Indianapolis businessman and philanthropist Louis Schwitzer, who is best known as the winner of the first automobile race at the Indianapolis Motor Speedway in 1909. Loewy's Frigidaire and Ralph Kruck's Westinghouse (above right) demonstrate the modernization of the form from designs that imitate furniture set on legs (above left), to a unified configuration that is still familiar today. –DAH

06

objects

EATING & DRINKING

WE MUST EAT AND DRINK TO SURVIVE, BUT IN THE MODERN WORLD, both acts are also defined by the rituals that accompany them and by the important social and cultural roles that they play in our everyday lives. We eat and drink everywhere: at home, in restaurants and bars, and outdoors. The complexity of ways in which we interact with our food and drink, in turn, requires a huge number of material accessories. The objects from the Kravis collection represented here cover a range of dining possibilities, from the most mundane to the luxurious. From the cloth that covers the table to the bowl that holds a salad to the pepper mill to the many containers that hold our beverages—they all serve to make our social interactions more lively and entertaining.

Russel Wright (1904–1976)

Tablecloth

Designed c. 1950

Cotton, rayon

68⅞ x 50⅜ in. (175 x 128 cm)

Produced by Simtex Mills, Division of Simmons Co., New York, New York

K2008.102

MARY AND RUSSEL WRIGHT CREATED INEXPENSIVE, MASS-PRODUCED GOODS TO ACHIEVE the goals described in their book, *Guide to Easier Living*, first published in 1950. To help consumers realize the casual lifestyle they encouraged, the Wrights developed a line of textiles to accompany their *Easier Living* furniture line and *American Modern* dinnerware, glassware, and spun aluminum pieces. These designs competed with the traditional lace and damask tablecloths that dominated the market at the time. Here, bands of bright color overlap in a plaid tablecloth, creating a complex pattern that could be combined with the other Wright wares. Wright worked for Simtex, one of several textile manufacturers he designed for, from 1949 to 1955. The back of the Simtex label reads, "Made Right in America," emphasizing the idea that the Wrights' line was distinctly American. A setting incorporating both the tablecloth and *American Modern* dinnerware is seen here —DAH

above left Reverse side of Simtex Mills label from this Russel Wright tablecloth.

above right Advertisement for *Simtex Modern* designs by Russel Wright, published in *Better Homes and Gardens*, May 1951.

left

Designer Unknown

Thermos Vacuum Jug, Model No. 65
Designed 1930–40
Plastic, glass, cork
10¼ x 5 x 4¼ in. (26 x 12.7 x 10.8 cm)
Produced by Thermos Ltd., London, England
K2010.176.1

right

Charles Horton (dates unknown)

Thermos Vacuum Jug, Model No. 140 Q
Designed c. 1925
Bakelite, glass, cork
11½ x 6¾ x 5¼ in. (29.2 x 17.1 x 13.3 cm)
Produced by Thermos Ltd., London, England
K2010.175

Advertisement for Thermos Brand Vacuum Ware, published in *The New Yorker*, November 30, 1940.

IN ADDITION TO INSULATED VACUUM PITCHERS AND TRAYS (PAGE 197), HENRY DREYFUSS ALSO DESIGNED vacuum bottles for the American Thermos Bottle Company, including the example in brown and yellow, faceted so it could not roll away, with a conical red cup/lid (page 183). Between the exterior housing and the interior reservoir of such containers is a sealed layer of open space that has been partially emptied of air—hence the name "vacuum bottle"—which greatly reduces the transfer of heat in or out and maintains the temperature of fluid, whether hot or cold, far longer than an uninsulated container. The range of designs by the American Thermos Bottle Company is seen in a 1940 advertisement from *The New Yorker*. The aluminum *Thermoses* from the Kravis Collection, with red plastic cups as tops (page 182), are similar to ones shown in the advertisement. The conical aluminum *Hollywood Set Thermos*, with horizontal lines and a black glass top, is named for its association with the glamour of Hollywood stars. According to contemporary advertising—"Drink with the Hollywood Set!"—the *Thermos* was originally accompanied by a conforming tray[1] and sold for $3.95. And it was promoted as "Ideal for Bedroom, Guest Room, Office Use . . . any place a cool, refreshing drink of water is wanted."[2] In contrast, the two British *Thermos* examples (opposite) have traditional tapering faceted forms with shaped handle, neck, and spout, suggesting a nineteenth-century pitcher, only transferred to the modern material of plastic. –DAH

See text page 181

left to right

Designer Unknown

Vacuum Bottle, Model No. 2484

Designed 1950–60

Aluminum, plastic, glass

13⅝ x 4⅝ x 3⅝ in. (34.6 x 11.8 x 9.4 cm)

Produced by the American Thermos Products Company, Norwich, Connecticut

K2011.58

Carl Bramming (1889–1978)

Thermo-King Vacuum Bottle,
Model No. 70-B Pint

Designed c. 1955 (utility patent 2,756,889)

Aluminum, plastic, glass

10⅛ x 4 x 3¼ in. (25.7 x 10 x 8.3 cm)

Produced by Aladdin Industries Inc., Nashville, Tennessee

K2011.54

Louis J. Darmstadt (1905–1986)
George W. Fuller (1915–1998)

Thermos Vacuum Bottle, Model No. M5219

Designed c. 1960 (utility patent 3,059,798)

Aluminum, plastic, glass

9⅜ x 3¾ x 3 in. (23.8 x 9.5 x 7.6 cm)

Produced by the American Thermos Products Company, Norwich, Connecticut

K2011.55

Henry Dreyfuss (1904–1972)

Thermos Vacuum Bottle, Model No. 370

Designed 1937 (D106,862)

Painted metal, aluminum, plastic

13⅞ x 4 x 4 in. (35.2 x 10 x 10 cm)

Produced by the American Thermos Bottle Company, Norwich, Connecticut

K2012.34.1

Attributed to
Henry Dreyfuss (1904–1972)

The Hollywood Set Thermos Vacuum Bottle

Aluminum, glass

Designed c. 1937

9½ x 4½ x 4½ in. (24.3 x 11.4 x 11.4 cm)

Produced by the American Thermos Bottle Company, Norwich, Connecticut

K2011.13

Finn Juhl (1912–1989)

Bowl

Designed c. 1950

Teak

6¾ x 14½ x 13½ in. (17.1 x 36.8 x 34.3 cm)

Produced by Magne Monsen for Kay Bojesen Modeller, Bredgade, Denmark

K2011.165

THE DANISH-BORN FINN JUHL BECAME ONE OF THE MOST IMPORTANT DESIGNERS OF FURNITURE, objects, and interiors in postwar Denmark. Though an architect by training, Juhl graduated from the Royal Danish Academy of Fine Arts in 1934 and focused primarily on design. Juhl debuted his first furniture collection in 1937—the result of a collaboration with cabinetmaker Niels Vodder that continued until 1959. Juhl's early work is characterized by a focus on artisanal craftsmanship rather than mass production, yet his designs often rejected traditional historicist styles. This carved bowl, made by lathe turning a single piece of solid teak, embodies both Juhl's design ideals and Danish craftsmanship in its simple elegance and integrity. Representing what may be considered a "softer" side of design of that period, Juhl held an oft-quoted belief that "one cannot create happiness with beautiful objects, but one can ruin quite a lot of happiness with bad ones." —PM

Jens Quistgaard (1919–2008)

Ice Bucket and Salt and Pepper Mills

Ice bucket, designed c. 1961:
18⅛ x 8½ x 8½ in. (46 x 21.6 x 21.6 cm)

Tall mill, designed 1975–80:
12¾ x 2¾ x 2¾ in. (32.4 x 7 x 7 cm)

Short mill, designed c. 1965:
5½ x 3⅛ x 2¾ in. (14 x 7.9 x 7 cm)

Palisander, teak

Produced by Richard Nissens, Langaa, Denmark, for Dansk Designs Ltd., Copenhagen, Denmark, and Dansk International Designs Ltd., Thailand

K2010.124.1–3

BROUGHT UP IN THE CRAFT TRADITION of Denmark, Jens Quistgaard learned woodcarving from his father, who was a sculptor, and was apprenticed at a young age at Georg Jensen Silversmithy. In 1954, the American engineer and entrepreneur Ted Nierenberg saw Quistgaard's work on view at the Danish Museum of Art and Design in Copenhagen and approached him to become the chief designer for a new Danish company, Dansk, headquartered in Mount Kisco, New York. Most of Dansk's designers were Scandinavian and emphasized the craft traditions. The staved construction of the ice bucket was inspired by the wooden hulls of Viking ships.[3] From his studio in Copenhagen, Quistgaard created a diverse range of products for Dansk through the mid-1980s.[4] The forms in wood recalled both traditional Scandinavian and Japanese designs. Quistgaard succeeded in transposing traditional crafts to mass production, yet retaining the look of hand-crafted objects. –DAH

Russel Wright (1904–1976)

Bun Warmer, Three Stacking Snack Trays with Lid, Lemonade Pitcher, and Two Cups

Designed 1930–1940

Aluminum, wood, wire

Bun warmer: 8¾ x 9¾ x 9½ in. (22.3 x 24.8 x 24.1 cm)
Stacking snack trays: 4¼ x 10⅝ x 10⅝ in. (10.8 x 27 x 27 cm)
Pitcher: 10⅛ x 10⅜ x 6 in. (25.7 x 26.4 x 15.2 cm)
Tall cup: 5¼ x 2½ x 2½ in. (13.3 x 6.4 x 6.4 cm)
Short cup: 2¼ x 2 x 2 in. (5.7 x 5.1. 5.1 cm)

Produced by Wright Accessories/Raymor, New York, New York

K2008.177, K2012.84.39, K2012.84.41, K2012.84.45, K2012.164

THE SIMPLE GEOMETRIC FORMS OF RUSSEL WRIGHT'S SPUN aluminum wares are evocative of the machine age, as seen especially in the spherical bun warmer and the circular stacking trays. In some pieces the shiny metal finish is softened by the addition of stained wood, rattan, or cork for handles and knobs. Wright later recalled that "aluminum was exciting for a young designer who wanted to design everything and had only a work bench and a lathe. The easy workability of the metal, its permanent integral coloring, inspired me to develop a variety of products."[5] During the Depression years, aluminum products were affordable and attractive for the informal lifestyle promoted by Mary and Russel Wright. The Wrights began producing aluminum wares around 1930, the same year Russel's aluminum cocktail shaker was shown in the annual exhibition of the American Union of Decorative Artists and Craftsmen (AUDAC).[6] They collaborated to produce Informal Serving Accessories, as they termed the aluminum ware, with Russel focusing on design and Mary on business and marketing. They used their carriage house on East 35th Street in New York as a workplace and residence, living on the upper floor and using the rest of the house to produce, ship, and sell their aluminum products.[7] Beginning in about 1935, Wright's aluminum wares were marketed by a cooperative venture called Russel Wright Associates, which was established between the Wrights and Irving Richards, who managed the business end. When the Wrights retired in 1940, Richards would go on to found Raymor, which became one of the largest wholesale distributors of modern design.[8] –DAH

right Illustration of Russel Wright aluminum ware, published in *Decorative Art 1939: The Studio Year Book*, 1940.

left

Designer Unknown

Artillery Shell Cocktail Shaker

Designed c. 1918

Silver-plated brass, brass, copper

11¾ x 3¼ x 3¼ in. (29.8 x 8.3 x 8.3 cm)

Produced by Gorham Silver Company, Providence, Rhode Island

K2012.114

center

Albert D. Mortrude

(1910–1979)

Konga Cocktail Shaker

Designed c. 1947 (utility patent 2,577,976)

Aluminum, Bakelite

11⅛ x 6⅛ x 3½ in. (28.4 x 15.6 x 8.9 cm)

Manufacturer Unknown

K2010.9

right

William Archibald Welden

(1900–1970)

Empire Cocktail Shaker and Two Cups

Designed c. 1938

Chromium-plated brass, Bakelite

Shaker: 12¼ x 7⅛ x 3 in.
(31 x 17.9 x 7.6 cm)

Cups: 2⅝ x 3¼ x 3¼ in.
(8.4 x 6.7 x 6.7 cm) each

Produced by Revere Copper and Brass Company, Rome, New York

K2010.205, K2011.209.1–2

COCKTAIL SHAKERS HAVE BEEN IN USE SINCE THE MID-nineteenth century, but the popularity of this form really took off during Prohibition in the 1920s, when these shining metal objects and the drinks that they enhanced became works of art. Companies like Tiffany & Co., Gorham, and Dunhill began making art deco shakers with a sense of fun—in the form of various animals, such as polar bears and penguins, or everyday objects like tanks, skyscrapers, bowling pins, and artillery shells. This 1919 Gorham *Artillery Shell*, a bullet-shaped shaker, was modeled on a World War I eighteen-pounder shrapnel shell. The top opens to reveal the strainer, and the base (not shown here) contains four shot glasses. The *Konga* shaker has a stainless steel cylinder body and red Bakelite rotating handles that allow cocktails to be made with a rolling rather than an up-and-down shaking motion, eschewing the notion of the traditional "shake and spill." The *Empire* cocktail shaker by Revere design head William Archibald Welden has a distinctive cylindrical finial, echoing the shape of the shaker itself, and a semicircular handle. –PM

Norman Bel Geddes (1893–1958)

Manhattan Cocktail Shaker, Six Goblets, and Tray

Designed c. 1934

Chromium-plated brass

Shaker: 13⅛ x 3¼ x 3¼ in. (33.3 x 8.3 x 8.3 cm)

Produced by Revere Copper and Brass Company, Rome, New York

K2011.4

Cooper Hewitt, Smithsonian Design Museum, promised gift of George R. Kravis II

IN 1934, CHARLES DALLAS, OWNER OF REVERE COPPER AND BRASS, a company originally established by the Boston silversmith Paul Revere, commissioned famous designers such as Walter Von Nessen, Lurelle Guild, Rockwell Kent, Russel Wright, and Norman Bel Geddes to design products for its giftware line. Bel Geddes & Co.'s *Manhattan* serving tray appeared in Revere's 1935 sales catalogue—along with sixteen other Bel Geddes designs—and was described as "a most individual tray which gains its effect by a frank use of straight lines, delicate flutings and subtle contrast of finishes."[9] The Bel Geddes cocktail shaker and goblets, which appeared in the 1936 Revere catalogue, were sold separately or as a set with the tray. The *Manhattan* cocktail shaker, a faceted cylinder towering over its goblets, represents the 1930s fascination with the high-rises, new at the time, that made the New York City skyline famous. —PM

Peter Müller-Munk (1904–1967)

Normandie Pitcher

Designed 1935

Chromium-plated brass

12 x 8¾ x 3 in. (30.5 x 22.2 x 7.6 cm)

Produced by Revere Copper and Brass Company, Rome, New York

K2012.82

WHEN THE *SS NORMANDIE* DOCKED in New York's harbor on its maiden transatlantic voyage on June 3, 1935, enthusiasm for the luxury ocean liner was seen in the throngs of visitors that greeted its arrival. The ship's interiors were richly decorated by leading French art deco designers, including remarkable *verre eglomisé* murals by Jean Dupas.[10] Inspired by this extraordinary ship, Revere Copper and Brass introduced the *Normandie* pitcher in August 1935,[11] which, in contrast to the luxurious interiors of the ship for which it was named, was mass produced and inexpensive. The Revere pitcher was streamlined modern, representing the new aerodynamics-based style sweeping across the United States, more comparable to the ocean liner's exterior: teardrop in plan and set at an angle, the form recalled the prow of the great French Line ship. In initial production, the joints of the exterior of the pitcher were seamless, but were later modified for stability with applied rims.[12] It was Peter Müller-Munk's most successful design aesthetically and reflected his transition from silversmith to industrial designer. —DAH

Tapio Wirkkala (1915–1985)
Two Bowls, Model No. TW 476
Designed 1974
Bronze
3 x 9¾ x 9¾ in. (7.6 x 24.8 x 24.8 cm)
3½ x 11¾ x 11¾ in. (8.9 x 29.8 x 29.8 cm)
Produced by Kultakeskus Oy, Hämeenlinna, Finland
K2013.69.1–2

FINNISH DESIGNER AND SCULPTOR TAPIO WIRKKALA WORKED IN A VARIETY OF MEDIA, including glass, ceramics, metal, plastic, and laminated plywood. Among his most famous works in glass are his bottle for Finlandia vodka, which was in production from 1970 until 1998, and Iittala's *Ultima Thule* set of drinking glasses. Both feature a "melting ice" look. The hand-hammered bronze fruit bowls, shown here, were designed around 1974 for the Finnish company Kultakeskus Oy, for which he created hundreds of beautiful products, including jewelry, vases, dishes, household accessories, and even animal figures in different metals such as bronze, silver, and gold. During the design process, Wirkkala carefully examined the end use of the features planned for each object, and molded and adapted them until the object was ready for production. The hammered bronze reflects the handmade craftsmanship, evidence that these bowls are not mass-produced—they hark back to the traditions of Scandinavian design. –PM

Jean G. Theobald (1873–1952)

below

Diament Tea Service

Designed c. 1928 (D76,434)

Silver-plated metal, painted wood

7¾ x 13½ x 6¾ in. (19.7 x 34.3 x 17.1 cm)

Produced by Wilcox Silver Plate Company, International Silver Company, Meriden, Connecticut

K2016.16

opposite top

Dinette Tea Service

Designed c. 1928

Silver-plated metal, plastic

4 x 13½ x 6¾ in. (10.2 x 34.3 x 17.1 cm)

Produced by Wilcox Silver Plate Company, International Silver Company, Meriden, Connecticut

K2016.15

opposite bottom

Dinette Tea Service, Model No. 7036

Designed 1928 (D76,564)

Silver-plated metal, painted wood

3½ x 8½ x 7 in. (8.9 x 21.6 x 17.8 cm)

Produced by Wilcox Silver Plate Company, International Silver Company, Meriden, Connecticut

K2010.65

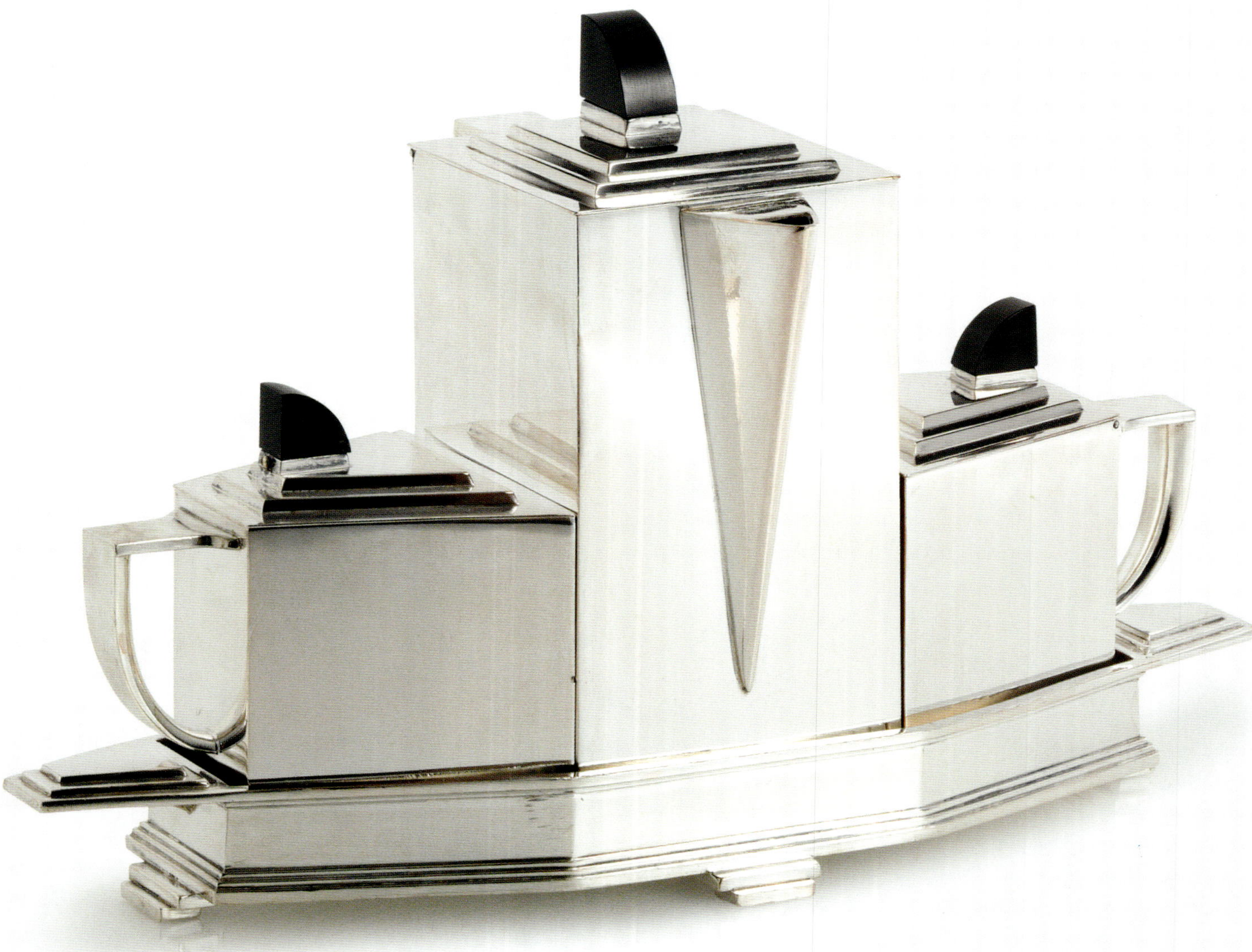

THESE THREE COMPACT TEA SERVICES, designed by Jean G. Theobald and styled by Virginia Hamill, were introduced in 1928. According to an advertisement of that year for the International Silver Company, which illustrates the *Diament* service, "These sets are perfect for modern apartments. They fit the smaller rooms we have today. They look charming on the little narrow tables."[13] The angled and stepped *Diament* service even echoed New York City's stepped architecture of the period. The aesthetic appeal of all these sets was in their tight, unified design; traditional services were comparatively awkward, with large trays and separate pieces. In 1928, the price of the *Diament* service was $75; the *Dinette* tea service with oblong tray was $35; and the circular *Dinette* tea service was $45.[14] Two were patented,[15] and all three were a collaboration of Theobald, who was a Wilcox staff designer, and Hamill. Trained at Parsons School of Design in New York, Hamill was a merchandising and styling consultant hired by International Silver Company in 1928 as a product stylist for modern hollowware.[16] –DAH

Peter Schlumbohm (1896–1962)

opposite left to right

Fahrenheitor Beverage Cooler
Designed c. 1945 (D142,795)
Cork, glass
14 x 6⅝ x 6⅝ in. (35.6 x 16.8 x 16.8 cm)
Produced by Chemex Corporation,
Pittsfield, Massachusetts
K2014.12

Fahrenheitor Cocktail Shaker
Designed c. 1946 (utility patent 2,532,373)
Aluminum, cork
10⅝ x 3⅞ x 3⅞ in. (27 x 10 x 10 cm)
Produced by Dr. Peter Schlumbohm,
New York, New York
K2012.196

Chemex Coffeepot
Designed c. 1941 (D137,943)
Pyrex glass, wood, leather
9⅜ x 6½ x 6⅛ in. (24 x 16.5 x 15.6 cm)
Produced by Chemex Corporation,
New York, New York
K2015.2

Promotional photo of Peter Schlumbohm with a selection of his inventions, published in *LIFE* magazine, April 1949.

FAMOUS FOR HIS MANY INVENTIONS, PETER SCHLUMBOHM, THE SON OF A CHEMICAL MANUFACTURER, graduated from Berlin University with a doctorate in chemistry. Schlumbohm first visited New York in 1931 and immigrated in 1936, "attracted by U.S. patent laws, which he found the world's best from an inventor's viewpoint."[17] Schlumbohm insisted that "no design was viable unless it was rooted in invention,"[18] and he held over three hundred patents for his many designs and innovations. Working in a loft studio at 41 Murray Street in Manhattan, he created a product line under the brand name *Fahrenheitor*. Among his inventions for the line was this beverage cooler designed to quickly chill a bottle of wine.[19] The cocktail shaker was another design in the *Fahrenheitor* line. The aluminum of its shaped body and the cork of its lid were materials the designer preferred. To improve on the many cocktail shakers of the day that were either apt to leak or nearly impossible to open once sealed, the utility patent stated, "the invention aims at a stopper for a cocktail shaker, which will be leakproof . . . the invention also aims additionally at a design which allows an easy removal of the lid."[20]

His most famous and successful invention was the *Chemex*,[21] an hourglass-shaped coffeemaker made of Corning's Pyrex glass, first marketed in 1941 and still in production. Based on forms found in laboratory glass, it was designed for use with filter papers for ease of clean-up and, according to Schlumbohm, filtered coffee was a better and healthier drink than coffee made by other methods because the "vile mixture of some fifty different chemicals are eliminated by filtering."[22] *Chemex* doubled the cone shape of an earlier Schlumbohm design called *Cinderella*, a small garbage pail intended as an elegant tabletop object. –DAH

Paul Schreckengost (1908–1983)
Teapot
Designed 1938
Glazed and painted earthenware
7¼ x 11 x 4½ in. (18.4 x 27.9 x 11.4 cm)
Produced by Gem Clay Forming Company, Sebring, Ohio
Cooper Hewitt, Smithsonian Design Museum, gift of George R. Kravis II, 2016-39-1

PAUL SCHRECKENGOST'S TEAPOT CONVEYS THE SPEED, ENERGY, AND CHIC OF THE dynamic streamlined style that was applied to "things that will never move."[23] Its forward thrust suggests the movement of a high-speed train; its body, curved handle, and spout, with its removable lid look like they would perform well in wind-tunnel testing. The body appears spherical in profile, but it is actually flattened, like two shallow domes pressed together. The downward sweep of the handle balances the curve of the spout. Horizontal speed lines at the base, spout, and bottom of the conforming handle, highlighted in platinum paint, enhance the sense of movement. Schreckengost came from a family of ceramicists, including his father and two brothers, Viktor and Don.[24] He worked as chief designer and mold maker for Gem Clay Forming Company, which produced a variety of household wares, and created this design as a Christmas gift for the firm's preferred customers in 1938. –DAH

Henry Dreyfuss (1904–1972)

Three *Thermos* Insulated Pitchers: Two Model No. 549, one Model No. 539 (with tray)

Designed 1935

Enameled steel, aluminum, glass, rubber

Large pitchers: 7 x 7⅞ x 5⅞ in. (17.8 x 20 x 14.9 cm) each
Small pitcher: 5¾ x 5⅞ x 4¾ in. (14.6 x 14.9 x 12.1 cm)
Tray: ½ x 9¼ x 6⅝ in. (1.3 x 23.5 x 16.8 cm)

Produced by the American Thermos Bottle Company, Norwich, Connecticut

TR2016.189–190, K2011.101.1–2

THIS CLASSIC AMERICAN THERMOS DESIGN OF 1935 FOR OFFICE OR HOME APPEARED IN various sizes and colors, with or without a tray. Dreyfuss later wrote that the form was inspired by an ancient Greek jug he saw in the Metropolitan Museum of Art.[25] Indeed, modernists had an affinity for the simplicity of classical designs of antiquity. However, this insulated pitcher also embodies principles of American streamlined design, seen in the emphasis on the horizontal speed lines at the base, the curved form, and the conforming handle. Prominently displayed on the bottom of the pitcher is the signature of Henry Dreyfuss in bold script, signifying the importance of the emerging industrial designer in the 1930s. This was an early step toward the kind of branding and name-recognized designers we are so familiar with today. Dreyfuss was among the first American industrial designers who convinced clients that improvements in design would lead to increased sales while the American economy was recovering from the Depression. –DAH

Russel Wright (1904–1976)

American Modern Dinnerware: Teacup with Saucer, Creamer, Celery Tray, Teapot, and Water Pitcher

Designed c. 1937

Glazed earthenware

Teacup with saucer: 2¼ x 6⅞ x 5⅞ in. (5.7 x 17.6 x 14.9 cm)
Creamer: 2⅞ x 7⅜ x 3¾ in. (7.5 x 18.7 x 9.5 cm)
Celery tray: 1⅛ x 13¼ x 3⅝ in. (2.9 x 33.7 x 9.2 cm)
Teapot: 5⅛ x 10⅛ x 6¼ in. (13 x 25.7 x 15.9 cm)
Water pitcher: 7⅝ x 7⅞ x 5⅞ in. (19.4 x 20 x 14.9 cm)

Produced by Steubenville Pottery Company, Steubenville, Ohio

K2008.104.1a,b; K2008.114.1c; K2008.173; K2008.174.1; TR2016.195.2

RUSSEL WRIGHT'S *AMERICAN MODERN* IS CONSIDERED TO BE THE BEST-selling dinnerware set ever produced in the United States. The annual retail sales were reported to be "$3,000,000, with demand much in excess of production."[26] In contrast to Wright's spun aluminum of the 1930s, which adhered to the aesthetic of geometric machine-age forms, the *American Modern* forms were undulating and unconventional. Unlike the bright colors of Frederick Hurten Rhead's *Fiesta* ware, Wright selected muted colors for *American Modern*, such as the sea-foam blue, chartreuse, and coral seen here. Customers were encouraged to mix the colors within settings. *House & Garden* praised the new pottery group, which could "take oven heat and refrigerator cold as well as grace the table."[27] The Wrights maintained that "a room used only for dining is a waste of space, steps, and labor" and was "becoming extinct."[28] In their world, the dining room was replaced by a more accessible arrangement that might include an eat-in kitchen. Even the living room could be utilized as a dining area, as seen in the illustration. –DAH

left Promotional photo of informal dining with *American Modern* dishes on Wright's *Easier Living* coffee table, mid-1950s.

Stefan Scholten (born 1972)
Carole Baijings (born 1973)

Paper Porcelain Settings: Mug, Espresso Cup and Saucer, Plate, Coffee Cup and Saucer

Designed 2009

Porcelain

Mug: 4 x 2¾ x 2¾ in. (10.2 x 7 x 7 cm)

Produced by HAY, Copenhagen, Denmark

K2016.1–4

THIS DISHWASHER-SAFE *PAPER PORCELAIN* COLLECTION, ISSUED BY DANISH BRAND HAY, IS CRAFTED IN fine porcelain but has a slightly speckled appearance to resemble coarse recycled paper. For the Amsterdam-based designers Stefan Scholten and Carole Baijings, design ideas emerge through the process of working with materials—what they call "thinking through making." "Our method is the atelier way of working, one that emphasizes constructive thinking," Baijings says. "It means making your own materials, your own colors, your own models, and looking at what's happening under your hands."[29]

Their initial approach to creating the *Paper Porcelain* collection was to hand make prototypes of cardboard scale models. For them, the scale model can make various problems visible and similarly produce a logical solution, and in this case, the resulting series of polygonal shapes appealed to them, and they decided to translate the angular shapes of the cardboard model into porcelain. They chose Japan's Arita Porcelain Lab to develop a special blend of porcelain containing tiny iron particles in order to translate the look and texture of the original gray recycled cardboard of the prototypes. The result is *Paper Porcelain*, a collection of fragile ceramic objects that replicate the delicacy of the cardboard models. Since Scholten and Baijings started collaborating officially in 2000, they have worked on a wide range of projects, from domestic objects to reimagining the Mini for BMW in 2012. –PM

George Sakier (1897–1988)

Three *Geometric* Serving Bowls

Designed c. 1929–33

Glass

Green: 3¼ x 12 x 11 in. (8.1 x 30.5 x 27.9 cm)
Pink and yellow: 2¾ x 8¾ x 8¾ in.
(7.1 x 22.2 x 22.2 cm) each

Produced by Fostoria Glass Company, Moundsville, West Virginia

K2008.244, K2008.417, K2012.167

GEORGE SAKIER BEGAN TO WORK FOR FOSTORIA GLASS COMPANY IN 1929 AND REMAINED there for fifty years, designing affordable glass in both modern and traditional styles. His repertoire included "geometric modern," seen in these examples, as well as "classic modern," and "romantic modern."[30] These angular, pastel bowls, number 2402 in the Fostoria sales catalogue,[31] are in the medium (pink and yellow) and large (green) sizes. In contrast to the fluted cylindrical shapes of Sakier's classic modern *Lotus* vases of the same year, these are strongly angled art deco designs. Sakier's motto was "To bring the most out of the machine," according to the Cheneys in *Art and the Machine*.[32] With the application of modern design standards' emphasis on mass production, the Cheneys posited, "Pressed glass, one of the cheapest and earliest products of American industry, takes on distinction, even when marketed in ten-cent stores."[33] Not surprisingly, Sakier's modern glassware—with a circular ashtray set and tall cylindrical drinking glasses, all of clear or transparent colored glass—was included in the 1934 *Machine Art* exhibition at MoMA.[34] A more adventurous step in Sakier's work, also in the MoMA show, was a prefabricated bathroom unit, ready for installation and produced by the American Radiator and Standard Sanitary Corporation in 1933. –DAH

Margarete Marks (1899–1990)
Coffee Service
Designed 1930
Glazed stoneware
Coffeepot: 5 x 10¼ x 6⅞ in. (12.5 x 26 x 17.5 cm)
Produced by Hael-Werkstätten für Künstlerische Keramic, Marwitz, Germany
K2008.261
Cooper Hewitt, Smithsonian Design Museum, promised gift of George R. Kravis II

MARGARETE MARKS STUDIED AT THE BAUHAUS FROM 1920 TO 1921 AND WAS unusual in that she studied ceramics rather than textiles, which was thought to be the more feminine profession at the time. In 1923 with her first husband, Gustav Loebenstein, she established the Hael-Werkstätten ceramic company, which became a successful business selling its pottery internationally until 1932. With the rise of Nazism, she immigrated to England in 1936 and continued to work as a ceramicist with mixed success. She subsequently married Harold Marks and worked primarily as a painter after the war. Marks is best remembered for the abstract pottery that she designed between the wars. She favored crisp geometric forms, especially circles and triangles, as exemplified in this coffee service. She often used vivid colors such as yellow and blue glazes. On occasion, she decorated her pieces with geometric patterns recalling the constructivist paintings of her Bauhaus colleague Wassily Kandinsky. Marks's contribution to modern design has only recently been accessed as museums in Europe and the United States have begun to collect her ceramics.

Donald Schreckengost (1911–2001)
Frank H. Sebring Jr. (1889–1934)

Tricorne Party Plate, Nut Dish, and Sandwich Plate

Designed 1934 (D94,245, plate shape)

Sandwich plate: 11½ x 11½ x ¾ in. (29.2 x 29.2 x 1.9 cm)

Vincent Broomhall (1906–1991)
Herbert A. Smith (1896–1979)

Streamline Teacup, Sugar Bowl, Creamer, and Coffeepot

Designed c. 1935 (D96,445, sugar bowl; D96,722, teacup)

Glazed earthenware

Produced by Salem China Company, Salem, Ohio

K2012.232.1,2,3,4,8,9

TRIANGULAR PLATES WERE NOT A NEW INVENTION in the 1930s—they are seen in traditional Japanese porcelain—but this geometric shape appealed to modern designers. Donald Schreckengost was from a family of artist-designers, including his brothers Viktor and Paul. In 1932, while still a student at the Cleveland School of Art, he became the design director of Salem China Company and remained there until 1935.[35] This service includes two Salem shapes: first, *Tricorne* plates and bowls, which are triangular with white circular centers, patented in 1934 by Schreckengost and Salem company director Frank H. Sebring Jr.;[36] second, the conical *Streamline* vessels, with angled handles patented by Vincent Broomhall and Herbert A. Smith in 1935.[37] The range of colors included Mandarin orange—the most popular variety—yellow, green, and blue. Decorated versions with decal designs, such as the *Sailing Ships*, were also available. –DAH

Advertisement for Salem China Company's new *Tricorne* set, published in the *Crockery and Glass Journal*, October 1934.

Ladislav Sutnar (1897–1976)
Teacup with Saucer, Teapot, and Sugar Bowl
Designed c. 1930s
Glass
Teapot: 5⅝ x 9⅞ x 7 in. (14.3 x 25.1 x 17.8 cm)
Produced by Družstevní Práce,
Prague, Czech Republic
K2011.148.2,3,6

THE DESIGNER LADISLAV SUTNAR WAS A LEADING MODERNIST IN CZECHOSLOVAKIA, now the Czech Republic, during the 1920s and 1930s. As art director of Družstevní Práce ("Co-operative Work"), a cultural organization and publishing company, Sutnar designed numerous books and objects, as well as a number of modernist household designs for the shop Krásná Jizba ("Beautiful Household") that were affordable and aimed at a growing middle class. Sutnar's glass tea set, for example, produced during the early 1930s for Družstevní Práce with heat-resistant glass by the Czech glass manufacturer Kavalier, came in several versions; the Bakelite handles of earlier models were later replaced with wooden ones. The functionalist design—comparable to Wilhelm Wagenfeld's heat-resistant glass tea service—was especially popular in Czechoslovakia. –PM

Russel Wright (1904–1976)
Eclipse Glassware
Designed 1957
Glass, decals
Tallest: 6⅞ x 3 x 3 in. (17.5 x 7.6 x 7.6 cm)
Produced by Bartlett-Collins Glass Company, Sapulpa, Oklahoma
K2008.245.1–4, K2016.27.2

NAMED FOR ITS PATTERN OF OVERLAPPING CIRCLES, THIS INEXPENSIVE AND MASS-produced glassware was made with gold rims paired with red, yellow, blue, or green dots. In addition to his designs in ceramic and aluminum, Russel Wright also designed several lines of glassware. Though the *Eclipse* glassware was designed in simple tapered forms, Wright created shaped forms for other lines, including undecorated, colored glassware for Imperial Glass in 1949 to accompany his *American Modern* dinnerware, and another line in 1951 for Morgantown Glass Guild. Oklahoma is not well known as a center for the glassmaking industry, but there were a number of factories established there in the early twentieth century because of good local resources for high-quality glass sand and natural gas. H. W Bartlett and George F. Collins organized the Bartlett-Collins Glass Company, in Sapulpa, Oklahoma, in 1914 to produce hand-pressed utilitarian glassware and lampshades,[38] and by mid-century Bartlett-Collins employed five hundred people.[39] –DAH

Shiro Kuramata (1934–1991)
Salt and Pepper Cellars
Designed 1980
Aluminum
3⅛ x 2¼ x 2⅛ in. (7.8 x 5.7 x 5.4 cm)
Produced by Sointu, Japan
K2013.124

SHIRO KURAMATA, ONE OF THE MOST IMPORTANT JAPANESE DESIGNERS OF the twentieth century, created work that was remarkable for its invention and tranquil beauty. In 1967, Kuramata saw Dan Flavin's and Donald Judd's minimalist works in a 1967 exhibition in Tokyo, which influenced his reductive designs.[40] This small salt and pepper shaker, produced with a shining aluminum or matte black finish, takes on monumentality because of its simple architectonic form: the cube that accommodates the sphere, combined in a harmonious composition for salt and pepper. In the 1970s and '80s, Japanese design was often characterized by high-tech, sometimes flashy, consumer products. Kuramata's work provided a different aesthetic—calm, lyrical, and spiritual. He explained, "If we simplify and go to the basic components that make up a thing, we can clarify the relationship between man and space, and grasp the true nature of function and rationality."[41] –DAH

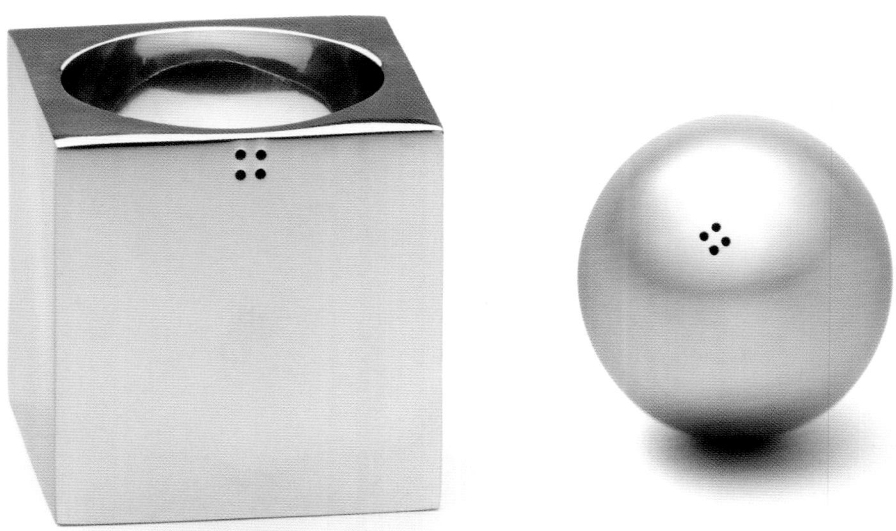

Zaha Hadid (1950–2016)
in collaboration with
Patrik Schumacher (born 1961)
Zaha Cutlery
Designed 2007
Stainless steel
Knife: 9⅝ in. (24.5 cm)
Produced by WMF, Württembergische Metallwarenfabrik AG, Greislingen, Germany
K2009.45.1a–e

ZAHA HADID WAS ONE OF THE MOST ACCLAIMED YOUNG EXPRESSIONIST DESIGNERS who emerged in the 1980s in reaction to postmodernism. Although she was best-known as an architect, Hadid was unusual in that she also designed largely hand-made, luxury products for companies such as Sawaya & Moroni in Milan and Established & Sons in England. Later in her career, which ended with her untimely death at 65, Hadid had begun to design pieces for mass production with long-established companies such as Alessi in Italy and WMF in Germany. As was typical of Hadid, this five-piece place setting—dessert fork, dinner fork, knife, soup spoon, and teaspoon—breaks all the rules. Rather than a straight handle with prongs, bowls, or blades, each piece of this asymmetrical service is a unique sculptural form. Made of solid stainless steel that is polished to a mirror finish, the form of each piece varies in thickness, producing an ergonomically balanced design, which was made possible by computer technology. When assembled, the set forms an elegant fan shape. –RCM

Nicholai Wiig Hansen (born 1979)
Geo Vacuum Bottle
Designed 2012
Plastic
7¾ x 6½ x 6½ in. (19.7 x 16.5 x 16.5 cm)
Produced by Normann Copenhagen,
Copenhagen, Denmark
K2014.11

WHEN NICHOLAI WIIG HANSEN WAS designing the *Geo* vacuum bottle, he sought a combination of forms "to give the flask weight and stability. Its shape is what gives *Geo* a sharp and graphic expression, and the masculine design perfectly counterbalances the lively color combinations."[42] With a minimalistic design and rigid geometric shapes, the highly portable hot-or-cold jug holds one liter (just over a quart) of fluid. Designed for Normann Copenhagen, this bottle is produced in six colors, including bold shades of red, blue, and purple, and won a Red Dot Design Award in 2013 and a German Design Award special mention in 2014. –PM

Massimo Vignelli (1931–2014)
Lella Vignelli (1934–2016)
Stacking Dinnerware
Designed 1964
Melamine resin
Cups: 4 x 3¾ x 3⅛ in. (10.3 x 9.5 x 8.1 cm) each
Plates: 1 x 9⅞ x 9⅞ in. (2.5 x 25.1 x 25.1 cm) each
Produced by Heller Designs, Inc., New York, New York
TR2016.191–194

THIS DINNERWARE, ORIGINALLY SOLD IN SETS OF TWENTY-FIVE PIECES, INCLUDED SIX DINNER PLATES, six salad plates, six bowls, a tray, and several serving dishes. In 1964, Massimo Vignelli convinced one of his Italian clients, Articoli Plastici Elettrici, which was making plastic tiles and Mickey Mouse ashtrays in melamine, to manufacture a very compact set of plastic dishes with interlocking units that would save space and achieve a clean shape as a final result. The set was immediately awarded the Compasso D'Oro for Good Design, but the manufacturer went out of business and the molds lay unused for several years. Alan Heller saw one of the sets on display at the Museum of Contemporary Crafts in New York City and, in 1967, imported them into the United States. A few years later, Heller added more colors, and finally the set was produced in the United States. The size of the plates, their edge-wall concept, their stackability, and the brilliant rainbow colors quickly captured consumers' attention, and the product became a symbol of contemporary housewares. From the initial set in melamine, many other objects followed. "In designing plastic objects, the limitations imposed by the material are very few—most are limitations from the molds," says Vignelli. "While glass and metal have low tolerance for sharpness, with plastics you can obtain very sharp corners, and ideal conditions for stackability. That explains why we have designed many objects with plastic materials."[43] –PM

Stephen Frykholm (born 1942)

Three Herman Miller Summer Picnic Posters
Watermelon, designed 1971
BBQ Chicken, designed 1979
Sweet Corn Festival, designed 1970

Screenprint with lacquer finish

39½ x 25 in. (100.3 x 63.5 cm) each

Produced by Continental Identification Products, Sparta, Michigan, for Herman Miller Company, Zeeland, Michigan

K2010.344.1,2,5

THE HERMAN MILLER COMPANY, PREMIER American manufacturer of modern furniture, was founded in 1938. The company's president, D. J. De Pree, ran it much like a family business, with paternalistic oversight, so it is not surprising that he held an annual picnic for employees as a natural way to bring the company family together. The De Prees would serve home-cooked food in their backyard or in a local park,[44] but in keeping with their Dutch Calvinist heritage, no liquor was served at company events. As the small company grew, the picnic became a tradition with more complex logistics. Stephen Frykholm, the company's head of graphic design, began to design posters in 1970 to commemorate the annual occasion. Each poster was a celebration of a culinary theme with flattened shapes and bold colors, as seen in these three examples that depict basics of the American summer picnic. For the Herman Miller Company, graphic design was an important part of the marketing of modern furniture, notably through the images created by Charles and Ray Eames. Even the company's annual reports, many designed by Frykholm, were graphically distinguished.[45] —DAH

left Herman Miller Company's annual employee picnic, Zeeland, Michigan, late 1970s.

INTERNATIONAL EXHIBITIONS AND INDUSTRIAL DESIGN

Juliet Kinchin

WE LIVE IN AN ERA MARKED BY GREAT EXHIBITIONS IN ALL THE MAJOR INDUSTRIALIZED countries of the world. Such events, driven by a competitive ethos and public appetite for spectacle, have helped to propagate new materials, technologies, and design ideas, and have been a testing ground for experimental products. At points throughout the twentieth century, international exhibitions presented visitors with both a stocktaking of achievements and an indication of things to come, marking moments of great collective transformation for the host city and the launches of popular interest in a particular designer, product, or technology. Despite the temporary nature of the installations, international exhibitions do not pass unnoticed. Catalogues, extensive media coverage, advertising, award systems, and souvenirs—all requiring specialist graphic design—form part of the larger phenomenon. Many of the pieces lauded in past exhibitions have been purchased by collectors and museums, or published in folios, magazines, and books that design students

FIG 1
Postcard of the Bastille station entrance to the Paris Métro, designed by Hector Guimard, 1900.

and manufacturers are encouraged to study. Other exhibits have subsequently been modified for commercial production or mentioned in companies' promotional materials.

Following the *Great Exhibition* held in London's Crystal Palace in 1851 the phenomenon of world exhibitions took off, with imperial powers and metropolitan centers vying to outdo each other in rapid succession. In 1900, Paris set the stage for the twentieth century with a world's fair, the largest to date, which attracted an estimated 48 million visitors. "Life seethes in this immense reservoir of energy," exclaimed De Vogüé about the fair.[1] At the time, the new "magic fluid," electricity, powered spectacular light shows including performances by American dance pioneer Loie Fuller. Cast iron and electrically lit entrances designed by French architect Hector Guimard marked points of descent into the Métro (fig. 1), a new network of underground trains, and their sprouting, organic imagery connected this new urban infrastructure to displays of art nouveau products and entire environments in the main exhibition pavilions. While French variants of the "new art" dominated, less politically powerful countries were also able to make their mark, presenting modern design rooted in national traditions as a form of cultural resistance to politically overbearing neighbors. The Hungarian section, for example, showcased colorful pyrogranite ceramics produced by the manufacturer Zsolnay, which embodied technical as well as aesthetic experimentation.

Two years later, the smaller *Prima Esposizione Internazionale d'Arte Decorativa Moderna* (First International Exhibition of Modern Decorative Art), held in Turin, Italy, broke with precedent by stipulating that all contributions were to be modern. The result was an even more comprehensive and coherent presentation of progressive design tendencies than at Paris. Scotland exhibited independently from England for the first time, with an installation designed by Charles Rennie Mackintosh that featured work in varied media by staff and students at the Glasgow School of Art, many of them women (fig. 2). It was not until the 1904 Louisiana Purchase Exposition, also known as the Saint Louis World's Fair, that American arts and crafts designers emerged on the international scene with a strong identity in furniture and ceramics. The fair was also notable for the presence of a substantial German

FIG 2
below left Rose Boudoir designed by Charles Rennie Mackintosh and Margaret Macdonald-Mackintosh, Turin Exposition of International Design, Italy, 1902.

FIG 3
below right Glass house designed by Bruno Taut, Werkbundaustellung, Cologne, Germany, 1914.

FIG 4
above left Workers' Club U.S.S.R. in the International Exhibition, Paris, 1925.

FIG 5
above right *Die Wohnung* ("The Apartment") poster for Werkbund Exhibition, Stuttgart, Germany, 1927.

contingent amid the displays of decorative arts as well as science, industry, and technology. Peter Behrens, who crafted a strong visual identity for the German section, exemplified the aim of the Deutscher Werkbund, established in 1907 to educate the domestic consumer and improve the reputation of German goods abroad. In 1912, more than 1,300 items by members of the Werkbund were assembled for a groundbreaking show, *Modern German Applied Arts*, at the Newark Museum in New Jersey, whose founding director, John Cotton Dana, was a passionate promoter of modern industrial design.

Exhibitions like *Modern German Applied Arts* were an important platform for the Werkbund in its efforts to stimulate German industry at home and abroad, as well as to address some of the larger social and economic concerns of the emerging modern state. The organization's exhibition held at Cologne in 1914 triggered a famous debate over the use of standardized "types" versus "individuality" in design. On display were standardized furniture and household objects, sleeping car interiors, and a model steel and glass factory designed by Walter Gropius. Architect Bruno Taut designed a showcase prismatic structure for the German glass industry to demonstrate the potential of building with glass blocks (fig. 3). The impact of the exhibition was curtailed, however, by the outbreak of World War I.

Branded as a former enemy of France and its allies, Germany was excluded from the first important exhibition after the war, the 1925 *Exposition des Arts Décoratifs, Industriels et Modernes* (Exhibition of Decorative, Industrial, and Modern Arts) in Paris, which did much to popularize the style subsequently known as art deco. The regulations, like those of the Turin 1902 exhibition, stipulated that "only articles of modern aspiration and real originality should be shown."[2] (The United States cited this as a reason for not participating.) Playing to France's traditional dominance of the international market for luxury goods, pride of place in the trade section was given to French firms like Sevres and Lalique and to famed Parisian department stores that projected a vision of chic glamour. Overall, the exhibition highlighted a gulf between sumptuous art deco, the dominant style, and the austere modernism of a few notable exceptions, such as Le Corbusier's Pavillon de L'Esprit Nouveau and the USSR pavilion (fig. 4), a radical constructivist design by

Konstantin Melnikov that was described by a British visitor as resembling "a dilapidated conservatory." Frantz Jourdain, a designer in favor of standardization, was among those critical of the exhibition for promoting a surface-thin modernism and failing to educate or offer suggestions for improvement.

By contrast, a 1927 international exhibition held in Stuttgart under the auspices of the Deutscher Werkbund, titled *Die Wohnung. Wie wohnen?* (The Dwelling: How Should We Live?) (fig. 5), highlighted issues of standardization, health, rational planning, and industrial production. The most famous element was the Weissenhof model housing estate, which was master-planned and supervised by Ludwig Mies van der Rohe and Lilly Reich and immediately recognized as a manifesto of a unified modernist architecture. Several international architects showcased their experimentation with new forms of tubular steel furniture, and some collaborated with German home-economics expert Erna Meyer on the efficient planning of the kitchens. Related displays in Stuttgart and at numerous international trade fairs of the period in centers such as Berlin, Brno, Leipzig, and Prague showed how industrial design was being significantly shaped by research into new materials and technologies. Aluminum and heat-resistant glass allowed for the development of exciting new products, and the increased availability of electricity and gas revolutionized appliance design.

One of numerous visitors to the Stuttgart exhibition was Swedish architect Gregor Paulsson, who was inspired to organize a similar event in Stockholm that materialized in 1930. The exhibition's stripped-down architecture and bright, airy model interiors offered visitors an attainable vision of modern living. Describing the exhibition for the Finnish press, architect Alvar Aalto wrote, "Exhibitions can and must serve as a social school for the correct use of technical opportunities, representing a very important step in our cultural development."[3] Naturally, Swedish craftsmen and companies, most notably the famous glass producer Orrefors, were given an opportunity to showcase their latest products, triggering a surge of international interest in Nordic design.

The Stuttgart and Stockholm exhibitions were models that Alfred Barr, the first director of the Museum of Modern Art in New York, and architect Philip Johnson had in mind

FIG 6
Installation view, *Machine Art*, the Museum of Modern Art, New York, 1934.

FIG 7
above left Installation view designed by Sonia and Robert Delaunay, *Palace of the Air, Exposition Internationale des Arts et Techniques dans la Vie Moderne*, Paris, 1937.

FIG 8
above right *Futurama*, installation view of a model of a future city designed by Norman Bel Geddes for the General Motors *Highways and Horizons* exhibition, New York World's Fair, 1939.

when conceiving *Machine Art*, a landmark exhibition held at the Museum of Modern Art (MoMA) that brought international attention to American industrial design (fig. 6). The selection of everyday objects—typewriter carriage springs, a self-aligning ball bearing, an outboard propeller, a toaster, a cash register, pots and pans, a microscope, a compass, and scientific flasks—and the way they were idealized through photography and the style of the installations, gave visual encouragement to people who felt uncertain about society's continuing progress during the Great Depression. Ultimately, the exhibition was more about proselytizing modern design than educating consumer choice and influencing buying habits. The museum's subsequent *Useful Objects* series in the 1940s and *Good Design* program of the 1950s changed the emphasis by providing more practical information about where and at what cost products could be purchased.

The ideal of international exhibitions as a medium for promoting world peace and harmony was stretched to breaking point at the last and largest of Paris's great world's fairs, the 1937 *Exposition Internationale des Arts et Techniques dans la Vie Moderne* (International Exhibition of Arts and Techniques in Modern Life). The Spanish pavilion, containing Pablo Picasso's monumental painting *Guernica*, drew attention to the unfolding horror of the Spanish Civil War and aerial bombardment of a civilian population. Also memorable was the architectural face-off between the Nazi German and Soviet Russian pavilions, which prefigured the military conflict between these two great ideological rivals. One of the most dynamic displays of industrial design was Sonia and Robert Delaunay's installation of suspended radial airplane engines and aluminum rings in the Palace of the Air (fig. 7). The Finnish pavilion, repeatedly singled out by critics, displayed organic shaped glassware and plywood furniture by Alvar Aalto and his wife, Aino. Bringing out the political dimension of the Finnish displays, Alvar commented, "Nature is, of course, freedom's symbol. Sometimes it is even Nature that creates and maintains the concept of freedom."[4]

The Aaltos' design of another Finnish pavilion was a highlight of the New York World's Fair in 1939–40. For most visitors, however, the fair was identified with its signature Trylon and Perisphere buildings, modernistic streamlining, and a more technologically determined

FIG 9
above left *Antelope* chairs designed by Ernest Race, *Festival of Britain* exhibition, 1951.

FIG 10
above right The Philips Pavilion designed by Le Corbusier and Iannis Xenakis, Brussels World's Fair, 1958.

view of the "World of Tomorrow" than at Paris. National pavilions were dwarfed by the ambitious contributions of big American corporations working with an up-and-coming generation of independent industrial designers including Walter Dorwin Teague, Norman Bel Geddes, and Raymond Loewy. General Motors was the force behind the most popular attraction, *Futurama*, which offered visitors a panoramic, cinematic, and car-centric look into the future (fig. 8).

Official initiatives to promote design as a tool of social and economic reconstruction after the traumas of World War II were behind a number of influential exhibitions. The 1951 *Festival of Britain*, for example, was conceived as a "tonic to the nation,"[5] an explosion of color and fun on London's South Bank and at sites around Britain. Science and technology featured strongly; in one of the pavilions, many Londoners saw their first-ever television pictures. The Festival Pattern Group brought together designers, manufacturers, and scientists to explore the potential of crystallography and atomic structures as a source of design inspiration. Aluminum and steel rod chairs designed by Ernest Race (fig. 9), ergonomic plywood furniture by Robin Day, and the abstract, colorful textiles of Day's wife, Lucienne, epitomized what became known as the "festival style." The Days were also recognized with the award of a gold medal at the 1951 Milan Triennale.

Occurring at three-year intervals and dedicated to crossovers between architecture, design, and craftsmanship, the Triennales of the postwar decades played an important role in shaping and promoting modern design culture in Italy and beyond. Boosted by investment delivered through the United States' Marshall Plan, Italian architect-designers were able to showcase their command over new plastics and electrical and lighting technologies, as well as their reinvigoration of Italy's traditional textile and glass industries. The Milan Triennales were an important shop window for companies such as Olivetti, Vespa, and Cassina to try out complex, sculptural prototypes produced in close collaboration with traditional craftspeople in the regions of Milan and Turin.

Increasingly, world fairs such as the ones in Brussels in 1958, New York in 1964–65, and Montreal in 1967 were framed by Cold War confrontations between the superpowers of the

FIG 11 AND 12
right and opposite *Seed Cathedral* designed by Heatherwick Studio, British Pavilion at *Expo 2010*, Shanghai.

United States and the USSR and their respective spheres of influence. Italian architect and cultural critic Bruno Zevi described the Brussels Expo as an "elephantine trade fair in which every building clamors for the attention of the public with the most crass propaganda."[6] The expo's icon was the giant model Atomium, a reminder of nuclear physics' destructive as well as peacetime potential. The Czechoslovakian, Polish, and Yugoslavian pavilions were each powerful expressions of pioneering postwar modernism and innovative display techniques derived from contemporary cinema. Also notable was the immersive Disney Circarama theater in the US section and the dissonant acoustic space of the Philips Pavilion, designed by Iannis Xenakis and Le Corbusier, in which an "electronic poem" by Edgar Varése was played back through 425 loudspeakers (fig. 10).

Officially sanctioned expos have continued to provide a vehicle for "nation branding," but at increasingly prohibitive costs to the participants. During the decades following the widespread student protests of 1968, many designers began to develop cross-cultural networks that functioned outside the traditional institutional and market forces of international exhibitions, and began to explore alternative ways of addressing diverse global audiences. In the late-twentieth century, the emergence of mass media and the rise of internet-based communication led many to believe that the great age of international exhibitions was over. Yet the first major expo of the twenty-first century, held in Shanghai, proved to be a record-breaking event that placed international exhibitions back at the center of innovative design, technology, and collective spectacle. *Expo 2010*, on the theme of "Better City—Better Life," confirmed Shanghai's status as "the next great world city," and broke all previous records for scale, expense, attendance, and number of participating countries.[7] Heatherwick Studio's extraordinary *Seed Cathedral* (fig. 11 and 12) was one of many pavilions to epitomize the contemporary focus on renewable energy and hi-tech, ecologically driven design. On this evidence it seems there is still life in the expo format as a tool of popular education, a catalyst of international debate, and an engine of product innovation.

ENDNOTES

1 De Vogüé, quoted in Debora Silverman, *Art Nouveau in Fin-de-siècle France: Politics, Psychology, and Style* (Berkeley and Los Angeles: University of California Press, 1989), 298.

2 Cited in Jonathan Woodham, *Twentieth Century Design* (Oxford, U.K.: Oxford University Press, 1997), 87.

3 Alvar Aalto, *Arkkitehti*, no. 8 (1930); reprinted in Goran Schildt, *Alvar Aalto in His Own Words* (Helsinki: Otava, 1997), 74.

4 Cited in Richard Weston, *Alvar Aalto* (London: Phaidon Press, 1995), 98.

5 A phrase coined by Gerald Barry, director-general of the Festival of Britain, and used as the title of a 1976 book about the Festival edited by Mary Banham and Bevis Hillier.

6 Bruno Zevi, "Bruxelles 1958: primi interrogativie," *L'architettura*, May 1958, 4.

7 Estimated attendance was around 72 million visitors. Adam Minter, "China Rules the World at Expo 2010," *The Atlantic*, April 29, 2010.

07
objects

LISTENING

IN THE MODERN ERA, TECHNOLOGY HAS ALLOWED US TO RECORD sounds and to listen to those recordings as one of our favorite leisuretime activities. In turn that was facilitated by the invention of a number of machines, among them the radio, the record player, and the hi-fi system. At first those machines were disguised as pieces of furniture, but later designers began to embrace the look of their technology, openly displaying their origins as machines in the context of the home. In more recent decades, objects that play music have moved closer to our bodies, turning this activity into one that could take place outside of the home, while on the move, thus bringing about a dramatic change in human behavior. The same applies to the telephone, which has evolved in the hands of designers from site-bound heavy chunks of plastic to small, light, multipurpose smartphones.

Daisuke Kajiwara (dates unknown)
Three *Toot-A-Loop* AM Radios, Model No. R-72
Designed c. 1971 (D229,347)
ABS plastic
6 x 6 x 2¾ in. (15.2 x 15.2 x 7 cm) (closed, each)
Produced by National Panasonic Radio, Osaka, Japan
K2008.92.2-4

IS IT A BRACELET, A RADIO, OR A WORK OF SCULPTURE? THE *TOOT-A-LOOP*, PATENTED by Japanese designer Daisuke Kajiwara,[1] is all three. Its sculptural form was a strong departure from the traditional rectangular-shaped portable radios, and its range of bold, bright colors was playful and fun. While Braun in Germany focused on minimal functionalism, Japanese designers, inspired by pop art, were commissioned to replace traditional forms with unconventional objects. According to an advertisement, the radio was as much fun to look at as to listen to: "Wind the Toot-A-Loop around your wrist and be the first on your block to have a bracelet that plays music."[2] When twisted, the swiveling joint on the thinner side rotates and the articulation on the wider side separates, and the two halves swing open to form an *S* shape that reveals a circular tuning dial set in the top. The speaker grille and earphone jack are on the exterior surface of the bottom half. The affordable *Toot-A-Loop* appealed to all ages. The radio was treated playfully even in a Panasonic sales brochure: "Our Craziest Crazy Color AM Portable Yet! Wear it. Swing it. Twist it. You can even listen to it!"[3] —DAH

Hans Gösta Thames (1916–2006)
Helge Edward Lindström (dates unknown)
Knut Hugo Blomberg (1897–1994)

Three *Ericofon* Telephones

Designed c. 1951 (utility patent 2,822,432)

SAN Plastic, rubber

8¼ x 4 x 5½ in. (21 x 10.2 x 14 cm) each

Produced by L. M. Ericsson Company, Stockholm, Sweden

K2008.15.1, K2016.29.3–4

THE ONE-PIECE *ERICOFON* TELEPHONE—"THE SOUND OF THE FUTURE"—WITH ITS all-in-one configuration anticipated both the cordless telephone and cellular phones of later decades. The range of colors was sold in both Europe and the United States. Sometimes called the "Cobra" due to its shape, the design was intended initially for institutional use, and hospitals were particularly interested in the phone for patients. Small and lightweight, the *Ericofon* accommodated the rotary dial on its base so the phone could not be left off the hook: when lifted off its base, a pressure switch released to engage the dial tone. Utility patents filed by Blomberg in 1941 and 1942 show the idea of the integrated handset and base,[4] but this specific case form appears on a patent filed in 1951. In the 1960s, Ericsson transferred the production to the North Electric Company in the United States. –DAH

left

Designer Unknown

Telephone, Model No. W48

Designed c. 1936

Bakelite, plastic, braided nylon

6 x 10 x 7½ in. (15.2 x 25.4 x 19.1 cm)

Produced by Siemens & Halske, Berlin, Germany

K2009.32

right

Gerard Kiljan (1891–1968)

Telephone

Designed 1962

Plastic, rubber, metal

5⅛ x 10 x 8 in. (13 x 25.4 x 20.3 cm)

Produced by Ericsson, Rijen, the Netherlands

K2008.628

IN 1900, ALEXANDER GRAHAM BELL FOUNDED AMERICAN TELEPHONE & TELEGRAPH—still known today as AT&T—which set up Bell Telephone Laboratories in 1925 as its research division. Its most famous telephone, the Dreyfuss Model 302, was the result of a competition that Bell Labs organized in the early 1930s to replace the 202 series of 1927. The Model 302 went into production in 1937 by Western Electric Manufacturing Company for Bell. The Dreyfuss design was aesthetically remarkable in form and function, with its square base providing a broader, more stable foundation than its dainty and precariously balanced predecessor's, and was a forerunner of the square shape that became the standard for the rest of the twentieth century. It was inspired by Jean Heiberg's 1931 design for the Swedish firm L. M. Ericsson Company.[5] The much later white Gerard Kiljan telephone also follows the Heiberg model. The German example—whose model number, W48, comes from the German word *Wählfernsprecher,* or "dialing telephone," and its production looks back to an earlier telephone model in its bulbous form. Its receiver is a pronounced arc resting on a prominent cradle. —DAH

Henry Dreyfuss (1904–1972)

Telephone, Model No. 302

Designed 1937

Enameled metal, steel, rubber

5½ x 9 x 7¼ in. (14 x 23 x 18.5 cm)

Produced by Western Electric Manufacturing Company for Bell Telephone Laboratories, New York, New York

K2008.325

Harold L. Van Doren (1895–1957)
John Gordon Rideout (1898–1951)

Air-King Skyscraper Radio

Designed 1933

Plaskon, metal, glass

11¾ x 8⅞ x 7½ in. (29.8 x 22.5 x 18.9 cm)

Produced by Air-King Products Company, Inc., Brooklyn, New York

K2008.734

DESIGNED BY HAROLD L. VAN DOREN AND John Gordon Rideout, the art-deco *Air-King Skyscraper* radio was made of a urea-formaldehyde resin plastic sold under the trade name Plaskon. The Air-King Products Company offered the radio with three different rectangular panels: a clock, a world map (as seen here), or an Egyptian panel with knobs that varied slightly from the other two versions. In its day, the *Air-King*, available in several colors, was the largest plastic radio on the market. –PM

A. M. Cassandre (1901–1968)

Miniwatt/Philips Radio Poster

Designed 1931

Offset lithograph

22½ x 15¼ in. (57.2 x 38.7 cm)

Printed by Alliance Graphique, Paris, France

K2011.96

Cooper-Hewitt, Smithsonian Design Museum, promised gift of George R. Kravis II

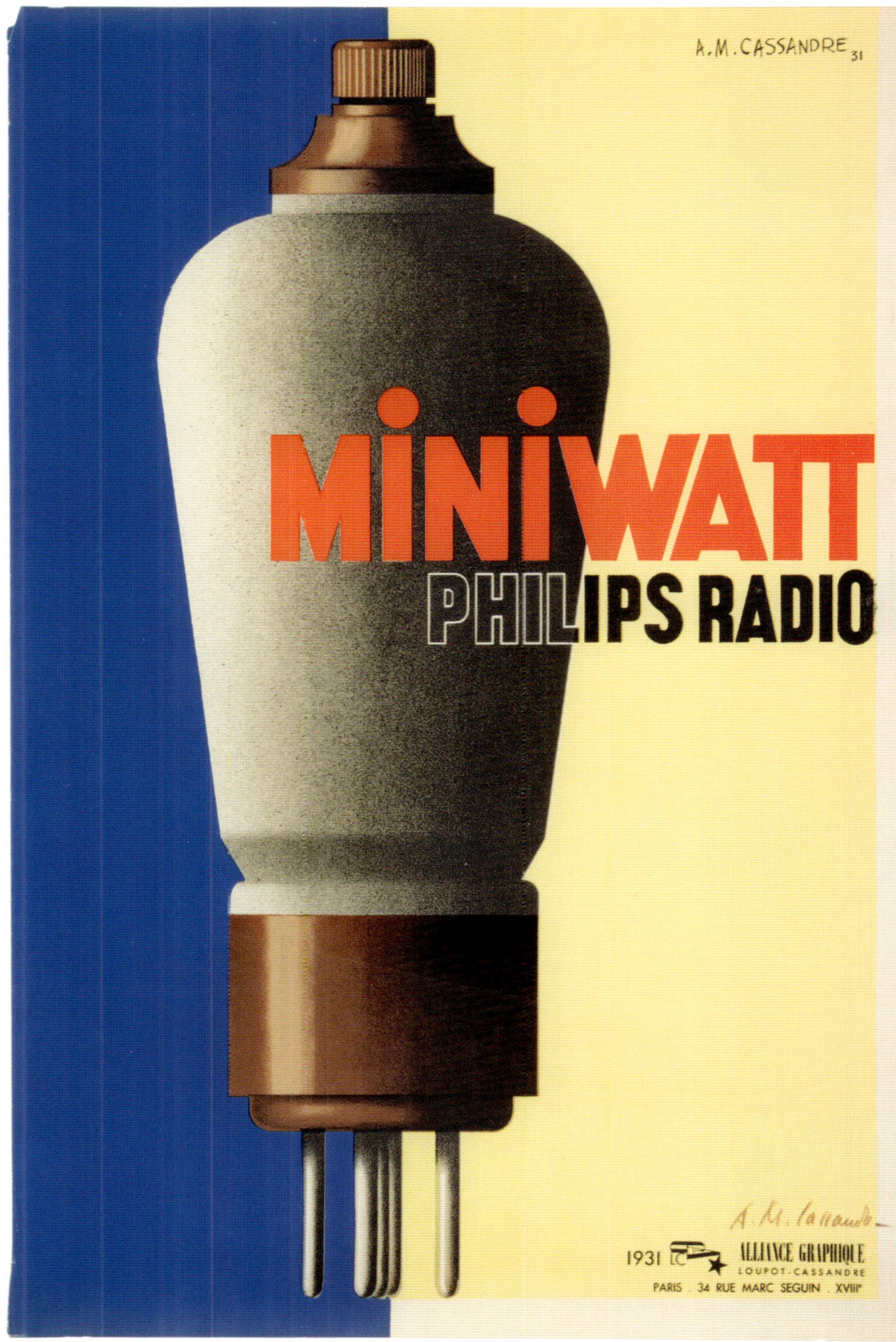

ANTICIPATING POP ART OF THE 1960S, A. M. Cassandre created a dramatic composition in this poster with the over-scaled Miniwatt radio tube, floating like a spaceship, balanced by the bold lettering. Increasing the size of this common functional object that would normally be hidden inside a radio casing was in keeping with pop art's penchant for celebrating the banal. It is reminiscent of the work of Claes Oldenburg, as seen in his 1970 *Giant Three-Way Plug* sculpture, which enlarges a standard electrical plug to gigantic scale. Cassandre designed a number of posters for Philips Radio, the Dutch light bulb and radio tube manufacturer. Russian born, Cassandre moved to Paris in 1915 and studied art at the Academie Julian. He was inspired by Cubism and Surrealism, both influences seen in this poster. A press release for his 1936 retrospective exhibition at MoMA claimed Cassandre was "the foremost poster designer in the world." –DAH

Wells Coates (1895–1958)
Ekco Radio, Model No. AD-65
Designed 1932–34
Bakelite, chromium-plated metal, woven fabric
15⅞ x 15½ x 8¼ in. (40.3 x 39.4 x 21 cm)
Produced by E. K. Cole Ltd., Southend, England
K2008.200
Museum of Modern Art,
promised gift of George R. Kravis II

Cover of Ekco Radio brochure for E. K. Cole Ltd., c. 1934.

RADIOS IN THE EARLY TWENTIETH CENTURY WERE USUALLY CONCEIVED AS PIECES OF WOOD furniture, often with such period details as cabriole legs and elaborate decoration. Canadian-born Wells Coates provided a new concept for the E. K. Cole company, using twentieth-century materials and technology. This radio's round shape was unprecedented and lent itself to production in plastic because of its simple design. Once tooling was set up, it was less expensive to reproduce than the more elaborate decorated products on the market. E. K. Cole rejected radios as antique furniture, and it developed new forms by such architects as Coates and Serge Chermayeff. Its bet on modern design during the Depression paid off, and the new *Ekco* radios became bestsellers, even in an age when electricity in homes was not yet universal: "Whether you have electric light or not there is an Ekco Set that will bring you 'Clear-cut reality' radio at a price to suit your pocket."[6] Sex appeal and chic fashion had also entered the market to help sell the new modern radios: the cover of the brochure places the Ekco radio on a pedestal like a work of art, while an elegant lady smoking a cigarette drapes her arm across the top of it. –DAH

MEDIUM WAVES
200/550 METRES
LONG WAVES
900/2000 METRES

John Vassos (1898–1985)

RCA Victor Special Portable Phonograph

Designed c. 1935

Aluminum, metal, plastic, velvet

Closed: 8 x 16½ x 18 in. (20.3 x 41.9 x 45.7 cm)

Produced by Radio Corporation of America (RCA), New York, New York

Cooper Hewitt, Smithsonian Design Museum, gift of George R. Kravis II, 2016-5-11

THIS SQUARE AND PORTABLE SUITCASE PHONOGRAPH, STYLED BY JOHN VASSOS, is streamlined and flashy. Powered by a hand crank to play 78 rpm records, it could be used indoors or outdoors. Its bright red detailing, including the turntable's velvet, contrasts with the shiny aluminum of the surround and exterior. It was also offered with black detailing. One could watch the records spin as reflected on the mirrored record-storage compartment set in the lid, for which record storage folders were provided; a protruding speaker grille was set in front of the turntable.[7] According to the utility patents, the mechanism of this portable phonograph was designed by engineers Selden T. Williams and Alfred Weiland.[8] Vassos was engaged by RCA Victor in 1933 while working on murals for WCAU radio station in Philadelphia. He was initially hired to work on its radio lines, but his design skills were soon used throughout the company to include graphics, corporate logo, sales displays, equipment such as microphones, and company products.[9] –DAH

John Vassos (1898–1985)

RCA Victor Record Player, 45 rpm, Model No. 9-JY

Designed c. 1949

Bakelite, plastic, metal, vinyl

6½ x 9¼ x 6⅞ in. (16.5 x 23.5 x 17.5 cm)

Produced by Radio Corporation of America (RCA), Camden, New Jersey

K2008.199

AMERICAN INDUSTRIAL DESIGNER AND ARTIST JOHN VASSOS DESIGNED THE FIRST MASS-produced television receiver, the TRK-12, which premiered at the 1939 New York World's Fair. He went on to make radios and TVs attractive and accessible to millions of Americans. While shellac 78 rpm records were fragile, had low fidelity, and were limited to ten minutes of playing time on each side, the seven-inch 45 rpm record produced higher fidelity, and in response, RCA developed the 9-JY model 45 rpm record player, an elegant square phonograph, with a red and gold-toned spindle, and elegant shaped arm. It was launched in 1949 and allowed the user to stack multiple discs for extensive listening. An advertisement that appeared in *LIFE* magazine extolled the Vassos-designed record player—the first in the industry to have an automatic record changer—as "the modern way to play records," and that 150 of the 7-inch "unbreakable" records fit in one foot of an ordinary bookshelf. As RCA branched out in new directions in the postwar period, Vassos went on to design everything from microphones, broadcast equipment, and transmitter buildings to RCA's first color television camera. –PM

Advertisement for RCA's *Victor 45* record player, c. 1949.

Isamu Noguchi (1904–1988)
Radio Nurse Receiver/Speaker
Designed c. 1937 (D108,837)
Bakelite
8¼ x 6½ x 6½ in. (20.9 x 16.5 x 16.5 cm)
Produced by Zenith Radio Corporation, Chicago, Illinois
K2011.200

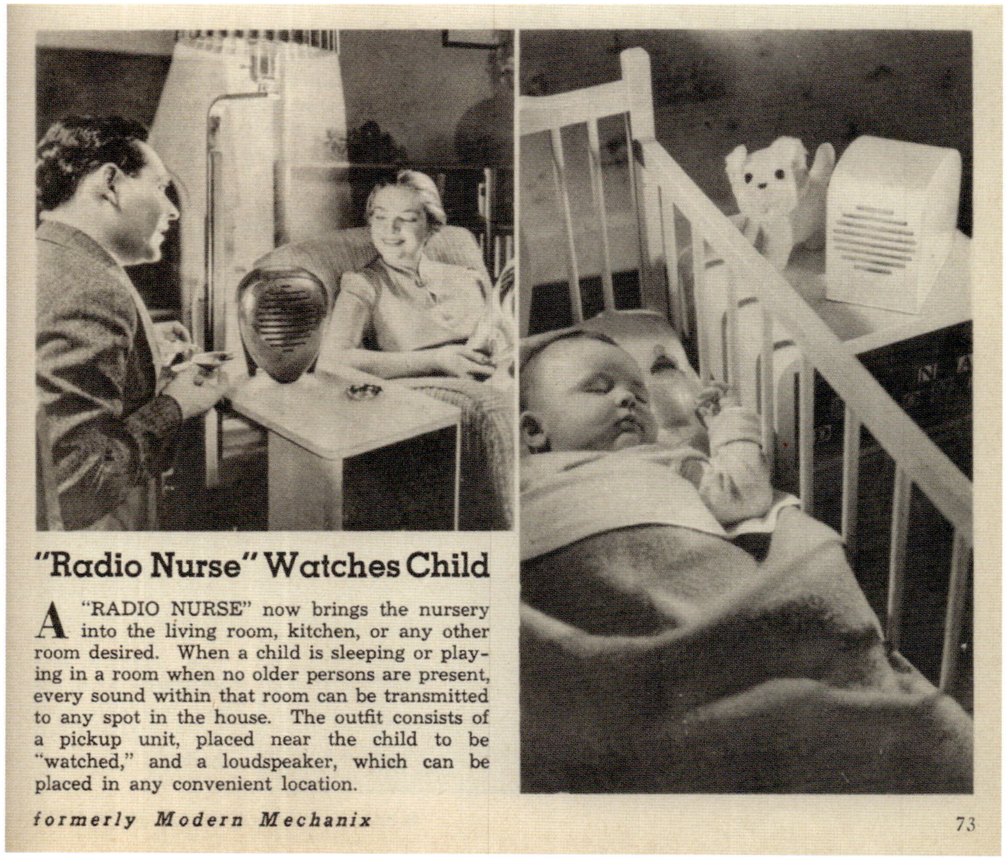

"Radio Nurse" Watches Child

A "RADIO NURSE" now brings the nursery into the living room, kitchen, or any other room desired. When a child is sleeping or playing in a room when no older persons are present, every sound within that room can be transmitted to any spot in the house. The outfit consists of a pickup unit, placed near the child to be "watched," and a loudspeaker, which can be placed in any convenient location.

formerly Modern Mechanix 73

"'Radio Nurse' Watches Child" article published in *Mechanix Illustrated*, July 1938.

ISAMU NOGUCHI'S *RADIO NURSE* IS AN EARLY BABY MONITOR, AN INTERCOM FOR THE home. Concern for children's safety had become widespread in the wake of the famous kidnapping of Charles Lindbergh's son in 1932, and Zenith commissioned Noguchi not long after to design a comforting form for families.[10] A brief article in *Mechanix Illustrated* describes its function: "When a child is sleeping or playing in a room when no older persons are present, every sound within that room can be transmitted to any spot in the house."[11] As seen in the illustration, the intercom set consists of a transmitter placed near the child—or any individual requiring constant supervision—and a receiver/speaker near the caregiver. The product, designed as the stylized head of a nurse, reflects the Japanese-American designer's training as a sculptor and his portrait heads of the 1930s, as well as the influence of Constantin Brancusi, for whom he worked in 1927 in Paris. The aerodynamic teardrop form with horizontal speed lines as the speaker cutouts also demonstrates aspects of American streamlining. The accompanying *Guardian Ear* transmitter in the collection retains its original box with the logos for both Zenith and the *Radio Nurse*. —DAH

Livio Castiglioni (1911–1979)
Pier Giacomo Castiglioni (1913–1968)
Luigi Caccia Dominioni (1913–2013)

547 Radio

Designed 1939–40

Bakelite, metal, plastic

8 x 9¾ x 10½ in. (20 x 25 x 26.5 cm)

Produced by Phonola Radio, Milan, Italy

K2012.87

Cooper Hewitt, Smithsonian Design Museum, promised gift of George R. Kravis II

THE *547* RADIO AND THE RR-126 STEREO (PAGE 247) ARE EMBLEMATIC OF THE IMPORTANT role that Italy played in twentieth-century design. Before World War II, Italy was a peripheral player in the industrial design field; by the 1960s, the country had become an international powerhouse. The Castiglioni brothers were central figures in this transformation. Livio and Pier Giacomo Castiglioni established a short-lived studio with Luigi Caccia Dominioni from 1938 to 1940. One of their most noted designs was the *547* radio, among the first Italian "plastic" radios, in this instance using Bakelite. The radio is, on the one hand, a late manifestation of the streamlined aesthetic that was so prevalent between the two world wars, but on the other hand, the form is also telling in its tentative expression of its function: a squarish base that holds the components, a rectangular control panel, and a circular speaker. A petite but dramatic design, the sleek red shell was in the end merely a streamlined casing for the complicated electronic components within. In the postwar period, the two Castiglionis would collaborate with their brother, Achille, to become one of the most renowned Italian design studios in the second half of the twentieth century. –RCM

David Ogle (1922–1962)

Seven Transistor Portable Radio, Model No. TR 82B

Designed c. 1959

Plastic, vinyl, metal

10½ x 13 x 3¾ in. (26.7 x 33 x 9.5 cm)

Produced by Bush Radio Ltd., Plymouth, England

K2012.1.1

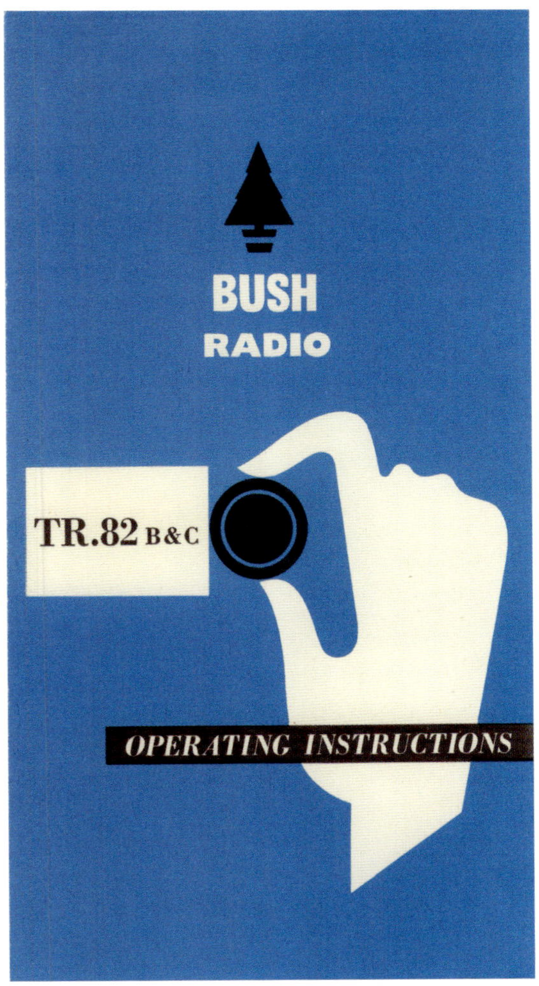

PORTABLE RADIOS BECAME INCREASINGLY POPULAR IN THE 1950S, WHEN CONSUMERS had more disposable income. The battery-powered Bush radios served this function for younger people, who regarded them as fashion accessories because they could be carried like a handbag or briefcase. This streamlined design, with curved edges and horizontal speed lines, is evidence of the continued popularity of the American consumer style after World War II, and stands in contrast to the minimal, rational designs of electronic goods produced by Braun in Germany.[12] The Bush company logo is prominently displayed in chromed letters on the front. The operating manual, with striking graphics illustrating the ease of tuning, explains that the radio had an internal antenna for use anywhere but could also be connected to a car's antenna for improved reception. Founded in 1932, Bush Radio expanded rapidly through the 1930s and 1940s, moving to larger factories in 1936 and 1949. This radio was designed by David Ogle, a British industrial designer specializing in automobiles. He first worked for Bush Radio in 1948 and continued to work on projects for Bush after he founded his own design firm in 1954.[13] The company relaunched this classic design in 1997. –DAH

left Cover of Bush Radio Operating Instructions, c. 1959.

CYARTS
DELUXE
STANDARD
BROADCAST
170 130 90 70 55
150 110 80 60
KILOCYCLES
METERS
120 100 85 75
140 65
160 60
175 55
550 175
500 200
400 325 250
FADA

Designers Unknown

rear
Cyarts Deluxe Radio
Designed c. 1946
Plexon, Lucite, woven fabric
7⅛ x 14 x 7⅜ in. (18.3 x 35.6 x 18.9 cm)
Produced by Cyarts Plastics, Inc., Bronx, New York
K2012.8

foreground
Bullet Radio, Model No. 115BA
Designed 1940
Catalin, plastic
6½ x 10⅜ x 6 in. (16.5 x 26.5 x 15.2 cm)
Produced by Fada Radio & Electronic Co. Inc.,
Long Island City, New York
K2008.470

right
Radio, Model No. 6AU-1
Designed 1941
Catalin, plastic, metal
7⅜ x 11½ x 6⅝ in. (18.7 x 29.2 x 16.8 cm)
Produced by Garod Radio Corporation, Brooklyn, New York
K2010.222

DURING THE 1930S, AS HOUSEHOLD GADGETS EVOLVED, manufacturers experimented with making radio housing out of different plastics, such as Bakelite, acrylic Lucite, and Catalin. The bullet-style radio made by Cyarts Plastics in 1946 has a cabinet with a panel that lights up and is hand-fabricated out of Plexon—an early plastic material known for its transparency that was also used for screens and lampshades—and Lucite grille and knobs made by Du Pont. The sole radio made by Cyarts Plastics, it originally sold for $69.95.

Catalin was the brand name for a thermosetting polymer, popular in the 1930s and 1940s, and used for many household objects, jewelry boxes, and radios. Examples include the *Bullet* radio, by the Fada Radio & Electronic Company, which employed a grille that was an integral part of the cabinet. Both the Fada and the Garod radios featured "drop" handles that lay flush when not in use and were adopted by other manufacturers. The Fada also featured a circular tuning dial, rather than the horizontal tuning indicators on the Cyarts and Garod radios. –FM

Designer Unknown

Dufono Intercom

Designed c. 1938

Bakelite, chromium-plated steel

5 x 6¾ x 3⅝ in. (12.7 x 17.1 x 9.2 cm)

Produced by Societa Radio Brevetti Ducati, Bologna, Italy

K2010.40

THE ONSET OF THE GREAT DEPRESSION IN 1929 DRAMATICALLY CHANGED THE DESIGN world. The art deco style of the 1920s with its emphasis on luxurious, handmade, and historicizing designs could no longer be sustained. Thus the 1930s saw the emergence of a new style—the moderne—and the industrial designer appeared, which led in turn to a dramatic increase in the number of new industrial products by manufacturers. Designers also sought to use new materials and technologies to create original forms for this new mass-produced merchandise. The moderne style had a streamlined aesthetic that was internationally embraced by designers, manufacturers, and consumers. The *Dufono* intercom is a striking example of this newly emerged approach. Made of Bakelite, the bipartite product features a sleek microphone poised on a tapered base. The design has often been attributed to the noted Italian designer Gio Ponti, but there is no data to date to support such an attribution. –RCM

John Vassos (1898–1985)

Microphone and Desk Stand, Model Nos. 77-C1 and 91-B

Designed c. 1938

Enameled metal, chromium-plated brass, plastic, silk

12⅝ x 4¼ x 9⅜ in. (32.1 x 11 x 23.8 cm)

Produced by Radio Corporation of America (RCA), Camden, New Jersey

K2012.15

JOHN VASSOS, WHO IMMIGRATED TO THE UNITED STATES FROM TURKEY in 1919, worked for RCA from 1933 to 1970. He designed many of its new products, such as RCA's first mass-produced television, which was launched at the New York World's Fair in 1939. He also redesigned all RCA in-house technological equipment, including a range of streamlined broadcast microphones. Here, the microphone is mounted on a Bakelite Model 91-B desk stand with metal banding. On the lower section of the capsule-shaped microphone, the RCA logo is centered in a metal band flanked by speed lines. According to RCA's 1941 sales catalogue—unlike the visually identical, uni-directional Model 77-B1—the Model 77-C1 is a three-way combination, or all-purpose, microphone that can be used uni-directionally, bi-directionally, or non-directionally for versatility in studio or auditorium environments, which "provided the station owner the equivalent of three microphones in a single unit."[14] –DAH

Dieter Rams (born 1932)

Compact Stereo, Type Audio 1 M

Designed 1962

Metal, acrylic, plastic, rubber

Closed: 6 x 25½ x 11 in.
(15.2 x 64.8 x 27.9 cm)

Produced by Braun AG,
Frankfurt, Germany

K2011.117

THE 1950S AND 1960S WERE DECADES OF RAPID CHANGE FOR INDUSTRIAL DESIGNERS, especially the engineers at Braun, the German company that had established itself as a pioneer in the electronics market. Following a period of technological advancement during World War II, Artur and Erwin Braun—sons of Max Braun, the company's founder—were ready to launch a new set of products. In 1962, Dieter Rams had the opportunity to update the Radio-Phonograph, Model SK 4/10, or SK4, with the latest in technology, replacing the bulky hot tubes required for clear sound with smaller transistors. This allowed Rams to create the Audio 1 M Compact Stereo, complete with an entire set of knobs, buttons, and dials next to the turntable. Rams's groundbreaking design for the Audio 1 M is said to have influenced the brushed aluminum unibody of the MacBook Pro. —PM

Dieter Rams (born 1932)
KH 1000 Headphones
Designed 1967
Metal, plastic
8⅝ x 10⅝ x 3¼ in. (21.9 x 27 x 8.3 cm)
Produced by Braun AG, Frankfurt, Germany
K2013.21

DIETER RAMS IS KNOWN FOR HIS minimal designs for Braun in the 1960s and for his ten principles for good design, which he began to develop in the 1970s. The Braun look—characterized by geometric forms, austere simplicity, and sober colors—appealed to a generation brought up on the principles of good design, and Rams's famous ideal of "less but better." These headphones are typical of Braun's minimal designs: the simple oval earphones are attached to a thin metal strip that is covered with a conforming plastic section to fit on the head. In the mid-1950s, Braun introduced a consistent austere aesthetic in its functional designs, becoming a leader in products, packaging, and advertising long before Apple. —DAH

David Gammon (1933–2005)

Reference Hydraulic Turntable

Designed 1964

Polished aluminum, brass, plywood, acrylic

7¼ x 17⅜ x 16½ in. (18.4 x 44.1 x 42.1 cm)

Produced by Transcriptors, Ltd.,
New York, New York

K2016.11

TRANSCRIPTORS, LTD., FOUNDED BY DAVID GAMMON IN 1960, WAS A LEADING PRODUCER of professional-grade record players and accessories into the early 1980s. Gammon invented the Transcriptor arm, around which this turntable was built to fulfill the request of a research lab in need of precision audio testing equipment. It was an innovation in the fields of sound technology and, with its visible mechanisms, audio equipment aesthetics. The distinctive turntable made many appearances in television and film, most notably one that Gammon gave to Stanley Kubrick, who used it in his films *A Clockwork Orange* and *2001: A Space Odyssey.* –PM

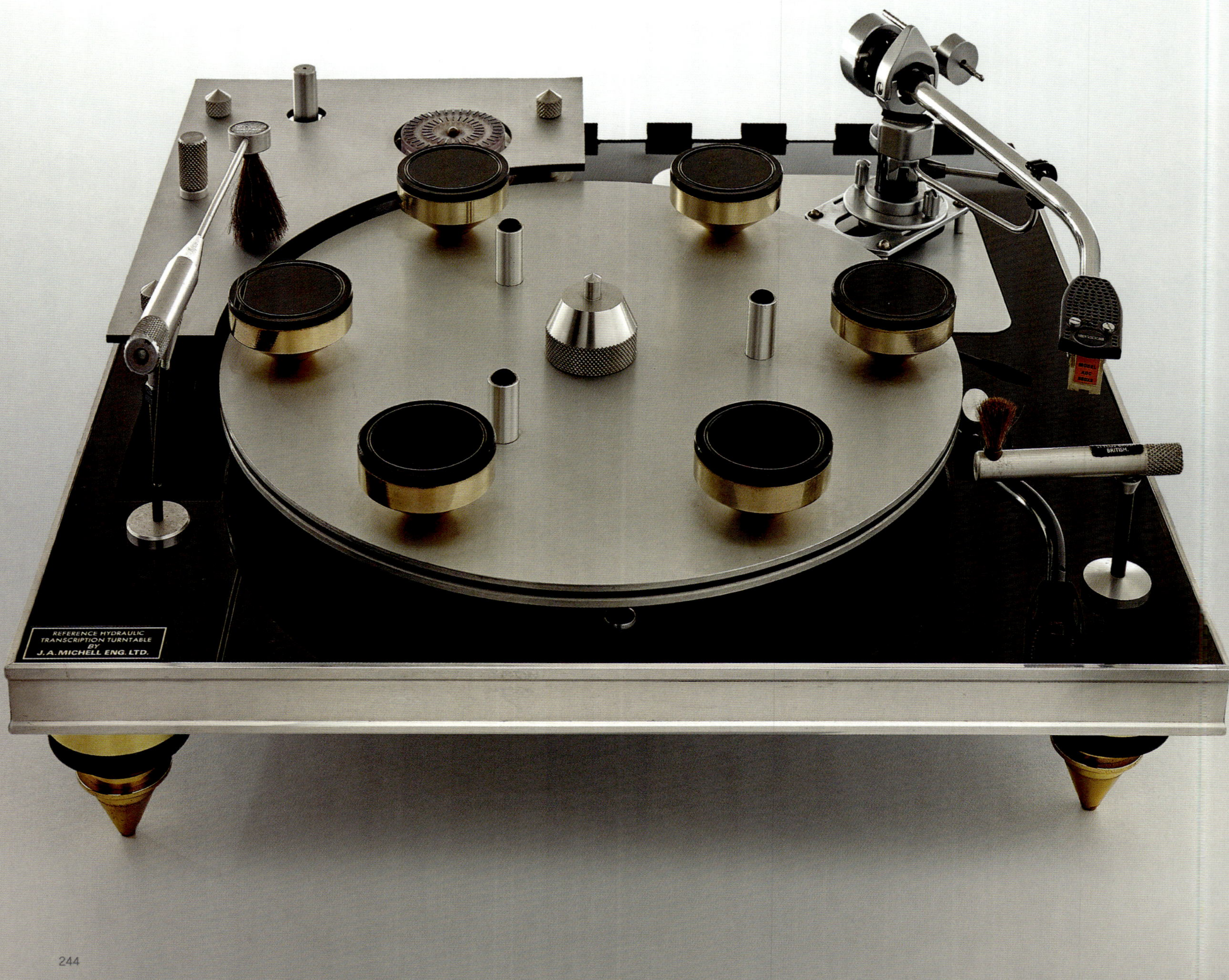

Mario Bellini (born 1935)
Dario Bellini (born 1937)

Totem Stereo System, Model No. RR-130

Designed 1970

Chipboard, plastic laminate, woven fabric

Closed: 20¾ x 20¼ x 20½ in.
(52.7 x 51.4 x 52.1 cm)

Produced by Brionvega, Milan, Italy

K2015.23

THE ARCHITECT AND INDUSTRIAL DESIGNER MARIO BELLINI BECAME ONE OF THE LEADING PROPONENTS of Italian design in the 1970s; twenty-five of his works are in the permanent collection of MoMA in New York. He has designed for major Italian manufacturers, including Olivetti, Cassina, B&B Italia, Flos, and Artemide. In designing this stereo system, Mario and his brother Dario Bellini decided not to emphasize the technical aspects of the object, choosing instead to create a supremely subtle hi-fi system that would fit easily into any domestic setting. This simple, compact cube organizes a complex, articulated technology. Despite the quiet simplicity of the form, an element of surprise is present when the hermetic cube is opened by rotating the removable speakers to reveal the controls, and a turntable that is hidden within the bottom half. In this case, the Bellinis made use of the feature of movement to highlight the complexity of this ingeniously designed system. –PM

BRIONVEGA

Achille Castiglioni (1918–2002)
Pier Giacomo Castiglioni (1913–1968)

Stereo Cabinet, Model No. RR-126

Designed 1965–66

Chipboard, plastic laminate, Perspex, aluminum, woven fabric

Closed: 36½ x 24¼ x 14⅜ in. (92.5 x 61.5 x 36.5 cm)

Produced by Brionvega, S.p.A., Milan, Italy

K2013.169

Cooper Hewitt, Smithsonian Design Museum, promised gift of George R. Kravis II

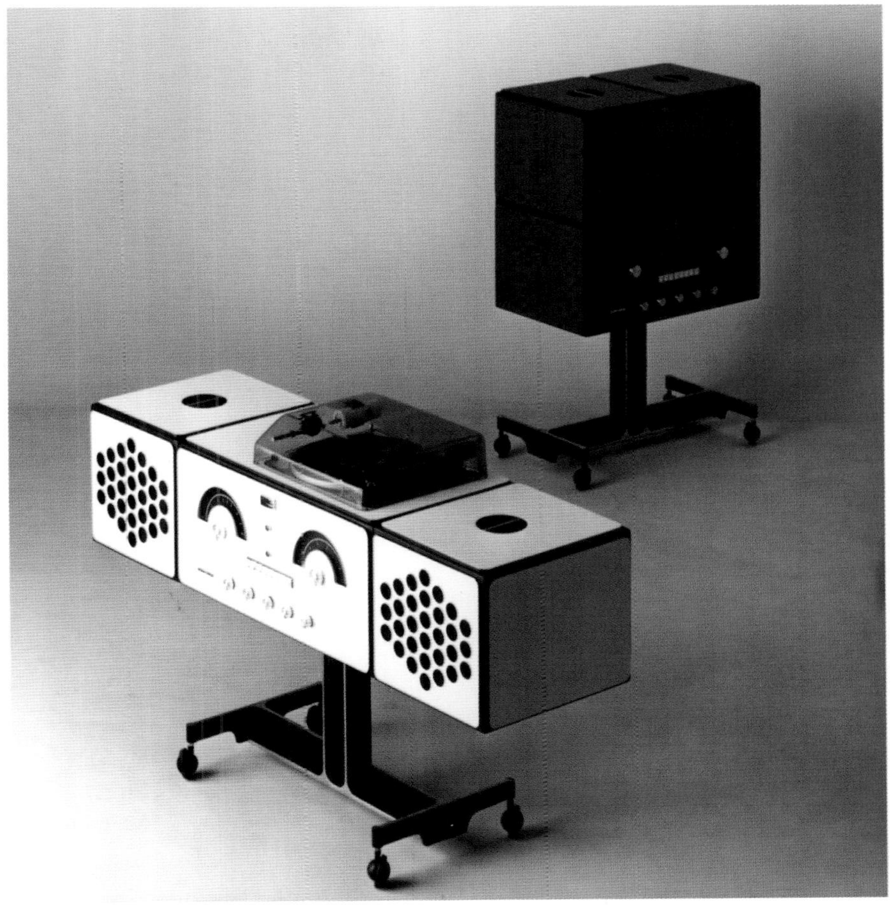

Promotional photo of RR-126 stereo cabinet, open (foreground) and closed.

BY THE 1960S, ITALIAN MANUFACTURERS SUCH AS BRIONVEGA BEGAN TO VIEW STREAMLINED forms as passé and instead embraced the idea that industrial design should be expressed in crisp geometric shapes. Brionvega increasingly turned to designers such as Marco Zanuso, Richard Sapper, and the Castiglioni brothers to create these new products. The 1960s also brought profound cultural and social changes that evolved into a new casual and more informal lifestyle. The Castiglioni's RR-126 stereo cabinet—a radio and phonograph—is a powerful exemplar of this change. First, it is not a static or fixed object; its sculpted aluminum pedestal base on casters recalls the furniture of Charles and Ray Eames. The form of the cabinet is also variable. The speakers can be adjusted into three different configurations: stacked on top of the controls to form a small cube; attached to the side for a dramatic cantilevered form; or used independently. The stereo unit was, moreover, made in multiple finishes, adding to the capricious quality of the design. The Castiglionis were noted for their attention to detail, as can be seen in the meticulous symmetrical layout of the knobs and controls, and for a playful, if not quixotic, aspect to their designs, which set a new standard in the field. –RCM

Sony Design Team

Sports FM Walkman Radio, Model No. SRF-5

Designed 1983

Plastic, metal

4 x 1½ x ½ in. (10 x 4 x 1.4 cm)

Produced by Sony Corporation,
Tokyo, Japan

TR2016.97

THE WALKMAN WAS EMBLEMATIC OF THE JAPANESE INNOVATIONS IN miniaturization and high-technology. When it was first released in the late 1970s, the Sony *Walkman*, a portable audio cassette/tape player, transformed people's listening habits by allowing them to carry whatever music they wanted, and to listen to it, through lightweight headphones anytime, anywhere. The owner could now effectively add their personal soundtrack to their surroundings wherever they were in the world, whether they were commuting to and from work, or simply going for a walk, bike riding, or running. This Sony SRF-5 *Sports FM Walkman* radio was designed for sports buffs (it was water resistant and came with a Velcro wristband) and was an ideal product for the health-crazed exercise aficionados of the early 1980s. Sony adapted its small invention over the years to play CDs, Mini-discs, MP3s, and finally, like the *iPod*, to stream music. –PM

Sony Design Team

TranSound Stereo Cassette-Corder, Model No. CFS-700

Designed c. 1981

Plastic, enameled metal

8⅝ x 21⅞ x 7 in. (21.9 x 55.6 x 17.6 cm)

Produced by Sony Corporation, Tokyo, Japan

K2008.869

THIS HIGH-PERFORMANCE PORTABLE BOOMBOX WAS CONTAINED IN A RUGGED, easy-to-carry box with a built-in stand and swiveling, detachable speakers. It also had the capability to connect with external speakers. When it debuted, the original *TranSound* was advertised as "the first Sony home stereo to leave home," for "a party, a picnic, your summer home . . . even to college."[15] The boombox was adopted by urban youth culture and advanced the rise of hip-hop music in America. –PM

left

Apple Design Team

First Generation *iPod* MP3 Player

Designed 2001

Plastic, aluminum, glass, stainless steel

4 x 2½ x ¾ in. (10.2 x 6.4 x 1.9 cm)

Produced by Apple Computer, Inc., Cupertino, California

K2013.167

right

Jonathan Ive (born 1967) and **Apple Design Team**

iPod Touch MP3 Player

Designed 2008

Polycarbonate, ABS plastic

4⅜ x 2½ x ½ in. (11.1 x 6.4 x 1.3 cm)

Produced by Apple, Inc., Cupertino, California

K2010.338

APPLE'S FIRST VERSION OF THE *IPOD*—WHICH WEIGHED A LITTLE OVER SIX OUNCES AND could store a thousand songs on its five-gigabyte hard drive—was introduced in October 2001 with a $400 price tag. It was not the smallest MP3 player on the market, nor did it have the most space, but it far outstripped the competition in storage capacity for its size. The *iPod* fulfilled a need in the market for digital music players, which at the time were generally badly designed or functioned poorly. The design team, including Jonathan Ive and Apple's hardware engineers Jon Rubinstein, Tony Fadell, and Michael Dhuey, was inspired by classic designs such as Dieter Rams's 1958 Braun T3 transistor radio, and Bang & Olufsen's BeoCom 6000 phone. It was visually striking; the shining white and steel gadget with a never-before-seen circular click wheel set on the face also introduced Apple's signature white-wired ear buds. In 2007, the *iPod Touch* was first introduced. It had a full touchscreen and looked much like an *iPhone*. Using Wi-Fi, it was possible to download audio files directly from Apple's iTunes store rather than connecting to a computer. The 8-GB *Touch* seen here is from the second release year. –PM

opposite Advertising poster of *iPod*, designed by Erik Miller and Chiat/Day team, 2007.

iPod

Dieter Rams (born 1932)

Tischsuper Radio, Type RT 20

Designed 1961

Beech veneer, powder-coated steel, plastic

10 x 19¾ x 7½ in. (25.4 x 50.2 x 19.1 cm)

Produced by Braun AG, Frankfurt, Germany

K2008.722

TRADITIONAL RADIOS MADE BEFORE WORLD WAR II WERE MEANT TO STAY IN ONE place in the home, like a piece of furniture to be shared by the family. This radio incorporates transistors, which allows users to take it anywhere with them, as well as making it more affordable. The aspect of portability made the radio popular with German Beatniks during the early 1960s. In the 1960s, some companies responded to youthful rebellion by making products with cheaper materials, such as plastic, that had a shorter lifespan. Braun designer Dieter Rams preferred to create objects that were long-lasting, well-designed, elegant, and useful. –PM

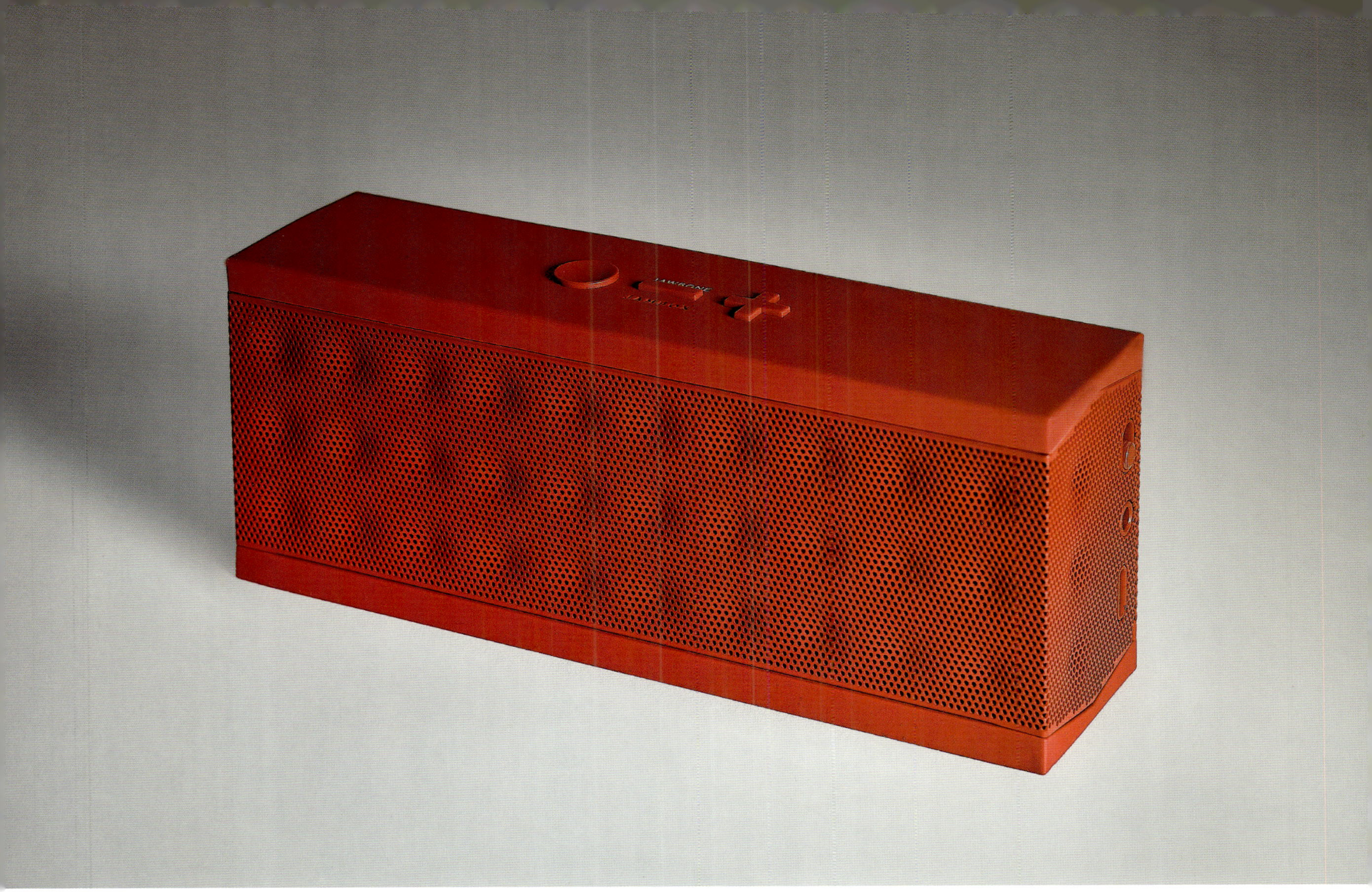

Yves Béhar (born 1967)
JAMBOX Wireless Speaker and Speakerphone
Designed 2010
Stainless steel, rubber
2¼ x 6 x 1⅝ in. (5.7 x 15.2 x 4.1 cm)
Produced by Jawbone, San Francisco, California
K2012.2

JAMBOX IS A TINY, PORTABLE, WIRELESS BLUETOOTH SPEAKER DESIGNED FOR THE consumer technology company Jawbone. It is half the size of a brick and allows the user to share music like an old boombox from the 1970s, but now in a more practical way. It is designed in different patterns—dots, diamonds, waves—that wrap around the speaker, both for shock resistance and for sound propagation. The top and bottom are two single pieces of molded rubber casing that render it shockproof and also dampen the vibration that comes from the mini-subwoofer and the two speakers. According to a company press release, "The two highly powerful acoustic drivers and ultra-small dome produce an audio spectrum almost as wide as the human ear itself. . . . *JAMBOX* is a sturdy, highly portable and versatile speaker system in a package that looks like minimalist art."[16] *JAMBOX* was an immediate standout for its high volume, broad sound range, and excellent sound quality in such a compact design. The speaker can be connected to other electronic devices, including mobile phones and iPads, and through the use of Bluetooth technology allows the user to seamlessly stream and share music, movies, and games, as well as make phone and conference calls. —PM

left Promotional photograph for the Jawbone *JAMBOX* speaker, c. 2010.

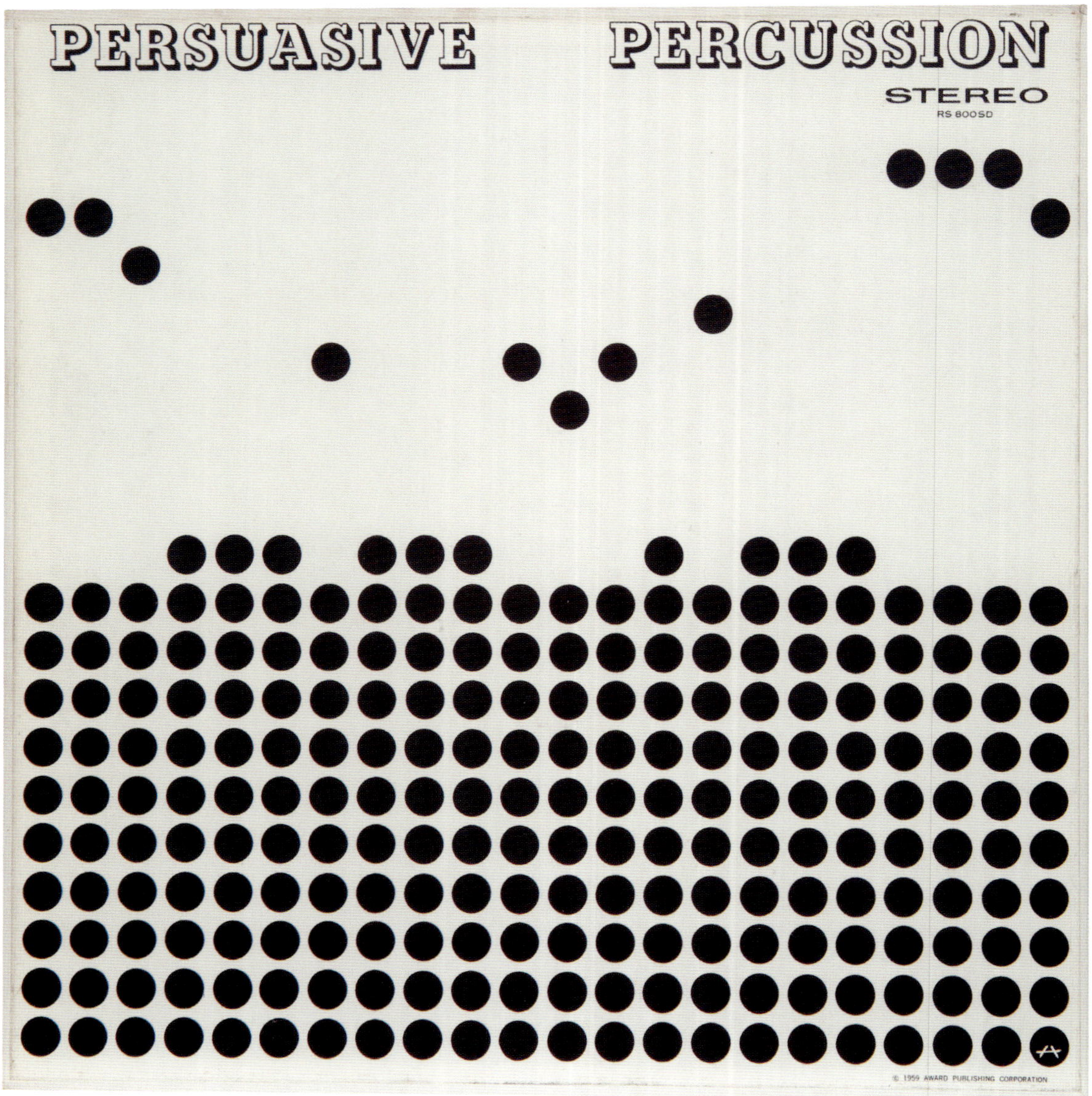

Josef Albers (1888–1976)
Persuasive Percussion Album Cover
Designed 1959
Offset lithograph on cardboard
12⅜ x 12⅜ in. (31.4 x 31.4 cm)
Produced by Command Records, New York, New York
MED.2016.1

THIS RECORD ALBUM SLEEVE WAS ONE OF A SET OF SEVEN ALBUM COVERS THAT pioneering modernist painter Josef Albers designed between 1959 and 1961 for Command Records. Albers was a student and then a teacher at the Bauhaus before immigrating to the United States to teach at Black Mountain College. Command Records founder Enoch Light commissioned him to create abstract designs that visually represented both the instruments and the qualities of sound in the modern music. This album showcases an innovative gatefold cover—first introduced by Light for Command Records—which doubled the surface area of the cover for liner notes and graphics. –PM

Alex Steinweiss (1917–2011)
Pal Joey Album Cover
Designed c. 1951
Offset lithograph on cardboard
12⅜ x 12⅜ in. (31.4 x 31.4 cm)
Produced by Columbia Masterworks, Columbia Records, Inc., New York, New York
MED.2016.8

ALEX STEINWEISS GRADUATED FROM PARSONS SCHOOL OF DESIGN IN 1934, AND worked for the Austrian poster designer Joseph Binder for three years. In 1939, at the age of 23, Steinweiss became the first art director of the recently formed Columbia Records, where he revolutionized the way in which records were packaged and marketed. According to Steinweiss, "The way records were sold was ridiculous. The covers were brown, tan, or green paper. They were not attractive, and lacked sales appeal."[17] Over his career, Steinweiss designed nearly nine hundred album covers as an artistic expression of the music they contained to appeal to consumers. His cover for *Pal Joey*—a 1940 musical that returned to Broadway in 1952—incorporates bold geometric text boxes and fonts that include his signature curly hand-drawn lettering, which is copyrighted as Steinweiss Scrawl. –PM

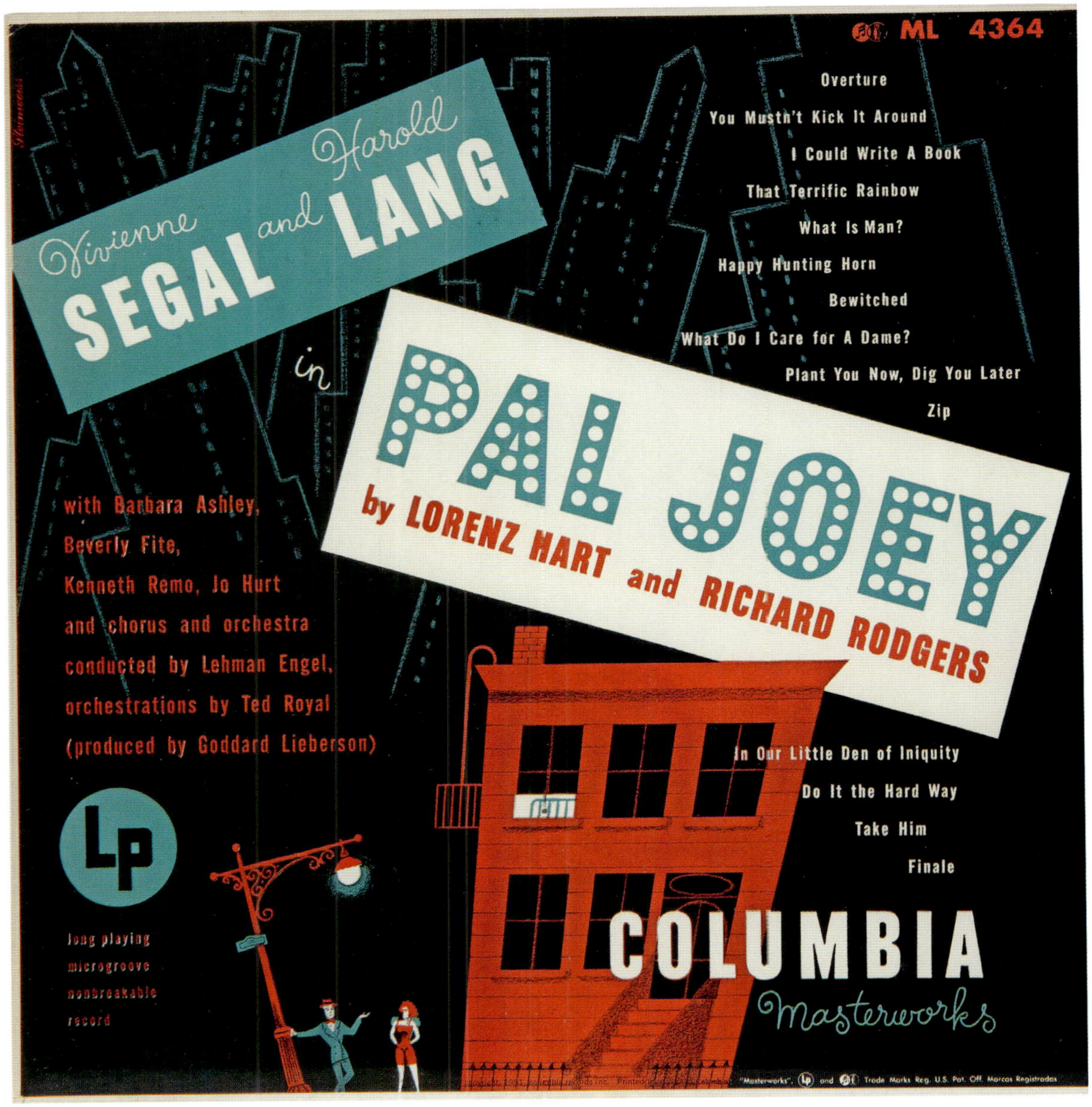

5
POLAROID
LAND CAMERA
B
MODEL 95

08
objects

VIEWING

FROM THE EARLY CAMERA OBSCURA TO OUR SMARTPHONES, WE obsessively look through lenses in order to make both still and moving visual records of our world. The cameras of all types that have made this possible have been overtly designed as functional objects of modern technology. In turn we have consumed these static and moving images with another set of artifacts—projectors—most of which have been designed to look like functional machines. Television sets on the other hand—among them Charbonneaux's elegantly tapering-legged television set from the 1950s—have inspired a more innovative approach and have become focal points in our homes.

left and rear

James Aitchison (1860–1911)
Thomas Bradley (dates unknown)

Two Sets of Collapsible Binoculars

Designed c. 1891 (utility patent 500,835)

Aluminum, glass

Open: 4½ x 4¼ x 2 in.
(11.3 x 11 x 5.2 cm) each

Produced by Aitchison & Co.,
London, England

K2010.186 and K2014.30

foreground

Designer Unknown

Opera Glasses

Designed c. 1950

Chromium-plated, enameled, and polished metal, glass

1⅝ x 3⅞ x 2 in. (4 x 9.8 x 4.9 cm)

Produced in Japan for Stellar Ware,
New York, New York

K2008.503

THE COLLAPSIBLE BINOCULARS ARE BASED ON THE SEVENTEENTH-CENTURY Galilean model, with a concave lens for the eyepiece and a convex objective lens at the opposite end. The Galilean construction is practical for use in inexpensive optics that need not offer strong magnification—for example, these Stellar Ware opera glasses with coated lens and focus adjustments. The Aitchison & Co. folding binoculars were designed by the London-based optician and inventor James Aitchison (with engineer Thomas Bradley), who took out thirteen patents between 1891 and 1906. His binoculars were first issued to British officers during the Boer War. They were advertised to the public as a "collapsible pocket binocular" that weighed about five ounces, and was described as "the lightest and most compact field glass in existence." —PM

MAKERS
AITCHISON'S
PATENT 1016
LONDON
ENGLAND

Walter Dorwin Teague (1883–1960)
Chester W. Crumrine (1889–1952)

Bantam Special Camera

Designed c. 1936 (D99,906)

Bakelite, enameled steel, stainless steel, glass

3⅛ x 5 x 1⅞ in. (7.9 x 12.7 x 4.8 cm)

Produced by Eastman Kodak Company, Rochester, New York

K2008.187

Advertisement for the Kodak *Bantam Special* camera, published in *Mechanix Illustrated*, July 1938.

THE INDUSTRIAL DESIGNER WALTER DORWIN TEAGUE, AN IMPORTANT PROPONENT OF streamlined design, was hired by Kodak to update its camera designs, including the popular *Brownie* and this 35-mm *Bantam Special* of 1936 (co-designed with Chester W. Crumrine). The smooth, shining housing of the compact *Bantam Special* gave the camera an aerodynamic appearance, so that even when the camera is stationary, the speed lines give the appearance that the camera is ready to move. The horizontal ribs were used not only for stylistic reasons, but also functioned to preserve the Bakelite housing, which was susceptible to cracking. Manufacturers like Kodak realized that hiring celebrity designers like Teague could help promote and sell their products. –PM

5
POLAROID
LAND CAMERA
MODEL 95

Edwin H. Land (1909–1991) (inventor)
Walter Dorwin Teague (1883–1960) (designer)
Harold R. Beal (dates unknown)
Robert H. Ensign (dates unknown)
Murray N. Fairbank (dates unknown)

Land Camera, Model No. 95

Designed c. 1948 (D152,229)
(utility patent 2,543,180)

Metal, plastic, glass, leather

5¼ x 9½ x 2½ in. (13.3 x 24.1 x 6.4 cm)

Produced by Polaroid Corporation,
Cambridge, Massachusetts

K2008.779

EDWIN H. LAND INTRODUCED HIS INSTANT CAMERA AT AN OPTICAL SOCIETY OF AMERICA meeting on February 21, 1947. At the presentation, he unveiled a one-step, one-minute process that both took a photograph and fully developed it, which was inspired by his young daughter's eagerness to see a photograph the moment that it had been taken. The following year, the first Polaroid instant camera, the four-pound Model 95 Land camera, and its film, Type 40 Land film, were offered for sale at a price of $89.75. The first run of fifty-seven cameras went on sale at Jordan Marsh department store in downtown Boston in time for the 1948 Christmas holiday shopping season. All fifty-seven cameras and film sold out on the first day, blindsiding Polaroid's marketers. A decade and a half later, Polaroid estimated that half of American households owned one of its products. By the time he retired from Polaroid in 1982, Land held 533 U.S. patents for his inventions.[1] –PM

Walter Dorwin Teague (1883–1960)

Gift Camera with original box and packaging, Model No. 1A

Designed 1930

Enameled and chromium-plated metal, leather, glass

Gift box: 2⅝ x 9¼ x 4¾ in. (6.7 x 23.5 x 12.1 cm)

Produced by Eastman Kodak Company, Rochester, New York

K2008.884

WALTER DORWIN TEAGUE, A PIONEER OF EARLY TWENTIETH-CENTURY INDUSTRIAL DESIGN, designed everything from cameras and radios to glassware for Steuben and a desk lamp for Polaroid. In 1928, he was hired by Eastman Kodak to create a new aesthetic for the company's cameras with an eye toward appealing to female customers. One tactic was to create a leather carrying case for a folding camera that was already in production in 1927, so that it could be carried like a purse. Kodak used words like "dainty" and "feminine" in its advertising. For this gift camera, Teague started with the existing *Pocket Kodak Junior* form and altered the exterior with brown leather and a geometric art deco pattern of enameled-metal inlay in brown and red on the front and the lens surround. The camera, with its matching gift box and packaging, was designed as a limited edition of ten thousand for the 1930 holiday shopping season, and sold for $15. Teague designed several of Kodak's well-known cameras of the era, including the *Bantam Special* and the *Baby Brownie*, examples of which are also in the Kravis Collection. –PM

BY EASTMAN KODAK CO
ROCHESTER N.Y.

Naoki Sakai (born 1947)
O-Product 35mm Camera
Designed 1988
Aluminum, plastic, glass, metal
Camera: 3¼ x 4¼ x 2 in. (8.3 x 10.8 x 5.1 cm)
Flash: 3½ x 1¾ x 2 in. (8.9 x 4.5 x 5.1 cm)
Produced by Olympus, Tokyo, Japan
K2008.242.1

THE OLYMPUS O-PRODUCT CAMERA—LAUNCHED TO MARK THE COMPANY'S seventieth anniversary—is a 35mm point-and-shoot compact camera with a detachable strobe unit that introduced a hard aluminum casing on personal cameras. Designed by Naoki Sakai and his studio, Water Design, the camera was an example of a style Sakai called "future retro"—more commonly called retro-futurism—which combines innovation with familiar forms from earlier eras. The unique, bold geometric form of circles and a square is reminiscent of designs from the 1930s and 1940s, particularly in the large, round external flash unit. Moving away from the standard-issue black plastic camera body, the *O-Product* camera helped popularize the brushed aluminum look in the late 1980s and 1990s. Issued as a limited edition of twenty thousand (ten thousand units were available in Japan, and ten thousand were sold overseas), the miniature import was a compact size and the detachable flash made it easily portable. The camera's mechanism was modeled after Olympus's point-and-shoot *Infinity Jr.* camera, with an autofocus 33 mm f/3.5 lens. —PM

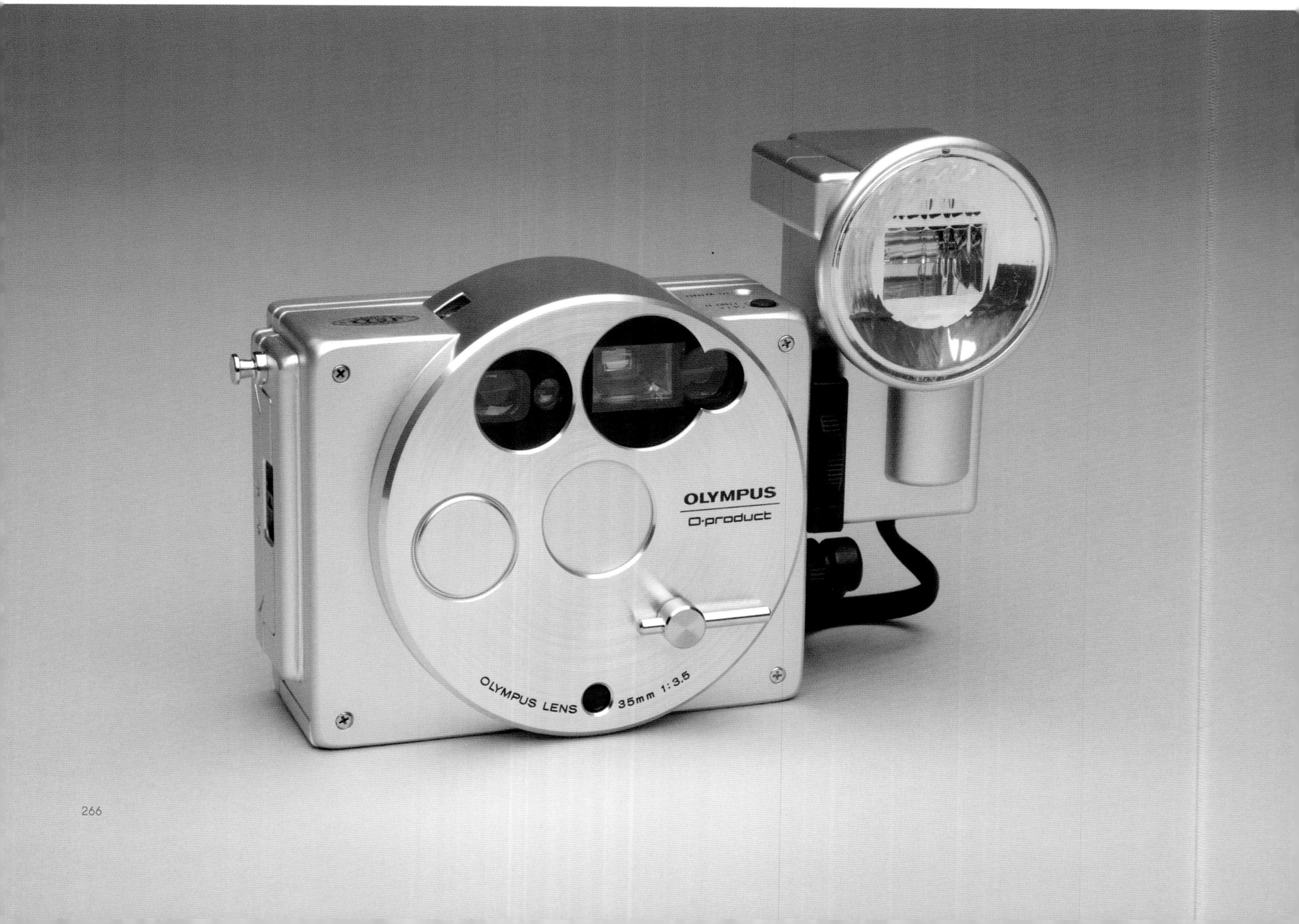

Dieter Rams (born 1932)
Slide Projector, Type D 10
Designed 1962
Painted metal, plastic, glass
6 x 6½ x 9 in. (15.2 x 16.5 x 22.9 cm)
Produced by Braun AG, Frankfurt, Germany
K2010.86

DIETER RAMS VALUED USABILITY FAR ABOVE CLEVERNESS, AND HIS goal for all products was straightforwardness and ease of use. Toward this end, he designed simplified and well-organized controls, sometimes with color coding. "I never trusted instruction manuals," he said. "We all know that most people don't read them."[2] The minimalist casing of this slide projector, Type D 10, reflects Rams's ten principles for good design. –PM

William B. Gruber (1903–1965)

View-Master Junior Projector

Designed 1949

Bakelite, metal, plastic

5 x 6 x 7½ in. (12.7 x 15.2 x 19.1 cm)

Produced by Sawyer's Inc.,
Portland, Oregon

K2008.489

WILLIAM B. GRUBER, AN ORGAN MAKER AND PHOTOGRAPHER TRAINED IN OPTICAL engineering, invented the *View-Master*, which made use of the idea to employ two cameras to simultaneously photograph an object from slightly different angles, then use a special viewer to optically overlap the two photos, causing the two to appear as a single three-dimensional image.

While Gruber was touring Oregon Caves National Park, a chance meeting with Harold Graves, the president of Sawyer's Inc., a Portland-based postcard manufacturing company that specialized in souvenir picture postcards, led to a partnership that put Gruber's invention into production. It would ultimately become *View-Masters* and reels, the thick cardboard piece set with miniature photo transparencies in tiny corresponding windows, that was inserted into the *View-Master* for viewing. During World War II, the US military purchased 100,000 viewers and 600,000 reels as part of its personnel training.

In December 1949, Sawyer's came out with a $9.95 *View-Master Junior* projector that used seven-scene picture reels and projected color prints up to 11 x 12 inches from a six-foot distance. At the time it was launched, more than three hundred *View-Master* picture reels were available. –PM

Designer Unknown

TV-22 Television

Designed 1950

Bakelite, glass

15 x 15½ x 16¼ in.
(38.1 x 39.4 x 41.3 cm)

Produced by Bush Radio,
London, England

K2010.106

TELEVISION FOUND A GROWING AUDIENCE IN ENGLAND AFTER THE CORONATION of Queen Elizabeth was televised in 1953. Bakelite, used extensively in radio cabinets, was also used in the new Bush Radio television sets, which were much easier to produce in plastic than in wood. Bakelite was perfect for the rounded corners and speed-line grooving seen at the base of this set, evidence of the influence of the American streamlined style in postwar British design. This 1950 *TV-22* model followed the *TV-12* set of 1949, which could receive only a single television channel, BBC, the only broadcaster at the time. The two models looked very similar, but this later model increased the accessibility as rival stations were established.[3] *TV-22*'s original design and advanced technology were signature Bush features. —DAH

BRAUN

Herbert Hirche (1910–2002)

FS 4 Television with Stand

Designed 1958

Wood, plastic, glass, metal; enameled-steel stand

Television: 22¾ x 24⅜ x 17 in. (57.8 x 61.9 x 43.2 cm)

Produced by Braun, AG, Frankfurt, Germany

K2012.231

Promotional photograph of the Braun *FS 4* television shown in a Knoll International model room, c. 1958.

HERBERT HIRCHE STUDIED UNDER WASSILY KANDINSKY AND MIES VAN DER ROHE AT THE Bauhaus in Dessau and Berlin between 1930 and 1933, and worked in the office of Mies van der Rohe from 1934 to 1938. From 1958 to 1963, he worked for Braun, designing radios and television sets. He was a trained architect and designed a house for Braun founder Erwin Braun.[4] His austere design for this wood-encased television is similar to his Braun *HF 1* television of the same year, which was further simplified by using a single material for the case and by placing its controls in a hidden panel at the top.[5] The boxy shapes, subdued color schemes, and simple thin-legged stands of both sets were typical of Braun's modern aesthetic, derived from Bauhaus principles. The warm wood of the *FS 4* harmonizes with the gray and white that surrounds the picture tube and the neat row of controls across the bottom, creating a simple, minimal whole. –DAH

Philippe Charbonneaux (1917–1998)
Téléavia Television, Model No. P111
Designed 1958
Painted wood, enameled metal, Perspex, glass, metal
54¼ x 25¾ x 29½ in. (137.8 x 65.4 x 74.9 cm)
Produced by Téléavia, France
K2013.144

ONE OF THE MOST STRIKING MID-CENTURY television sets—the *Téléavia*—was created by French industrial designer Philippe Charbonneaux, who was known primarily for his automobile designs, including the 1962 Renault 8. The forceful presence of his television was in contrast to the austere televisions being produced by the German company Braun around the same period, as seen on page 270. The *Téléavia* P111 provided functional advantages by having casters on the feet for easy movability and an adjustable 54-cm panoramic screen. The screen could also be aimed at the viewer—tilted vertically or turned on its axis from side to side.[6] In contrast to the heavy rectangular wood cabinets that normally housed American televisions, the *Téléavia* rests on a control console section supported by tapering crossed legs. A portable television with the same model number was designed in 1966 by Roger Tallon. –DAH

Marco Zanuso (1916–2001)
Richard Sapper (1932–2015)
Black ST 201 Television
Designed 1969
Acrylic, metal, plastic
10 x 12¼ x 11⅝ in. (25.4 x 31.12 x 29.53 cm)
Produced by Brionvega, Milan, Italy
K2015.22

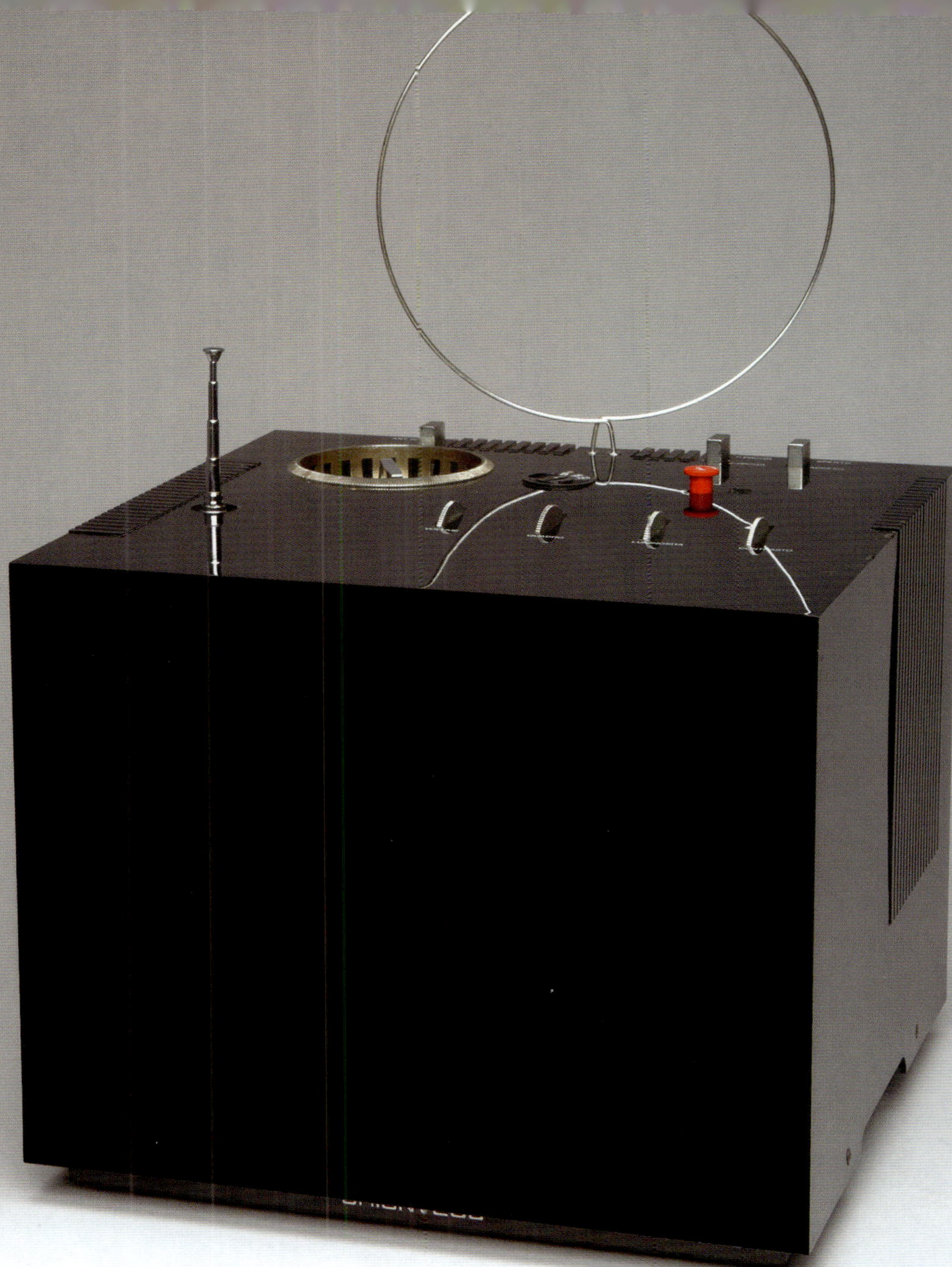

IN 1959, THE ARCHITECT AND DESIGNER MARCO ZANUSO JOINED FORCES WITH industrial designer Richard Sapper to create a formidable design partnership. According to MoMA, the products of that partnership embody "what Zanuso has called the 'metaphoric quality which objects have for those who use them.' It is the exploitation of this metaphoric quality that endows their products with a level of emotional and intellectual substance usually outside the realm of product design."[7]

The *Black ST 201* television was designed for Brionvega, an electronics company that during the 1960s hired designers such as Mario Bellini, as well as Achille and Pier Giacomo Castiglioni. To create this minimalist black cube of a television set, which embodies the very nature of its acrylic plastic material, the designers created a stripped-down version of the classic television set, with a simple cube and a casing that hides and protects the cathode tube and other mechanisms. Zanuso said of the television: "When switched off, the object is an abstract thing that holds you at a distance; when on, it vanishes, and is replaced only by an image."[8] –PM

Designer Unknown

Videosphere Television, Model No. 3240

Designed 1970

Plastic, metal, chromium-plated metal

14¼ x 9½ x 9½ in. (32 x 27 x 27 cm)

Produced by JVC, the Victor Company of Japan, Ltd., Yokohama, Japan

K2008.196

Cooper Hewitt, Smithsonian Design Museum, promised gift of George R. Kravis II

INTRODUCED IN 1970, THE *VIDEOSPHERE* WAS A POPULAR PORTABLE JVC TELEVISION THAT was shaped in the form of a space helmet, with a separate raised square platform. It came in a variety of colors including white, red, black, and orange (a gray model was launched in the 1980s). Some models were available with an alarm clock and radio. The retro-futuristic design was specifically targeted at a young audience that was fascinated by all things space: "Have a Ball! JVC Reshapes the World of Television with its 'mod-pod' Videosphere" was one of its advertising slogans. While its novel plastic casing gave it a space-age design, viewers would watch an almost flat TV screen through the plastic sphere, making it somewhat difficult to watch. However, the swivel mount allowed it to be rotated a full 360 degrees, and the viewer could watch it from any position. —PM

Yoshinobu Nameki (born 1955)
Shin Miyashita (dates unknown)
SONY KV-4000 "Essen Design" Television
Designed c. 1980 (utility patent 4,354,206)
Cast aluminum, glass, chromium-plated metal, plastic
Closed: 4⅞ x 4¾ x 14 in. (12.4 x 12.1 x 35.6 cm)
Produced by Sony Corporation, Tokyo, Japan
Cooper Hewitt, Smithsonian Design Museum,
gift of George R. Kravis II, 2016-5-24

INTRODUCED IN THE UNITED STATES IN APRIL 1980, THE *KV-4000* WAS SONY'S highest quality small-screen color TV using the high-resolution Trinitron color CRT, with a 3.7-inch screen. A wonder of miniaturization (at the time, it was the smallest portable color TV in the world), the compact design had such distinctive features as concealed knobs and a screen that was movable in height and angle. It was encased in platinum-tone brushed aluminum and had a beveled glass screen. A button on the side released the cabinet to tip it up for better viewing. The screen could also be tilted. Weighing under seven pounds, the TV took eight C batteries and could easily be carried. Its unusually high picture resolution and brightness made it a favorite of TV station engineers. –PM

Philmore F. Sperry (1890–1942)

Film Projector, Model No. P-90

Designed c. 1939 (D115,771)

Enameled steel, aluminum, glass, plastic

15¾ x 11 x 9¼ in. (40 x 28 x 23.5 cm)

Produced by Revere Camera Company, Chicago, Illinois

K2008.139

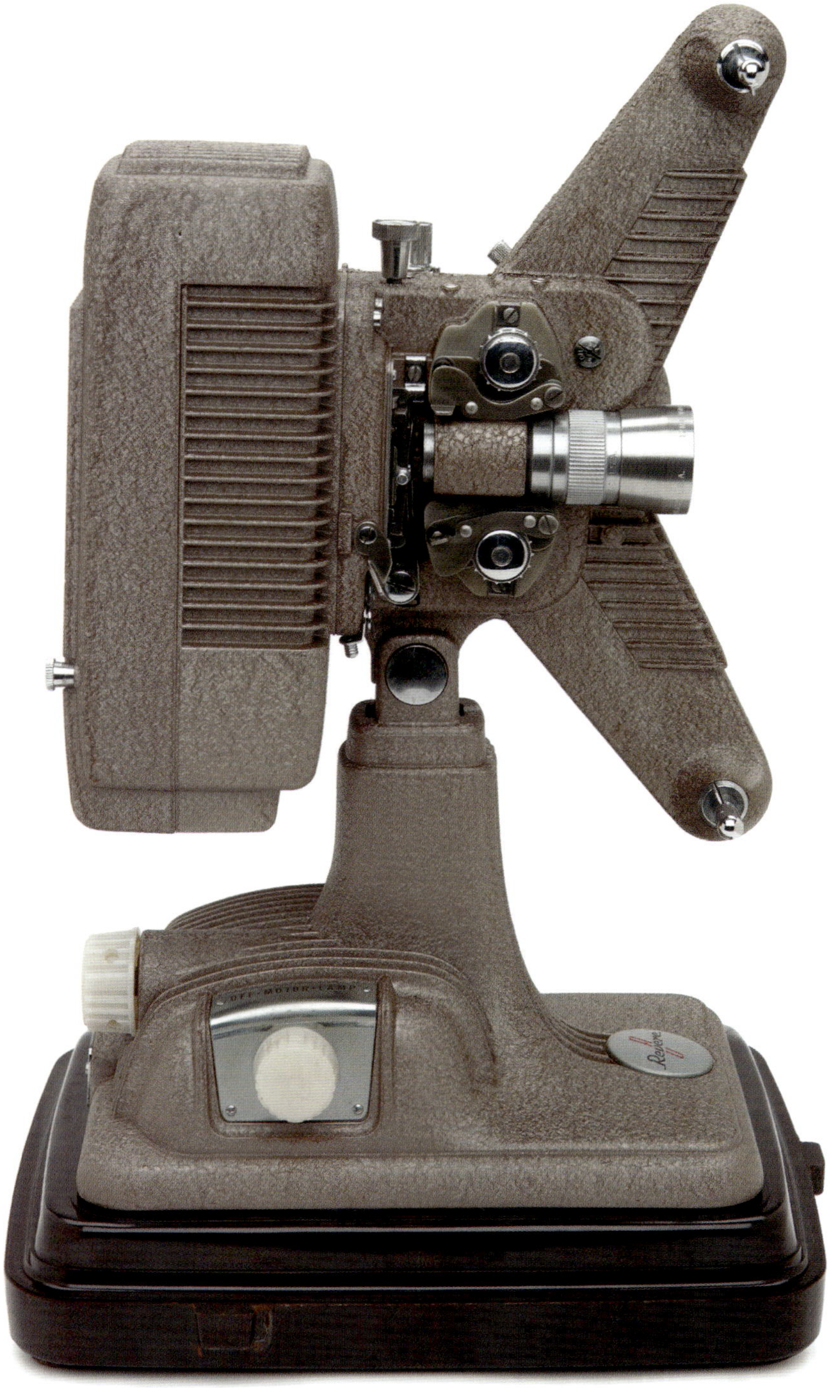

THE 16MM PROJECTOR IS STREAMLINED WITH speed lines that adorn all its parts, giving it the appearance of a space-age robot. It suggests the ever-increasing speeds of the twentieth century and the growing appetite for movies during the Great Depression. Transporting the heavy projector was possible with its handy plastic carrying case, which echoed the speed lines at its base. Film reels could be stored in a compartment in the base. The original operating manual promised a combination of "operating ease with 'theatre quality' projection to an unexcelled degree."[9] Revere Camera Company was founded in Chicago as an outgrowth of the Excel Auto Radiator Company and in 1939 began to produce movie cameras and projectors. –DAH

above *Owners Operating Manual* for Revere 16mm projector, published for Revere Camera Company, Chicago, 1948.

left

Designer Unknown

35mm Motion Picture Film Reel
Designed 1930–50
Aluminum
15 x 15 x 1⅞ in. (38.1 x 38.1 x 4.8 cm)
Produced by Goldberg Brothers, Denver, Colorado
K2010.199

right

Morris Goldberg (1903–1994)

Self-threading Motion Picture Film Reel
Designed c. 1968 (utility patent 3,450,366)
Aluminum
24½ x 24½ x 1⅞ in. (62.2 x 62.2 x 4.8 cm)
Produced by Goldberg Brothers, Denver, Colorado
K2011.89

FILM REELS ARE AS OLD AS THE MOVIE INDUSTRY. THESE EXAMPLES ARE VARIATIONS OF a standard design, one of several popular designs for reels in the twentieth century. Four holes in the center accommodate sprockets to mount it on a projector. Goldberg Brothers, a hardware business established in Denver in 1897, began to make film reels in 1913 to supply the silent film industry. The firm also obtained a license to manufacture film reels under the Eastman Kodak brand, which led to them becoming a leading source for the product. Still in business today, Goldberg has transitioned to the digital age with new products for film-related hardware for the motion picture industry as well as returning to its original business in hardware.[10] These sleek film reels recall the golden age of movies with their appeal as machine art. Indeed, such pieces would have fit well in MoMA's *Machine Art* exhibition of 1934, in the company of other machine parts such as ball bearings, propellers, and springs, which were admired for their abstract beauty. –DAH

H16
REFLEX
Bolex
paillard

Designer Unknown

H16 Reflex 16mm Movie Camera

Designed c. 1956

Enameled and chromium-plated metal, faux leather, metal, glass, plastic

8½ x 9⅝ x 5½ in. (21.6 x 24.5 x 14 cm)

Produced by Bolex Paillard, Sainte-Croix, Switzerland

K2012.41

THE USER MANUAL FOR BOLEX PAILLARD'S *H16 REFLEX* MOVIE CAMERA gives detailed instructions for the use of the camera and its numerous accessories, and also praises its many positive attributes: "Beauty of line—technical perfection. . . . The designers of the Bolex *H16 Reflex* left nothing to chance. Every part of its intricate mechanism has been studied with the most meticulous care, only selected, high-class raw materials are used throughout, and all components are painstakingly inspected at each stage of assembly."[11] Technically brilliant, through Swiss ingenuity and inventiveness, the Bolex *H16* is also aesthetically masterful: the three lenses of varying sizes are arranged like a still life. –DAH

opposite Cover of user manual for the *H16 Reflex* 16mm movie camera and accessories, published for Bolex Paillard, c. 1956.

A SHORT HISTORY OF PRODUCT DESIGN EDUCATION

Catherine McDermott

A DEFINING MOMENT IN PRODUCT DESIGN EDUCATION WAS THE 1851 GREAT EXHIBITION, which took place in London's Hyde Park, opposite to where the Royal College of Art now stands. There, six million visitors experienced what was at that time a modern global world, marveling at objects such as Colt's repeating gun in the American section and the exceptional quality of Shanghai silk from China.

By 1851, manufacturing had developed at an unrivaled pace, and the new generation of entrepreneurs began to question how designers might be educated for industry. The long-held tradition of the European guilds, in which apprentices trained alongside craftsmen, was being abandoned in favor of an education system that separated design and making. The debates surrounding this monumental evolution, which revolved around the needs of industry, society, individual creativity, and art school independence, continue to this day.

The first challenge for design education in Victorian Britain was one of scale. In Glasgow, the demand for places to study design was so high that the Glasgow School of Art offered classes in shifts—early morning, day, and evening. Another solution was a centralized curriculum that was rolled out to manufacturing cities not only in the United Kingdom, but also to industrial centers throughout the British Empire. The work of designers at the Sir J. J. School of Art, established in 1857 in Bombay, and the National Art School in Sydney, Australia, was shipped to South Kensington in Great Britain to be graded and returned months later. Also widespread was the influence of British design school architecture; institutions like Pratt Institute in New York and Bezalel Academy of Arts and Design in Jerusalem resembled their counterparts in London.

The key to South Kensington's design curriculum was eye-to-hand skills, practiced through technical drawing based on nature and historical examples, in large studio groups. These skills were not expected to produce copies, but to act as a starting point for creativity and, indeed, connectivity to the modern world. Only a few designers, such as Dr. Christopher Dresser, who studied and taught design at South Kensington, brought this disciplined method to life. British design education also pioneered an international market for books on teaching design. In the collection of Beijing's Central Academy of Fine Art's

FIG 1
above left View from the southwest of the Bauhaus building, Dessau, Germany, c. 1926.

FIG 2
above right View of the metal workshop at Bauhaus showing Marianne Brandt (left) and Hin Bredendieck working at the drawing tables, Dessau, Germany, c. 1927.

there is a slim 1906 British design manual, translated into Mandarin and complete with illustrations. Such early textbooks were the start of the global academic discipline of design practice research, which is now established in universities across the world.

South Kensington also shaped design education by using museum collections as a teaching tool. By 1900, the European cities of Vienna, St. Petersburg, and Stockholm had built their own versions of Victoria and Albert's museum and design school model. The value of museum partnerships for design education carries on in the twenty-first century; Mario Botta's museum for Tsinghua University in Beijing opened in 2016 and the Bauhaus Collection for Hangzhou's Academy of Art both continue this tradition.

At the turn of the twentieth century, however, design education was ready for change. In 1900, when Archibald Knox was dismissed as design director at Kingston School of Art for refusing to adhere to the South Kensington curriculum, it was more a badge of honor than a disgrace. New thinking from other disciplines was starting to have an impact on design education, such as Friedrich Fröbel's work on the value of play in children's learning and Sigmund Freud's theories on the subconscious in the creative process.

In the twentieth century, those changes came to a head in a small German design institution founded by Walter Gropius: the Bauhaus. When the school moved to Dessau in 1926 (fig. 1), its buildings, education, and industry partnerships caused an immediate sensation and set an agenda for design education that still survives today. For the Bauhaus, design education was no longer only about mastering technical skills in the service of industry. Instead, the school pioneered ways to develop individual creativity, starting with the famous foundation year, which exposed students to specialist areas of practice and guided them through a curriculum offering broad cultural references.

Importantly, the Bauhaus also offered design education case studies on how to operate in the modern world, and industrial sponsorship was a case in point. Gropius's Dessau school was funded by the Junkers company, which at the time was manufacturing the most modern central heating systems in Europe. Gropius also explored marketing and communication strategies well ahead of their time, through publications, posters, and displays. Visitors to the Bauhaus's 1926 exhibition were offered a tram ride to the Torten Estate, with its Marcel

FIG 3
View of exhibition of work by students and lecturers at the Ulm School of Design, Ulm, Germany, 1958.

Breuer kitchens and Marianne Brandt lights, and the villas where the teachers, or masters, lived were open for tours to help people learn about the modern home.

Design education at the Bauhaus aspired to influence not only new ways of learning, but also new ways of living. The spaces were designed to be communal, flowing from the entrance to the well-known School Hall, which staged Oskar Schlemmer's theater and the famous parties devised by Gropius to promote another enduring Bauhaus principle: interdisciplinary practice (fig. 2). Gropius paid for a Berlin tutor to teach students modernist dance, the Charleston, and connected them to new German developments in science. The brief for the Metallisches Fest ball required students to dress as a "radioactive substance." The sexual, social, and creative freedoms encouraged at the Bauhaus would later reemerge in the 1960s, when art and design schools became sites for activism and social change.

When the Nazi party came to power in 1933 the Bauhaus was forced to disband, but it went on to influence design education in many different yet often interconnected ways. Gropius and Mies van der Rohe moved to the United States and taught at Harvard University and the Art Institute of Chicago, shaping design education not only in the US, but also through their students who went on to teach at leading design schools across Mexico, Brazil, and Chile. Creative education experiments flourished in the US, such as the Cranbrook Academy of Art, which reconnected industry and crafts through the work of one of its most renowned students, Charles Eames. In 1960, Ray and Charles Eames went on to develop India's flagship design school, the National Institute of Design in Ahmedabad.

In postwar Europe, one of the influential design schools came from a generation that had experienced the Nazi regime, the Ulm School of Design. The ambition of Max Bill, Ulm's director, was for design education to use and learn from disciplines such as semiotics and aesthetics (fig. 3). His vision influenced other educators like Bruce Archer at the Royal College of Art. Ulm's modernist legacy was kept alive by Dieter Rams and passed on to the best-known product designer of his generation, Jonathan Ive of Apple. In 2016, the technological success and confidence of the United States has ensured that over half the world's top twenty design schools are American.

European design schools continue to be ranked highly. The collapse of the Soviet Union reestablished national Russian and Eastern European centers for high-level design education. Scandinavia's commitment to sustainability continues in Helsinki's Aalto University, while the Politecnico di Milano and the Domus Academy remain flagships for Italian design expertise. The Academy of Fine Arts Vienna, Design Academy Eindhoven in the Netherlands, and the Royal College of Art in London are examples of small postgraduate design schools that lead international creative practice. Their leadership in critical design has moved education toward conceptual thinking, writing, and curatorial practice.

Demand for design education has also increased in Africa and Asia. The emerging Nigerian film and fashion sectors and the creative hub cities of Johannesburg and Cape Town in South Africa have seen the expansion of design education. Since 2000, India has opened eight new design schools in and around Mumbai and New Delhi. South Korea is now a global provider of design education, with Seoul's Hongik University one of many institutions leading the way.

Since the 1990s, however, design education in China has been at an entirely different level of growth. China is set to become the largest economy in the world by 2030 and, like Britain in the nineteenth century, it needs trained designers. The expansion of design education in China has focused on building the design capacity of high-ranking universities such as Tsinghua, Tongji, and Shanghai Jiao Tong, alongside smaller specialist art schools like Central Academy of Fine Arts in Beijing and China Academy of Art in Hangzhou. China has also massively invested in regional training schools offering technical skills-based design courses. Previously seen as a country that makes, as opposed to designs, products, China now understands that design education will be critical to its new role on the global stage.

In the twenty-first century, design education continues to grow. There is now widespread recognition of the benefit of interdisciplinary teams and a growing desire to integrate designers and design thinking into the very core of all types of companies and organizations. Governments and corporations alike are no longer seeing design and design education as merely nice additions to their agenda, but as necessary and vital tools to help build a robust creative economy that will protect and ensure the future of our planet.

09

objects

WORKING

MOST WORK TAKES PLACE IN THE PUBLIC SPHERE, WHETHER IN A factory, an office, a workshop, a studio, or an endless range of sites, including, increasingly, our homes. One kind of workplace features a variety of hand tools—hammers, saws, pliers, drills—which became powered with the arrival of electricity. Designers of these objects valued efficiency and effectiveness over elegance, although many find beauty in their utilitarian forms and materials. Artifacts destined for the office are quite different. In addition to the task-oriented furniture for clerks, managers, and administrators, a range of tools and machines, from the typewriter to the calculator to the computer, emerged that made office work more efficient. Heavy, desk-bound machines were gradually replaced by more compact ones that denoted fun rather than work—Ettore Sottsass's *Valentine* typewriter, for example—while, with the help of Jonathan Ive, Apple's computers fit easily in every environment, whether in the workplace, at home, or in a pocket.

left to right

Designer Unknown
Wrench
Designed 19th–20th century
Steel
6½ x 2⅜ x ½ in.
(16.4 x 6 x 1.3 cm)
Manufacturer Unknown
K2012.150

Designer Unknown
Claw Hammer
Designed 19th–20th century
Steel
8⅜ x 3 x ¾ in.
(21.3 x 7.6 x 1.9 cm)
Manufacturer Unknown
Produced in Germany
K2012.147

Designer Unknown
Tin Snips
Designed 19th–20th century
Steel
8⅞ x 2 x 1⅛ in.
(22.5 x 4.9 x 2.9 cm)
Produced by H. Weiss & Co., New York, New York, and J. Jones Co., Germany
K2012.152

John Anderson
(1861–1941)
Pipe-Expanding Pliers
Designed c. 1896 (utility patent 559,763)
Steel
11⅛ x 3½ x 2⅜ in.
(28.3 x 9.1 x 5.9 cm)
Manufacturer Unknown
K2012.148

Designer Unknown
Pipe Cutter
Designed 19th–20th century
Steel
8⅞ x 2⅝ x ⅝ in.
(22.5 x 6.7 x 1.8 cm)
Manufacturer Unknown
K2012.149

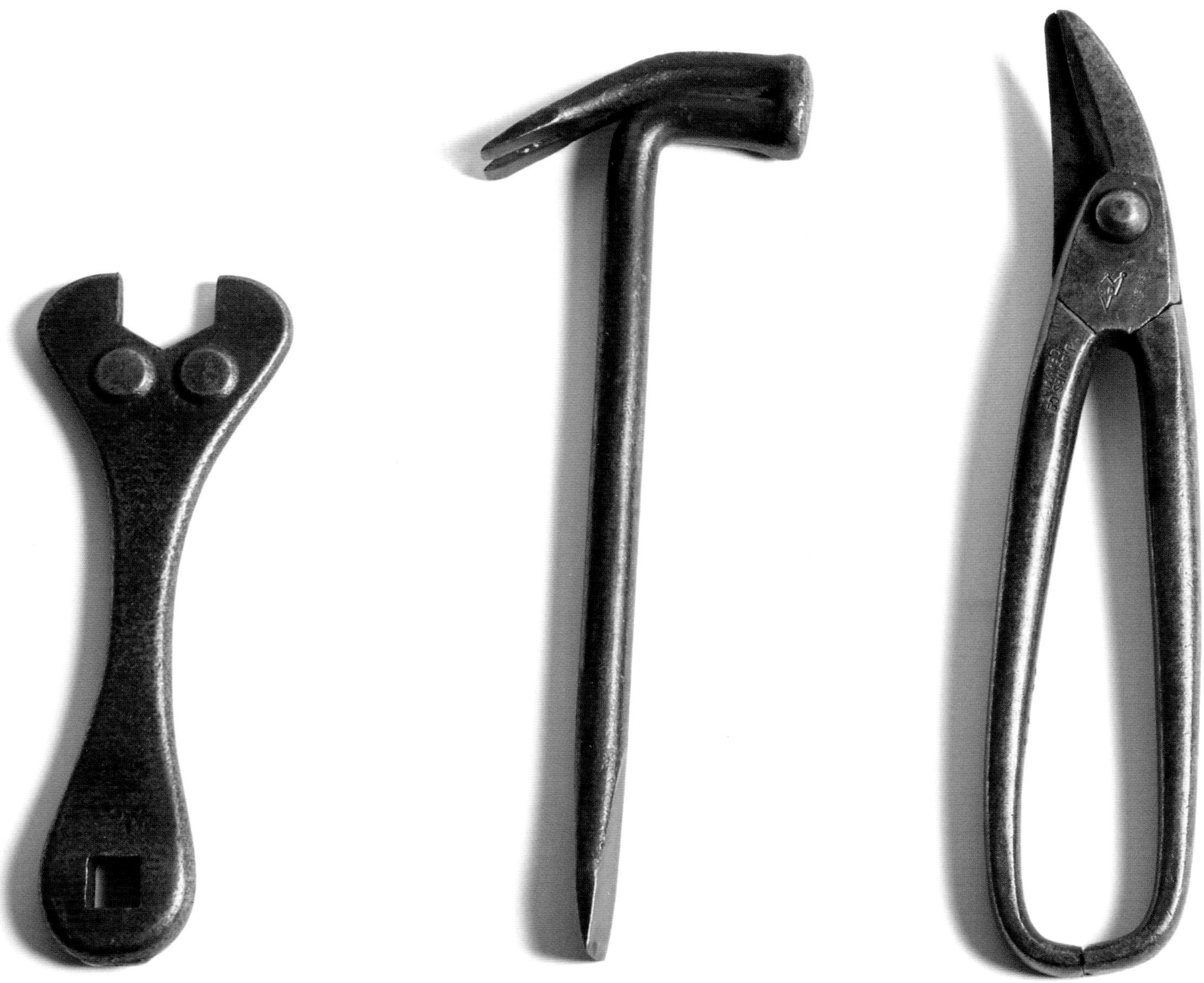

ONE'S TOOLS ARE PERSONAL. THEIR SHAPES ARE DETERMINED BY THEIR ASSIGNED TASKS and the user's needs. As an extension of the human hand, their job is to make certain tasks easier to perform. The pipe cutter, for example, allows a quick, smooth, controlled cut through thick metal tubing and is essential to plumbing work. The claw hammer's primary purpose is the insertion and removal of nails, but of course it can be applied to a variety of pounding or pulling jobs. This particular hammer also has a flattened end on the bottom of the handle, probably for other kinds of prying, scraping, or screwdriving activities. Tin snips, just as they appear, are to sheet metal what scissors are to paper. This wrench is a combined open-end and square-box-end wrench, though the exact intended purpose for the square punch out at the bottom is uncertain. At six and a half inches long, it is quite short for a wrench of that width by today's standards, and user strength would have to make up for its limited leverage. The pipe-expanding pliers could be inserted into the end of a pipe and, with a squeeze of the handles, the two sides of the split-cone top would be pushed apart, forcing the opening of the pipe wider. –PM

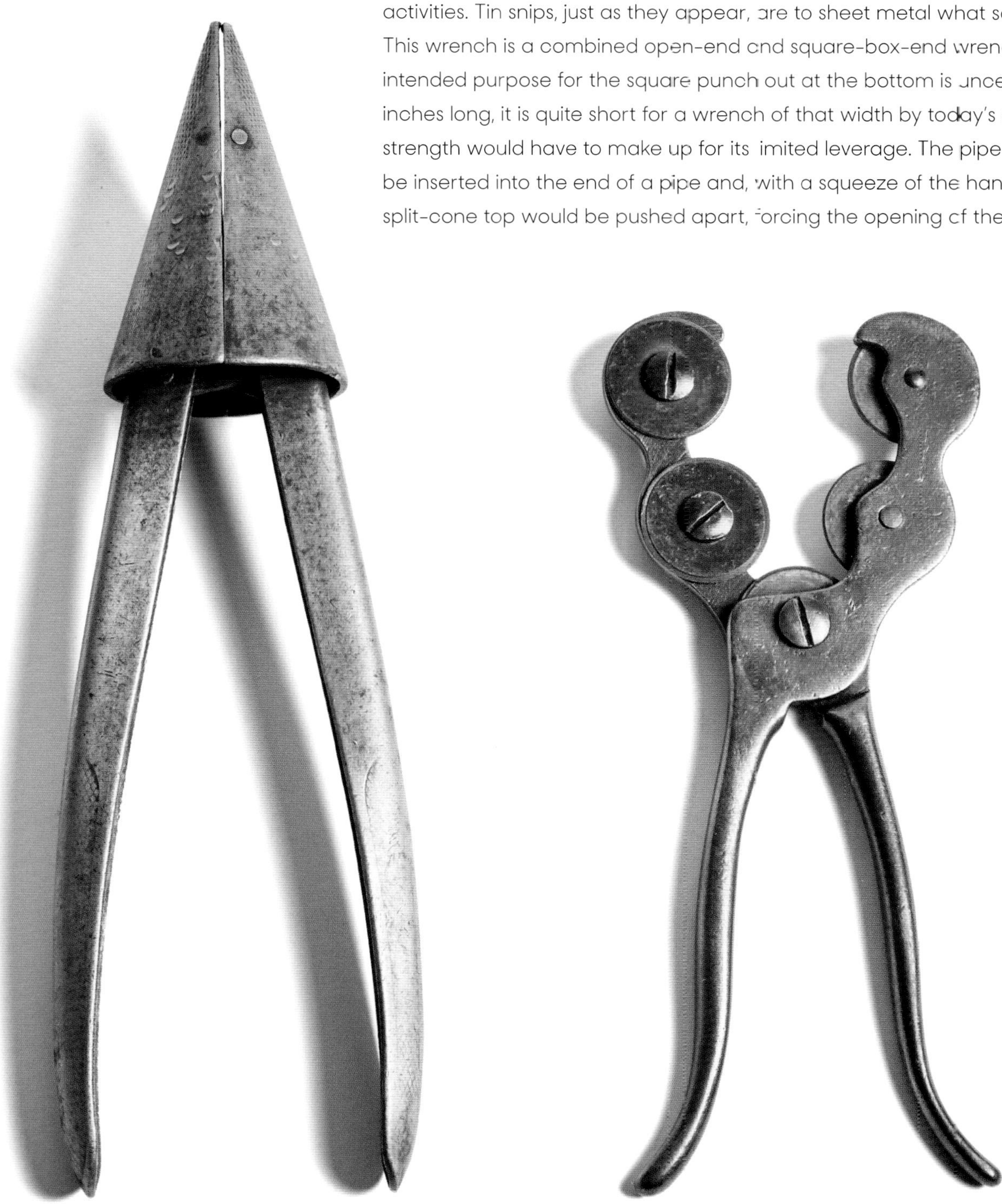

Designer Unknown

Elcona Sewing Machine
Designed c. 1950
Enameled metal, metal, plastic
9⅜ x 16¾ x 7⅛ in. (23.8 x 42.6 x 18.1 cm)
Produced by Zündapp-Werke GmbH, Nuremberg, Germany
K2010.300

FOUNDED BY FRITZ NEUMEYER IN 1917, ZÜNDAPP-WERKE WAS KNOWN FOR MAKING MOTORCYCLES and scooters. After World War II, Zündapp was no longer allowed to manufacture motorcycles, so the factory turned to producing sewing machines.[1] The electric sewing machine was commonly used in households through much of the twentieth century, replacing laborious hand sewing and mechanical pedal-powered machines. The basic form of the sewing machine evolved from awkward, angular designs in the nineteenth century that exposed the complex, and potentially dangerous, mechanical workings to smoother covered forms in the 1920s and 1930s. The green exterior of the *Elcona* reveals characteristics of streamlining, as seen in the curved shape, continuous sheathing, and use of speed lines on the top, base, and inner side of the bobbin compartment. Streamlining flourished in Europe after the war. –DAH

opposite Promotional photograph of the *Elcona* sewing machine, with storage case at the side, Germany, c. 1950.

Designer Unknown

Craftsman Hedge Trimmer,
Model No. 283,25652

Designed c. 1948

Steel, aluminum, brass, rubber

6⅝ x 11⅞ x 11⅝ in. (16.8 x 30.2 x 29.5 cm)

Produced for Sears, Roebuck,
and Company, Chicago, Illinois

K2010.59

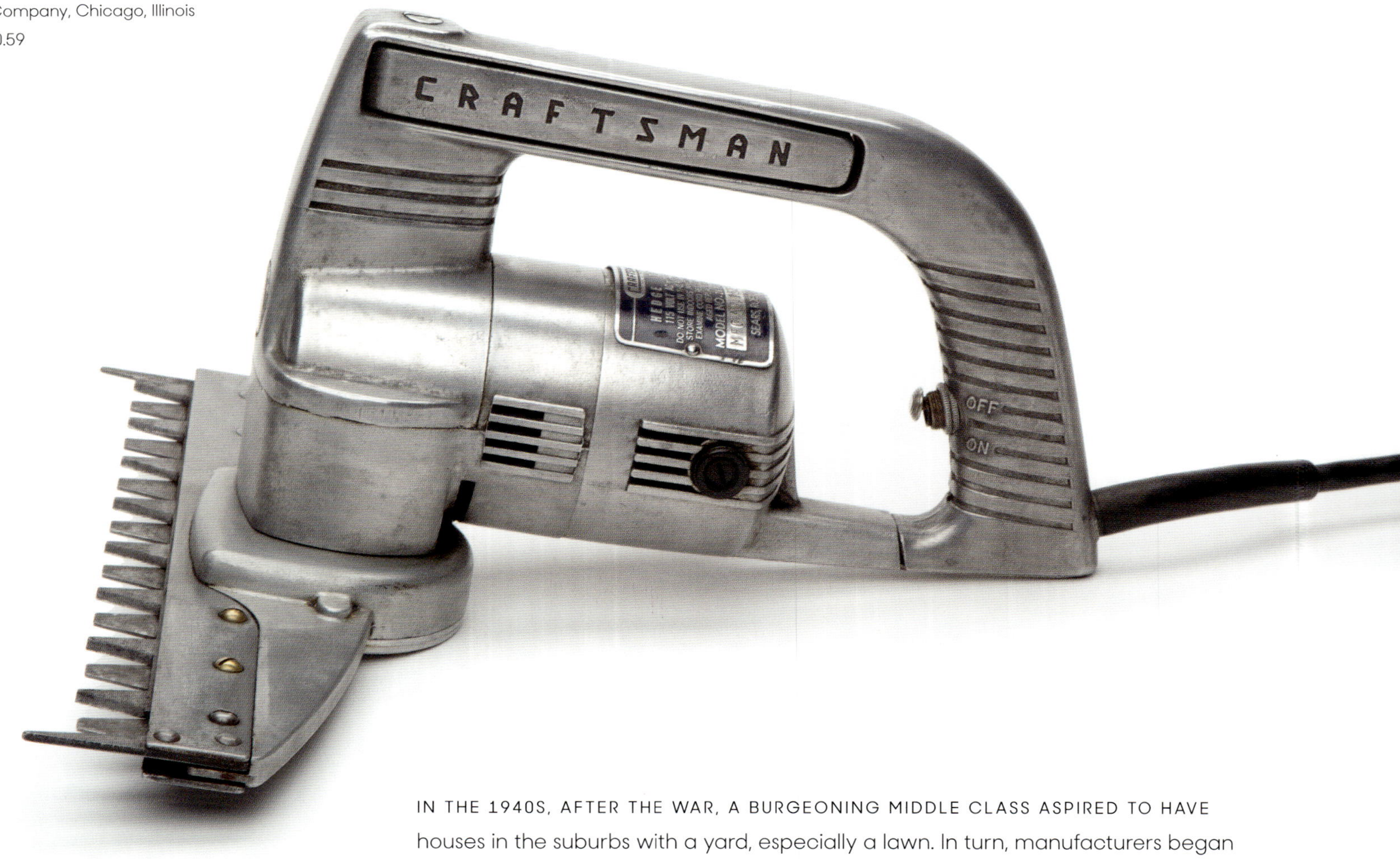

IN THE 1940S, AFTER THE WAR, A BURGEONING MIDDLE CLASS ASPIRED TO HAVE houses in the suburbs with a yard, especially a lawn. In turn, manufacturers began to develop tools to lighten yard work from novel gadgets to practical gear, like flower cutters, pruners, sprinklers, lawn mowers, and hedge trimmers. While indoor household appliances that lightened the drudgery of domestic chores were marketed to housewives, tools and other timesaving gadgets were promoted to the men of the family. Mail-order catalogue companies such as Sears, Roebuck, and Company made tools and equipment available to a mass market. This *Craftsman* electric hedge trimmer—with its small motor attached to the front of the handle that could be plugged into any socket and operated with an extension cord—was just one tool among many that Sears marketed to handymen. Electric trimmers made short work of a job that took tedious hours when done manually with hand shears, and saved the cost of hired labor. The blades are long and fast cutting. The Craftsman label, which appears three times on the trimmer, has a nostalgic name that harkens back to a simpler time. The trimmer originally sold for $29.50, making it affordable to almost anyone at the time. –PM

John R. Morgan (1903–1986)

Craftsman Portable Air Compressor

Designed 1935–40

Steel, aluminum

10 x 14 x 7⅝ inches (25.4 x 35.6 x 19.4 cm)

Produced for Sears, Roebuck, and Company, Chicago, Illinois

K2008.990

JOHN R. MORGAN LIVED IN VARIOUS LOCATIONS THROUGHOUT THE WESTERN HEMISPHERE before establishing a studio in Wheeling, Illinois, and exhibiting works in cities across the United States. After studying at the Detroit Technical Institute, he went on to work with C. G. Spring and Bumper Company and General Motors before becoming the chief product designer at Sears, Roebuck in Chicago in 1934. The wraparound horizontal grooving for the vents of this air compressor, characteristic of streamlining, is echoed in the handle above. Craftsman also sold matching paint sprayer guns to be used with its compressors. Morgan went on to design other Sears products such as vacuum cleaners, kitchen cabinet systems, and a number of appliances. He left the company in 1943, and the following year established his own office in Chicago under the moniker Jack Morgan Associates and continued to design industrial objects. –PM

De Witt Clausen (1897–1953)

Floor Polisher, Model No. P-12

Designed c. 1946 (D147,890)

Aluminum, chromium-plated steel, wood, straw bristles, plastic

51 x 11½ x 12⅜ in. (129.5 x 29.2 x 31.4 cm)

Produced by Clarke Sanding Machine Co., Muskegon, Michigan

K2012.43

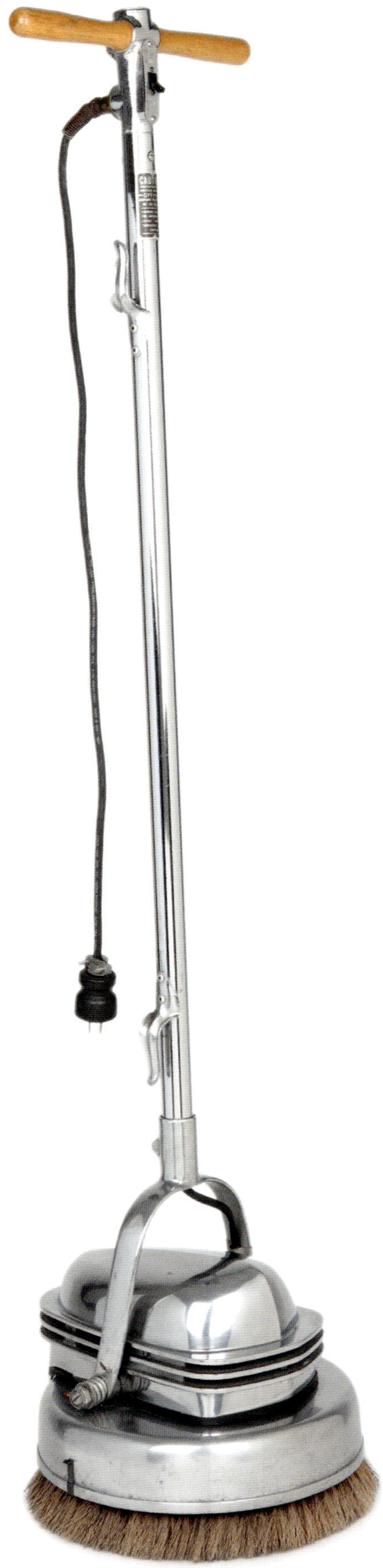

THE P-12 FLOOR POLISHER HAS SIGNATURE ELEMENTS OF STREAMLINING: the motor is sheathed in a tear-drop casing, with louvres serving as speed lines. The motor rests, like a space craft, on a round aluminum covered base above the bristles. The long vertical neck is attached to the body with a *U*-shaped prong that straddles the motor and is surmounted by a short horizontal wooden handle. The Clarke Sanding Machine Company was started by Alex A. Clarke in Dallas, Texas. The initial use for Clarke sanders was to refinish bar tops, which then led to refinishing school desks manufactured in Chicago. In 1922, Clarke moved his company to Chicago, producing table-top sanders as well as floor versions, with the addition of the long handle. In 1929, Clarke moved the company to Muskegon, Michigan, where it remained until 1985.[2] –DAH

Designer Unknown

Case for *Nuggets* Wrench Set, No. 49-K
Designed 1930–45
Enameled metal, metal
8½ x 22⅞ x 7⅛ in. (21.8 x 58.1 x 18.1 cm)
Produced by Blackhawk Manufacturing Co., Milwaukee, Wisconsin
K2008.680

FIRST MARKETED TO PROFESSIONAL MECHANICS, THIS BULLET-SHAPED CASE WAS designed to carry seven wrench handles, four attachments, and thirty-eight sockets for general work. It had a "scoop bottom" for easy withdrawal of the tools and a hinged lid to serve as a work tray for nuts, bolts, and parts. The portable chest had wheels on one end for easy transport. It advertised that customers paid for fewer wrenches, thus fewer to carry: the 49-K series of *Nuggets* wrenches weighed in at only 62% of the wrenches it claimed to replace. –PM

Miguel de la Torre (dates unknown)

Vibro-Tool Engraver

Designed c. 1945 (utility patent 2,480,451)

Plastic, metal

3 x 7⅜ x 1½ in. (7.6 x 18.7 x 3.8 cm)

Produced by Burgess Battery Company, Chicago, Illinois

K2010.56

IN 1917, CHARLES FREDERICK BURGESS OPENED THE BURGESS BATTERY COMPANY, ONE OF the early battery producers that later became part of Duracell. During the 1930s and 1940s, small hand tools like this *Vibro-Tool* engraver appealed to the middle-class "do-it-yourselfer" who would use it for countless home projects. An ad in the October 1945 issue of *Popular Mechanics* demonstrated that the *Vibro-Tool* would "add power to your craft skill," allowing the user to write on steel; engrave on glass or plastics; slice into materials such as wood, linoleum, cardboard, and cloth; and tool leather.[3] Two models were available: one with twenty ordinary engraving needles and another with an extra set of attachments. The vertical black stripes on the circular aluminum label background are typical of art deco motifs of the 1920s and 1930s. The spherical design at the end of the unit provided a comfortable ergonomic grip. –PM

Designer Unknown

Radiator Cleaning Tool

Designed 1930–50

Steel, brass

5½ x 11 x 3½ in. (14 x 27.9 x 8.9 cm)

Produced by Miller Tool and Manufacturing Company, Detroit, Michigan

K2011.142

THE APPEAL OF THIS TOOL IS THE RINGED TAPERING NOZZLE, WHICH GIVES IT a futuristic laser-gun-like appearance. A water or compressed-air hose can be attached to the threaded socket at the back of the gun, which would be used to flush out a radiator. The lever on the side controls a valve for adjusting or closing off pressure. Although the designer is not known, the manufacturer's name is clearly stated in block letters within an impressed lozenge on the flared handle. Born in Ontario in 1884, expert toolmaker A. L. Miller immigrated to the United States in 1909 and, with his brother R. H. Miller, organized Miller Tool and Manufacturing Company in Detroit in 1914.[4] –DAH

left

Glenn Calvin Wilhide (1896–1954)

Black & Decker, Junior, ¼" Electric Drill

Designed c. 1940 (D129,046)

Aluminum, steel, plastic

5⅛ x 11⅜ x 3¼ in. (13 x 28.9 x 8.3 cm)

Produced by Black & Decker Manufacturing Co., Towson, Maryland

K2013.183

center

Frithiof Petrus Forss (1882–1950)

Thor Electric Drill

Designed c. 1936 (utility patent 2,072,551)

Steel

5⅝ x 8⅛ x 2½ in. (14.3 x 20.6 x 6.4 cm)

Produced by Independent Pneumatic Tool Co., Aurora, Illinois

K2011.50

right

Designer Unknown

Electric Drill, Model No. ¼"

Designed 1940–50

Aluminum, steel

4⅜ x 11⅛ x 3⅛ in. (11.1 x 28.3 x 7.9 cm)

Produced by Cummins Portable Tools, Division of C. B. M. Corp., Chicago, Illinois

K2013.181

THIS TYPE OF HANDHELD DRILL WAS INVENTED AND PATENTED BY SAMUEL BLACK AND Alonzo Decker. According to the 1914 patent application, the configuration of handle and controls eliminated confusion about how to operate it: "There is nothing more to do than merely pull a trigger after the fashion of firing a revolver."[5] The Black & Decker drill seen here has a large "easy-grip" handle, a sliding thumb switch, a "husky" aluminum casing, *Compo* oil-less bearings, and a powerful universal motor. A 1935 ad in *Popular Mechanics* called it "the kind of fast-working tool that every practical man wants handy—ready to speed through a thousand and one different metal and wood drilling jobs around the shop, home, or garage."[6] The Cummins electric drill had a 2200-rpm motor and was frequently sold with attachments such as a polisher-sander or saw bits. The *Thor* electric drill was made by the Independent Pneumatic Tool Co., which originally produced a variety of products such as a pneumatic hammer and motorcycles. The company adopted the trade name in its new corporation name, Thor Power Tool Co., in 1953. A 1957 ad for the *Thor* speed drill stated that it was "the only drill ever engineered in advance to handle a complete selection of attachments." —PM

Designer Unknown

Welder's Mask

Designed 1925–30

Coated cardboard, glass, Bakelite, metal

10¼ x 8¾ x 7 in. (26 x 22.2 x 17.6 cm)

Produced by American Optical Corporation, Southbridge, Massachusetts

K2011.143

THIS MASK'S ODD CIRCULAR EYEHOLES AND PROMINENT NOSE EVOKE THE HUMAN-like features of an art deco android. Created by an unknown designer, its design predates the now-standard oval welder's mask with a single ocular pane. In addition to shielding the wearer from heat, sparks, and debris, this welder's mask blocks the blinding ultraviolet and infrared light emitted by a welder's electric arc. The modern welding helmet, still used today, was first introduced in 1937 by Wilson Products. Prior to that, it was more common for welders to use a combination of face masks and goggles. –PM

left

Joseph E. DuBois (dates unknown)

Allsafe Mono Goggle

Designed c. 1943 (D136,378–379)

Plastic, elastic

2⅞ x 5⅜ x 2½ in. (7.3 x 13.7 x 6.4 cm)

Produced by American Allsafe Company Inc., Buffalo, New York

K2014.39

center

Harvey S. Cover (1867–1961)

Nod and Shake Goggles

Designed c. 1922 (utility patent 1,433,676)

Rubber, glass, elastic

2½ x 7 x 1 in. (6.4 x 17.8 x 2.5 cm)

Produced by Cover, South Bend, Indiana

K2014.42

right

George R. Hoffmaster (1915–1997)

Flip-top Welding Goggles

Designed c. 1953 (D171,852)

Plastic, glass, rubber

3⅛ x 7 x 3¼ in. (7.9 x 17.8 x 8.3 cm)

Produced by Wilson Products, Inc., Reading, Pennsylvania

K2014.43

WORK GOGGLES APPEAL TO OUR MODERN SENSIBILITY AND, RATHER LIKE SUNGLASSES, CONVEY A SENSE of high fashion, whether in spite of or because of their association with gritty labor. Most are anonymous designs, but the *Nod and Shake* goggles have a patent that identifies the designer as Harvey S. Cover of South Bend, Indiana. According to the patent specifications, his eye goggles were to protect workers against the "dense fumes, gasses, poisonous dust, lime, spattering fluids, pathogenic bacteria, viruses and other deleterious substances that injuriously affect the eyes and call for complete protection and good vision."[7] Cover's inventions focused on designing and producing such work goggles, and his legal effort to protect his designs from copies led to a Supreme Court decision in his favor. In 1907, he patented eye guards for early automobiles without windshield protection. He adapted his goggle designs to the particular function required.[8] Less is known about the other two goggles. However, the wraparound *Allsafe Mono* Goggle, an improvement on Cover's design, is still in production today. The patent numbers listed in the accompanying instructions reveal that these goggles were designed by Joseph E. DuBois of Detroit in 1943.[9] –DAH

Clifford H. Flanigan (1905–1996)

Lawn-Do-All Mower

Designed c. 1948 (D158,847)

Aluminum, steel, chromium-plated steel, plastic

32 x 19½ x 48 in. (81.3 x 49.5 x 121.9 cm)

Produced by Universal Auto Parts,
New Albany, Indiana

K2011.47

Detail of the U.S. design patent D158,847 for the *Lawn-Do-All* mower, showing its handle folded for storage, 1948.

CLIFFORD H. FLANIGAN'S STURDY ELECTRIC-POWERED MOWER IS A CLEVER DESIGN. It included an electric light under the top of the handle for lawn mowing at night, and its hinged handle folds over the machine for ease of storage, as seen in the patent drawing.[10] The control knob in the center of the motor housing adjusted the height of the cutting blade. The instructions printed on the lightbulb housing are marked copyright 1958, and explain how to avoid running over the power cord—an all-too-common mishap that would bring an abrupt end to the day's mowing. Universal Auto Parts was a small company, established in 1932 in New Albany, Indiana, in a former Universalist church building. An automobile equipment and supply store, the company probably made the lawn mower as a side line to its regular production. Following the war, aluminum was abundant, and it was used generously in the *Lawn-Do-All.* –DAH

left

Designer Unknown

Lawn Sprinkler

Designed 1935–45

Cast iron

3½ x 7 x 6¾ in. (9.1 x 17.6 x 17.1 cm)

Produced in the United States

K2010.36

center

Designer Unknown

Rain Dial Sprinkler, Model No. R

Designed 1963–70

Enameled metal, metal, plastic

6¼ x 8⅝ x 10⅞ in. (16 x 21.8 x 27.8 cm)

Produced by the Gates Rubber Co., Denver, Colorado

K2010.196

right

Designer Unknown

Pride-o-Lawn Sprinkler

Designed 1940–46

Enameled metal, metal

3½ x 5 x 7⅝ in. (8.9 x 12.5 x 19.2 cm)

Manufacturer Unknown

K2010.195

FREDERICK LAW OLMSTED IS OFTEN CREDITED WITH INSPIRING THE RAGE FOR GRASS lawns in the yards of homes across America. Olmsted not only designed public spaces such as New York's Central Park, but also created a new vernacular for the green spaces of private residences. His design for the Chicago suburb of Riverside, developed with Calvert Vaux in 1868–70, included a lawn for each home. In 1871, Joseph Lessler of Buffalo, New York, patented a lawn sprinkler that could "be made to slowly revolve . . . by means of a clock-work mechanism."[11] After World War II, the demand for affordable housing (with lawns) in America skyrocketed with the return of GIs from the war. Time-saving sprinklers proliferated on the market and came in many shapes, sizes, and prices. Some sprinklers could be attached to the end of a garden hose, and a spike at its base would be driven into the ground. Many sprinklers could be adjusted to produce a range of water streams, from a fine mist to a heavier droplet, and some included glass containers that allowed the user to add fertilizer, lawn weed killer, or insecticide. Other sprinklers were designed in novelty shapes, like a frog or an atomic-era sputnik. –PM

Designer Unknown

Hat Form

Designed 1935–1965

Aluminum

14¼ x 9¼ x 7⅞ in. (36 x 23.5 x 20 cm)

Manufacturer Unknown

K2012.213

THIS HAT FORM—OR HAT BLOCK—WAS USED BY a milliner to shape and stiffen hats. Hat blocks were traditionally carved from soft wood into the shape of a hat, taking into consideration the fashion and style of a hat design, how a hat is balanced, and how it should sit on the wearer's head. During the twentieth century, milliners began using hat forms made of other more durable and long-lasting materials such as aluminum, like the one seen here, which were used with a hydraulic press or a stamping machine. Metal forms, however, were primarily used for men's hats, because they were of shapes that remained standard over time, whereas wooden hat blocks were more practical for the quickly changing fashions of women's hats. —PM

Designers Unknown

Flashlights

Chromium-plated and enameled metal, plastic

Produced by Eveready Battery Company, Inc., St. Louis, Missouri

left to right

Flashlight

Designed 1950s

9⅞ x 3¾ x 3¾ in. (25.1 x 9.5 x 9.5 cm)

K2012.133

Eveready Masterlite Flashlight, Model No. 7231P

Designed 1954

6⅛ x 2 x 2 in. (15.6 x 5.1 x 5.1 cm)

K2012.137

Flashlight, Model No. 3351

Designed 1956

9⅞ x 2 x 2 in. (25.1 x 5.1 x 5.1 cm)

K2012.136

Eveready Masterlite Spot-Flood Flashlight, Model No. 7253

Designed 1953

7 x 2¼ x 2¼ in. (17.8 x 5.7 x 5.7 cm)

K2012.135

Eveready Masterlite Flashlight, Model No. 7351P

Designed 1954

9½ x 2¼ x 2¼ in. (24.1 x 5.7 x 5.7 cm)

K2012.134

THE NEED FOR A PORTABLE LIGHT SOURCE IS AS OLD AS THE CAVEMAN. TORCHES, CANDLES, oil, and kerosene lamps were all predecessors of the far less hazardous modern flashlight. The battery-powered flashlight (or torch, as it is known in England) was invented at the turn of the twentieth century, and the basic form—a long cylindrical body with the flared light source on one end, with a small bulb set within a reflective cone to brighten the light—has changed little over time. Other forms of flashlights, such as the lantern-shaped version, were alternatives to the classic cylinder seen here. According to their markings, all of these flashlights were produced by Eveready. Founded by Conrad Hubert, a Russian inventor who immigrated to the U.S. in 1890, the company was originally the American Electrical Novelty and Manufacturing Company, selling flashlights and batteries under the trademark Ever Ready. In 1905 Hubert changed the name to The American Ever Ready Company, later condensed to Eveready. –DAH

George Nelson & Associates

(active 1947–83)

Pronto Posters

Designed 1964

Screenprint on Masonite

23 x 17 in. (58.4 x 43.2 cm) each

Produced by Howard Miller Clock Company, Zeeland, Michigan

K2012.117.1–3

THESE POSTERS WERE ORIGINALLY DESIGNED FOR THE CHRYSLER PAVILION AT THE 1964–65 World's Fair in New York. The Chrysler Corporation commissioned the firm George Nelson & Co. to design its display, which would compete with General Motors' *Futurama* and Ford's pavilion, designed by Walt Disney. "All of the Chrysler attractions were designed for children, so I kept the safety posters simple and to the point by designing a graphic factory worker to demonstrate the consequences of factory accidents," says Lance Wyman, one of the designers who worked on the project at Nelson. "Safety hazards were familiar after working three summers in factories during my college years."[12] The architectural historian Vincent Scully praised Nelson's Chrysler exhibit: "It is pop art at its best, and presents Detroit with welcome wit and irony. There is a 'working' engine you can walk into and a demented mock-up of a car of the future you can examine from below—the bucket seats are buckets."[13]

The posters were eventually produced by Howard Miller and marketed around 1965 as safety signage for factories: "They communicate fast. And with style and wit."[14] Twenty-four posters addressed various workplace safety rules, telling workers to "Wear Your Helmet" or to "Think" (illustrated by a pantless man). "They add flair and humor to the surroundings, improve morale, reduce accidents, and aid efficiency," promised Howard Miller.[15] —FM

Studio Architetti BBPR (founded 1932)
Spazio Desk
Designed c. 1961
Enameled steel, vinyl, plastic
30¼ x 55⅛ x 28¾ in. (76.8 x 140 x 73 cm)
Produced by Olivetti, Ivrea, Italy
K2013.19

THE *SPAZIO* MODULAR DESK SYSTEM WAS DESIGNED FOR OLIVETTI IN 1961 BY the modernist Milan-based Studio BBPR, an architectural partnership founded in 1932 by Gianluigi Banfi, Lodovico Barbiano di Belgiojoso, Enrico Peressutti, and Ernesto Nathan Rogers. After the war, they went on to design projects such as the American Pavilion at the Milan Triennale in 1951 and the Torre Velasca in Milan in 1958. This flexible desk system is composed of multiple basic elements such as desk units and shelving, with multipurpose connectors and hardware for creating many possible configurations. Users could choose from four color options. The Olivetti *Spazio* line of office furniture was awarded the 1962 Compasso d'Oro. –PM

William Plumb (1932–2017)
for Eliot Noyes and Associates (active 1948–1977)
Dictating Machine Stand
Designed 1960
Walnut, chromium-plated steel, lacquered Masonite, rubber
24 x 17 x 17 in. (61 x 43.2 x 43.2 x cm)
Produced by International Business Machines Corporation (IBM), Armonk, New York
K2012.50.1

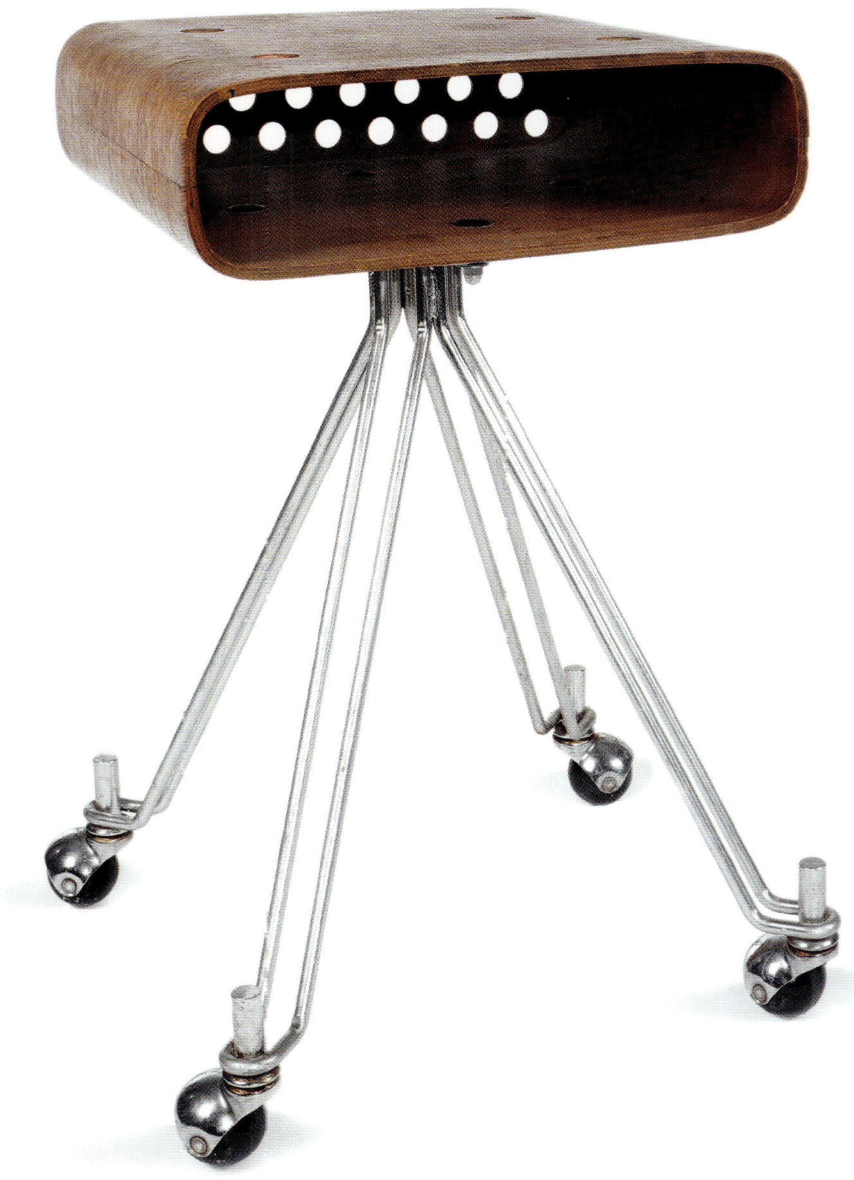

KNOWN AS THE FATHER OF CORPORATE DESIGN identity in America during the 1950s and 1960s, the architect and designer Eliot Noyes—first with IBM in 1956, then with Westinghouse, Mobil Oil, and Cummins Engines—revolutionized branding in modern business. At IBM, he brought in a team of consultants (including Paul Rand, George Nelson, and the Eameses) who together were responsible for all visual expressions of the brand—everything from visual identity to architecture, publications, advertising, and product design. "In a sense, a corporation should be like a good painting," Noyes said. "Everything visible should contribute to the correct total statement; nothing visible should detract."[16] Noyes brought on a team of talented in-house architects and designers, including William Plumb, who designed this dictation machine stand. Plumb was a Cornell-trained industrial designer who went on to pursue postgraduate studies at the Politecnico di Milano in Italy, finally working in the studio of Gio Ponti, the noted Milanese architect and designer. At Noyes's office, he worked on a wide variety of activities from brand identity to the design and development of a number of successful IBM products. –PM

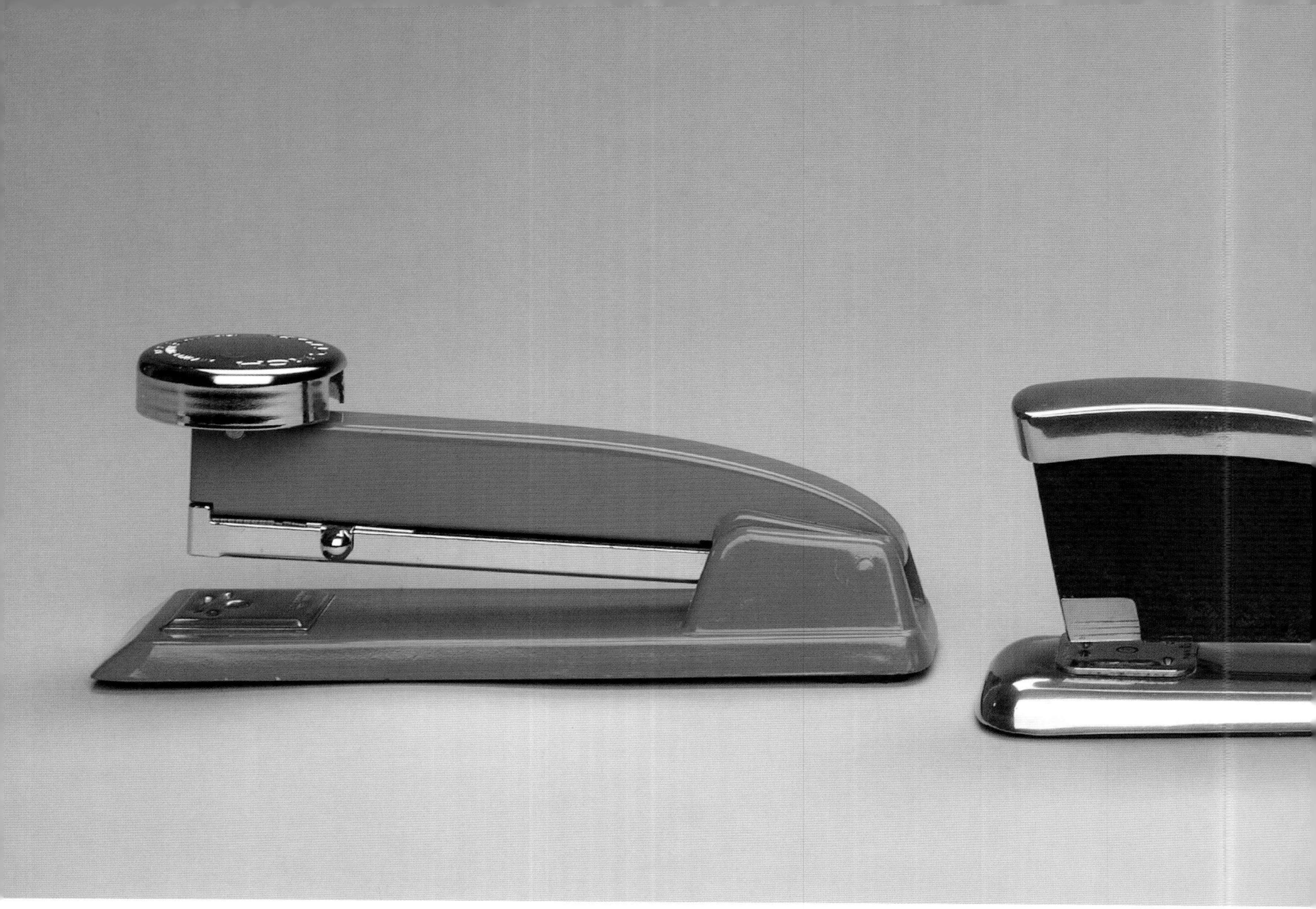

left

Curt E. Nyberg (1902–1970)

Monarch Stapler

Designed 1949 (D158,401)

Enameled and chromium-plated metal

2⅞ x 8¼ x 2¼ in. (7.3 x 21 x 5.7 cm)

Produced by Vail Manufacturing Co., Chicago, Illinois

K2011.185

center

Robert Heller (1899–1973)

Zephyr Stapler Model No. S101

Designed c. 1938 (D117,144)

Steel, enameled metal

3⅛ x 8⅛ x 2⅛ in. (7.9 x 20.5 x 5.4 cm)

Produced by Star Paper Fastener Company, Norwalk, Connecticut

K2012.29

right

Herbert W. Marano (1907–2001)

Tatum Aluminum Stapler

Designed c. 1949 (utility patent 2,680,245)

Aluminum, chromium-plated steel, plastic

4 x 7¼ x 1¾ in. (10.2 x 18.3 x 4.6 cm)

Produced by Wilson-Jones Company, Chicago, Illinois

K2012.36

THE FOUR-WAY PAPER STAPLER THAT IS MOST COMMONLY USED TODAY HAD BEEN DEVELOPED by the 1940s. The enameled and chromium-plated *Monarch* stapler produced by Vail Manufacturing Company is a typical early example. It was designed by Curt E. Nyberg, a designer who had at least seventeen other patents for everyday household items like flashlights, staplers, staple removers, stepladders, and vacuum cleaners. The design of the *Zephyr* stapler for the Star Paper Fastener Company was patented in 1939 by Robert Heller, a pioneer of American industrial design best known for his *Airflow* fan of 1937 (pages 108–109). The three utility patent numbers listed on this stapler are for its mechanism, designed for Hotchkiss by Fridolin Polzer. The *Zephyr*, a streamlined stapler—with its nickel-plated and black-enamel painted "wrinkle" finish, on a heavy steel body—loads from the front. The patent application for the *Tatum Aluminum* stapler—made of Alcoa aluminum—was filed in 1949 by Herbert W. Marano,[17] who held nearly two dozen patents for designs for the Wilson-Jones Company. –PM

Egmont Arens (1888–1966)
National Package Sealer, Model No. 208
Designed 1941 (D132,802)
Enameled and chromium-plated steel, plastic, enameled metal, rubber
6½ x 12⅝ x 4½ in. (16.5 x 32.1 x 11.4 cm)
Produced by Nashua Package Sealing Division of Nashua Corporation, Nashua, New Hampshire
K2010.46

EGMONT ARENS, A LEADING AMERICAN INDUSTRIAL DESIGNER, HELPED PROMOTE the concept of the "consumer engineer," using design innovations and consumer psychology to increase object appeal and sales. He founded his own industrial design firm in 1935. His products were often streamlined, as seen in this dispenser for gummed paper tape. It is sheathed in a sleek form that resembles a train engine and hides the functional mechanism that simultaneously moistens and dispenses a length of tape. According to a contemporary account, "Old style tape dispensers have usually been kept in the back room, but this new model, owing to its good style, is finding a place right on the counter of retail stores."[18] The gray enameled body contrasts with the bright chromium-plated lever at the side and the polished band on the curving front of the top, where the word "Prince" appears in red letters. A mark on the underside of the dispenser identifies it as a "National Package Sealer." On most of the *National* tape dispensers, the National logo appears on the top instead of "Prince," but the designation of these differing brand names on the same sweeping form is unclear. The design patent for this "Strip Server" was filed in January 1942,[19] and correspondence supports the claim that the design was completed in 1941.[20] Like many products of this time, however, production was delayed until after the war ended in 1945, when metals became available for civilian use. Despite the association of streamlining with planned obsolescence, Nashua Corporation has continued to produce an adaptation of the original design to the present. –DAH

Willis A. Kropp (1906–1999)

Adding Machine

Designed c. 1941 (D130,070)

Bakelite, plastic, metal

6⅞ x 9 x 12⅝ in. (17.5 x 22.9 x 32.2 cm)

Produced by Victor Adding Machine Co.,
Chicago, Illinios

K2010.307

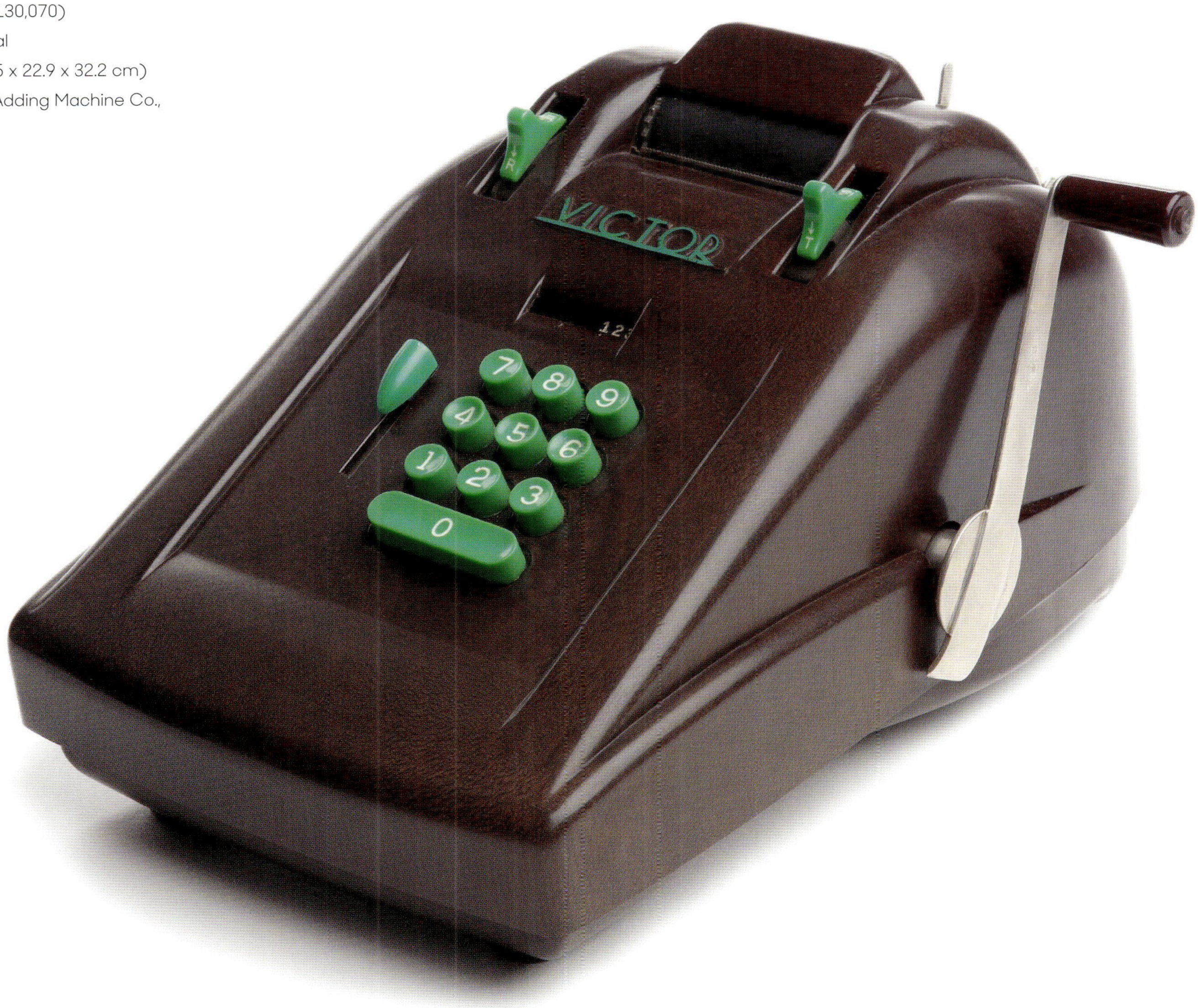

ADDING MACHINES WERE FIRST USED BY BANKS AND INSURANCE COMPANIES AND LATER, by World War II, they were ubiquitous in most medium-size firms. New uses for adding machines also emerged at this time, including calculation of payroll deductions, weekly labor and work summaries, managing petty cash expenditures, and tracking and calculating weekly travel expenses. Following the death of founder Oliver D. Johantgen in 1932, Victor Adding Machine Company hired the engineer Thomas O. Mehan, who designed the mechanism inside this lightweight, streamlined *Victor* machine,[21] which had only one-third the parts of the older models. The green number keys with white numerals are presented in three rows of three, followed by a fourth row for the wider zero key—an arrangement that was first introduced by Oscar J. Sundstrand in 1914 and became an industry standard for mechanical calculators. –PM

Ettore Sottsass (1917–2007)
Hans von Klier (1934–2000)

Praxis 48 Electric Typewriter

Designed 1964 (D205,971)

Enameled and chromium-plated metal, plastic, rubber

6¼ x 17½ x 13¼ in. (15.9 x 44.3 x 33.7 cm)

Produced by Olivetti, Ivrea, Italy

K2013.79

Advertisement for the Olivetti *Praxis 48* typewriter, 1965.

ENGINEER ADRIANO OLIVETTI LED HIS NAMESAKE COMPANY IN DEVELOPING A STRONG brand identity based on the idea of giving "happy products to society." The design of all aspects of the company's image, from graphics and logos to products and architecture, projected a strong cohesiveness and unity. It was also important that the function of design was controlled directly by the management and not, as was more common, by the marketing department. The *Praxis 48* electric typewriter has distinctive green-capped ivory keys and a wide, conforming spacebar set in a keyboard that projects straight forward from the square-cornered carriage housing and hangs suspended over the tabletop. —PM

olivetti
PRAXIS 48
TAB +
MIN
TAB −
MAX

Marianne Brandt (1893–1983)
Paper Tray
Designed c. 1931
Enameled steel, chromium-plated steel
1⅝ x 3⅞ x 5⅛ in. (4.1 x 9.8 x 13 cm)
Produced by Ruppel Mehrfach Geschützt, Gotha, Germany
K2010.76

MARIANNE BRANDT STUDIED WITH JOSEF ALBERS, LÁSZLÓ MOHOLY-NAGY, WASSILY Kandinsky, and Paul Klee at the Bauhaus. She worked in the metal workshop instead of the weaving workshop, where women had traditionally been expected to study. Despite her exceptional metalwork, she only achieved the status of "acting master." She wrote, "There was no place for a woman in a metal workshop, they felt. They admitted this to me later on and meanwhile expressed their displeasure by giving me all sorts of dull, dreary work. . . . Later things settled down, and we got along well together."[22] Today, the modern geometric designs she created between 1924 and 1933 are still admired. This paper tray includes her signature spherical forms in the supporting feet and the terminal for the arm, which secures a stack of note paper. True to Bauhaus functional minimalism, only two sides were provided to support the stack of note paper, along with a simple curved pen rest. –DAH

Edward C. Hoffmann (1879–1950)
Electro-Pointer XI Pencil Sharpener
Designed c. 1939 (D117,901)
Plaskon, plastic, rubber
5¾ x 8¼ x 3¾ in. (14.5 x 21 x 9.5 cm)
Produced by Swingline, Inc.,
Long Island City, New York
K2010.143.2

RESEMBLING AN AIRPLANE PROPELLER HOUSING, THIS AERODYNAMIC PENCIL SHARPENER was designed for the office of the future. Plaskon, a new plastic that replaced Bakelite, allowed for the continuous curves and sculptural form. Electric pencil sharpeners first became available in the late 1910s, but they did not come into common use until the 1940s. The more ubiquitous mechanical pencil sharpeners were made of iron or steel—heavy and sometimes awkward machines with a hand crank to turn internal blades for sharpening. For Edward C. Hoffmann's modern, labor-saving design of 1939, a manually adjustable ring allowed for a range of pencil sizes to be inserted, which activated the electrified sharpening device inside, and a drawer of clear plastic caught the shavings below. It was first manufactured by Triple "E" Products Company in St. Louis, but Swingline started to produce the *Electro-Pointer* after 1951, evidence of the ongoing popularity of the streamline style.[23] According to the patent drawing, the pencil sharpener was designed by Edward C. Hoffmann of University City, in suburban St. Louis.[24] –DAH

Mario Bellini (born 1935)
Divisumma 18 Electronic Printing Calculator
Designed 1972 (D233,350)
ABS plastic, synthetic rubber, melamine resin
4¾ x 12 x 2⅛ in. (12.1 x 30.6 x 5.4 cm)
Produced by Olivetti, Milan, Italy
K2009.8

DURING THE 1960S AND 1970S, MINIATURIZATION AND ELECTRONICS improvements meant that every object in daily use could be made smaller, lighter, and faster. In 1963, at the age of only twenty-eight, Mario Bellini was hired as Olivetti's chief industrial design consultant. His design for the portable *Divisumma 18* calculator was a modern update of the large, clunky machines of old. Bellini covered the key panel with a rubber skin, turning what were normally independent buttons into raised bumps beneath a sealed, protective surface. The underside of the device is molded in a way that recalls the organic shape of the human hand. With the exception of the division symbol and one of the equals signs in black (perhaps connected to the name *Divisumma*), the numbers, letters, and symbols on the keyboard are expressed as white inserts embedded in the yellow rubber. The sensual contact between fingers and the smooth rubber sheathing is one of the most appealing and successful features of this calculator. –PM

right Advertisement for the *Divisumma 18* electronic printing calculator, 1974.

Dieter Rams (born 1932)
Dietrich Lubs (born 1938)
Control Calculator, Model No. ET 22, Type 4955
Designed 1976
Plastic
5¾ x 3⅛ x ⅞ in. (14.6 x 7.9 x 2.2 cm)
Produced by Braun AG, Frankfurt, Germany
K2010.305

DIETER RAMS CREATED A HIGHLY RECOGNIZABLE AND ELEGANT aesthetic identity for Braun's products that is embodied in this calculator and continues to have a profound impact on today's industrial design. Early in his forty-year tenure at Braun, Rams established an in-house design department filled with talented designers and fostered fruitful collaborations with some of Europe's most recognized names in product design, including Otl Aicher, Hans Gugelot, Wilhelm Wagenfeld, Gerd Alfred Müller, Reinhold Weiss, Dietrich Lubs, Jürgen Greubel, and Robert Oberheim. This calculator is the second in a series that Rams and Lubs designed together. The small round buttons—looking like rows of M&M candies—are color-coded by category. All numerals are white-on-black, all but one of the functions are brown with white markings (including the smaller, oblong function keys and the switch above), and the equal sign is black on bold yellow. –PM

olivetti
valentine

Ettore Sottsass (1917–2007)
Valentine Typewriter
Designed 1969 (utility patent 3,711,175)
Plastic, enameled metal, metal
4⅛ x 12⅞ x 13¼ in. (10.5 x 32.7 x 33.7 cm)
Produced by Olivetti, Barcelona, Spain
K2011.133

Advertisement for the *Valentine* typewriter, showing case below, 1969.

THE ITALIAN ARCHITECT AND DESIGNER ETTORE SOTTSASS'S *VALENTINE* TYPEWRITER WAS part of Olivetti's corporate image-making program that was both visually and materially innovative. In 1956, Sottsass became a consultant to Olivetti, which was in the process of developing a computer. Following his work on the *Elea 9003* computer, he went on to design a series of typewriters for Olivetti into the 1960s, among them the *Tekne 3*, the *Praxis 48* (pages 312–313), the *Lettera de Luxe*, and the *Dora*. Although the *Valentine* is a radical design, it is a revamp of an earlier Olivetti typewriter—the *Lettera 32*. Its bright red color contrasts sharply with the black keys and carriage and with the white of the letters and numbers. The typewriter is also emblematic of 1960s pop art design, which connected to the youth culture and challenged the principles of modernism, aligning it instead with the values of instant gratification. –PM

Yves Béhar (born 1967) and
fuseproject
Nicholas Negroponte (born 1943)
One Laptop Per Child XO Laptop Computer
Designed c. 2007 (D609,703)
PC/ABS plastic, rubber
Closed: 1⅜ x 9 x 9½ in. (3.5 x 22.9 x 24.1 cm)
Produced by Quanta Computer Inc., Taiwan
Cooper Hewitt, Smithsonian Design Museum, gift of George R. Kravis II, 2015-5-8

THE *ONE LAPTOP PER CHILD* PROJECT IS AN EXAMPLE OF "SOCIAL DESIGN," WHERE DESIGN is used as a force for positive social and environmental change. Nicholas Negroponte, co-founder of MIT Media Lab, initiated the idea of putting laptop computers into the hands of impoverished children around the world at a cost of only $100 per laptop. He then urged the computer industry to figure out a way to make it possible, thereby helping to revolutionize education. Yves Béhar and his company fuseproject designed the laptop with Negroponte and others in consultation with technological and educational experts. It is a sturdy machine, designed to operate in a wide variety of environmental conditions, and it can be repaired easily if needed. The laptop uses flash memory instead of a hard drive to make it extremely energy efficient, which is an important factor in developing countries where electricity is often unreliable. —PM

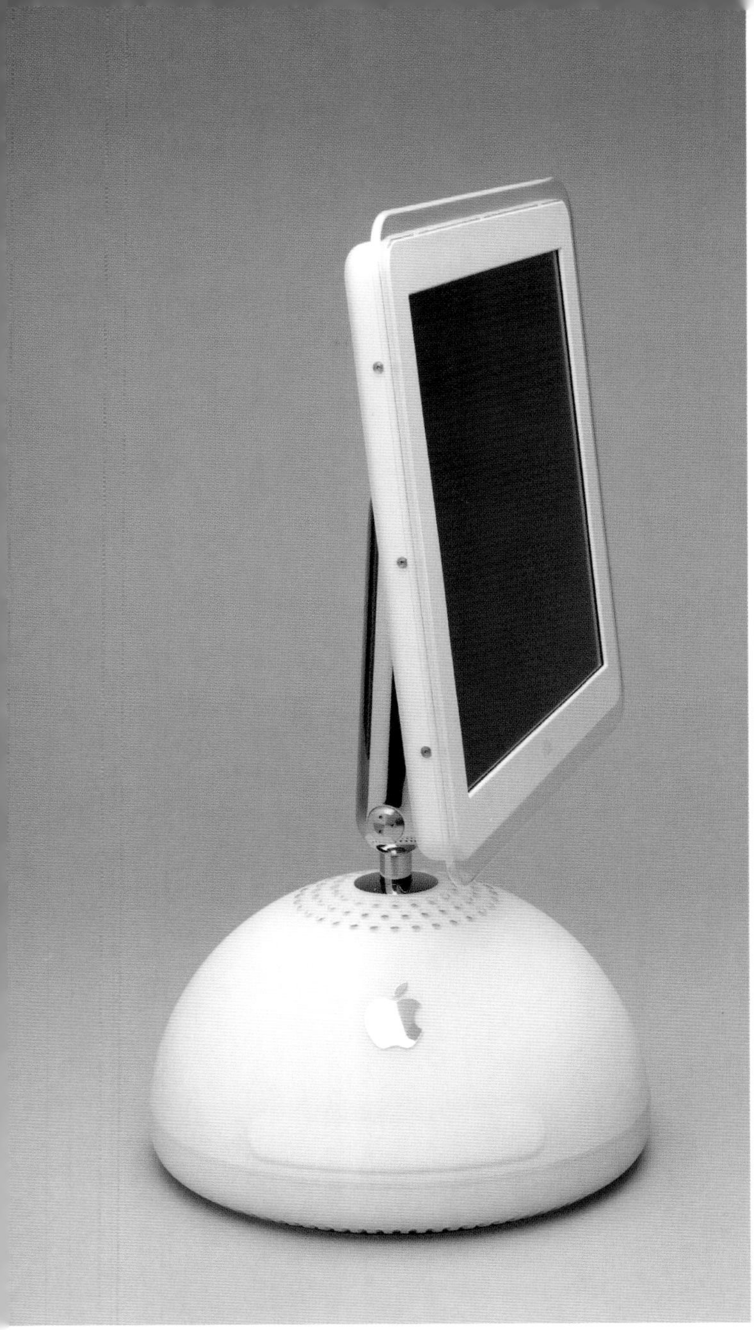

Apple Design Team

iMac Flat Panel Computer

Designed 2001

Plastic, chromium-plated metal, acrylic

19¾ x 15⅛ x 11 in. (50.2 x 38.4 x 27.9 cm)

Produced by Apple Computer, Inc., Cupertino, California

K2008.28

ANNOUNCED IN JANUARY 2002, THE *IMAC FLAT PANEL* COMPUTER WAS A COMPLETE transformation from the brightly colored biomorphic form of the *iMacs* of the previous four years. The flat screen that could be rotated and change angle was mounted to a domed white base—inside of which were housed the computer's internal components—on a hinged arm in shining chrome. Joints at the top and bottom of the arm allowed an exceptionally broad range of motion for optimal positioning of its 15-inch LCD screen. This *iMac* was the first to come with Apple's signature all-white keyboard and mouse. Unifying all the elements into a one-piece, self-contained computer reduced the need for extra cables for a separate monitor and speakers, resulting in a graceful design with a flat screen that appears to float in the air on a chrome elbow. –PM

10
objects

PLAYING

PLAY IS AN ESSENTIAL PART OF LIFE. FOR CHILDREN IT IS THE MEANS through which they first engage with the material world, often in miniature form. With the aid of a doll's house or a toy car, and any number of small props, children's play often involves fantasy and invention, a form, that is, of scenario creation that often mimics adulthood. Children play organized games that involve objects to facilitate them, from packs of cards to LEGO sets. Adults engage in games and sport as a means of relaxing and distancing themselves from day-to-day life. The Kravis collection contains a variety of "play" objects for both children and adults, among them old wooden toys for children, as well as bicycles, skis, and an archery bow for adults. Their design reflects their function and aims at maximum efficiency and performance.

Ko Verzuu (1901–1971)
ADO Toys
Produced by ADO, Berg en Bosc, the Netherlands

far left
ADO Wybert Trailer with 17 Wooden Blocks
Designed 1930s
Painted wood, rubber, metal
7¼ x 9½ x 6 in.
(18.4 x 24.1 x 15.2 cm)
K2013.138

front, center
ADO 390 Tractor and Trailer
Designed 1950s
Painted wood, metal
3⅞ x 17⅞ x 4 in.
(9.8 x 45.4 x 10.2 cm)
K2013.151

rear, center
Large Trailer with 36 *ADO 456* Wooden Blocks
Designed 1950s
Painted wood, metal
6¼ x 16⅜ x 10 in.
(15.9 x 41.6 x 25.4 cm)
K2013.149

far right
ADO 954 Covered Trailer
Designed 1930s
Painted wood, linen, metal
7⅝ x 11⅞ x 6¾ in.
(19.4 x 30.2 x 17.4 cm)
K2013.136

KO VERZUU WAS A FOLLOWER OF THE DUTCH DE STIJL GROUP OF AVANT-GARDE architects and artists from the 1920s, which is reflected in the bright primary colors of his geometric toy designs. Verzuu was particularly influenced by Dutch architect and de Stijl member Gerrit Rietveld, who likely developed an interest in children's furniture when he became a father, and designed a child's wheelbarrow in 1923.[1] According to Juliet Kinchin, "The synthesis of aesthetic and educational reform principles was a feature of several interwar initiatives in the Netherlands." ADO's designs, as well as designs of educational toys by Rietveld and others, were promoted in support of progressive ideas about teaching children.[2] –DAH

Designer Unknown

LEGO Minifigure Model
Designed 2000
Plastic
19¾ x 10⅞ x 7¼ in. (50.2 x 27.6 x 18.4 cm)
Produced by LEGO Group, Billund, Denmark
K2013.36

THESE COLORFUL, INTERLOCKING PLASTIC bricks have been in continuous production from 1958 to the present day and are popular with every age group. Once connected and constructed, LEGO blocks can be disassembled and used again and again. The LEGO Group originated in the workshop of a Danish carpenter, Ole Kirk Christiansen, who began to produce wooden toys in 1932. In 1934, the company was named LEGO, from the Danish *leg godt*, which means "play well." LEGO, and other similar companies, considered play an important part of a child's educational development. In 1947, LEGO began to produce plastic toys, and, in 1949, the interlocking "automatic building bricks" were introduced. However, according to company leader Godtfred Kirk Christiansen, the final basic form we know today was created in 1958: "That was when we invented the brick with the studs on top and tubes on the inside underneath. It was the tubes that gave the product its versatility and building stability."[3] The priorities for this final design included "unlimited play potential" for kids of all ages to inspire "development, imagination, creativity."[4] This new structural integrity created potential for design configurations limited only by the imagination. The creative, sculptural design of the *LEGO Minifigure* model, used for store display purposes, is only one example of the infinite possibilities.[5] –DAH

Victor Vasarely (1906–1997)
Vasarely Planetary Folklore Participations No. 1 Magnetic Puzzle
Designed 1969
Polystyrene, magnets, metal
20 x 20 x 1 in. (50.8 x 50.8 x 2.5 cm)
Published by Éditions William Wise, Lacoste, France, and Éditions Pyra, Zürich, Switzerland, in collaboration with Éditions Denise René, Paris, France
D0122

Cover of instruction booklet for *Vasarely Planetary Folklore Participations No. 1*, showing one possible pattern for the puzzle, 1969.

HUNGARIAN-BORN FRENCH ARTIST VICTOR VASARELY DEVISED THIS PUZZLE FOR PARTICIPANTS OF ALL ages to create their own art based on his original work. Vasarely was a leading advocate for op art in the 1960s, a movement characterized by its use of geometric patterns to create optical illusions. In 1965 Vasarely's work was included in MoMA's pioneering op art exhibition, *The Responsive Eye*, and in 1970 the museum's annual Christmas Shop sold *Vasarely Planetary Folklore Participations No. 1* as a limited-edition art object for $500.[6] This set is signed and numbered in ink on the back and produced in an edition of three thousand. Using nineteen "highly nuanced" colors in circles and squares, Vasarely created an assembly that is visually appealing and engages the user through its interactive elements. The puzzle offers creative play in the same way that the educational toy products produced by Creative Playthings did in the 1960s; their goal was to teach the use of color as a stimulus for creativity and the imagination. This puzzle set included a total of 195 square magnetic pieces—some containing smaller squares, some containing circles—to fill the 100 spaces in the steel frame, allowing the user to produce virtually endless configurations. Also included were "programmation sheets" that codified the color/shape pattern of Vasarely's original and a second preset pattern, as well as a few blank forms. Vasarely advised in the instructions that accompanied the set: "Make each of your compositions freely, seeking that which satisfies you the most. Then copy the codes on a blank programmation sheet."[7] —DAH

below top

Designer Unknown

Codeg Model Sports Car
Designed 1940–50

Bakelite, metal, plastic

3¾ x 13½ x 5½ in. (9.5 x 34.3 x 14 cm)

Produced E. K. Cole, Southend, England

K2001.1

below bottom

Designer Unknown

Toy Midget Racer

Designed 1935–45

Aluminum, rubber

2¼ x 9⅝ x 3½ in. (5.9 x 24.5 x 8.9 cm)

Manufacturer Unknown

K2011.170

opposite

Carl Swenson (1921–1975)

NY-LINT Toy Car, Model No. 600

Designed 1945

Steel, enameled steel, rubber

4⅝ x 14 x 5⅜ in. (11.8 x 35.6 x 13.7 cm)

Produced by Ny-Lint Tool and
Manufacturing Company,
Rockford, Illinois

K2008.11

SINCE THE EARLY DAYS OF THE AUTOMOBILE, TOY CARS HAVE BEEN POPULAR WITH BOTH children and adults. The Toy Midget Racer, designed sometime between the two world wars, is a replica of the very small, very fast American midget racer cars of the 1930s.

The *Codeg* model sports car, made of Bakelite, was probably designed during the 1940s, and was likely inspired by the 1939 Citroën Traction Avant Cabriolet. *Codeg* was the trade name of Cowan, de Groot Ltd., a London-based importer of toys and other goods that went on to become one of England's biggest wholesalers of toys.

The Ny-Lint Tool and Manufacturing Company of Rockford, Illinois, which began in 1937 as a tool shop, entered the toy-making business in late 1945 and became known for its high-quality toy transportation and construction vehicles. The *NY-LINT* model seen here was a large, automatic spring-wind car, which incorporated three different movements (winding up the spring motor allowed the car to turn around, turn corners, or park itself), all governed by a key mechanism. The car, designed by Carl Swenson, was based on the Chrysler *Airflow* of 1934, which was one of the earliest examples of streamlining in mass-produced automobiles. Finished in bright red (the most popular color), blue, or green (the rarest color), with black rubber bumpers and tires, the toy car, made of "baked enamel," was a hit at the 1946 Toy Fair in New York City, and the press praised its ability to drive itself. –PM

Viktor Schreckengost (1906–2008)
Torpedo Child's Pedal Car, Model No. J-840
Designed c. 1950
Enameled and chromium-plated metal, plastic, rubber
20⅛ x 39¼ x 17 in. (51.1 x 99.7 x 43.2 cm)
Produced by Murray Ohio Mfg. Co., Cleveland, Ohio
K2009.41

Promotional photo of Viktor Schreckengost working on a clay model for the *Torpedo* pedal car, 1948.

VIKTOR SCHRECKENGOST CAME FROM A FAMILY OF CERAMICISTS, BUT AFTER STUDYING design at the Cleveland School of Art and at the Kunstgewerbeschule in Vienna, Austria, he became a prolific industrial designer, creating everything from tableware and children's toys to flashlights, furniture, and even an artificial limb. During the Depression, he designed streamlined objects for mass production by such companies as Sears and American Limoges, many of which were patented. In addition to many lines of ceramics, Schreckengost became known for designing children's bicycles—especially the banana-seat bicycle—and pedal cars, including this *Torpedo* child's pedal car, manufactured by Murray. He would first sculpt the pedal cars in clay, and from those would create working models. He designed numerous bicycles that were put into production for Sears, using a combination of basic parts and modifying elements such as handlebars and chain guards. "If we sold 600,000 of something, I felt I was on the right track," he once said in an interview published in *Metropolis* magazine. –PM

torpedo

Designer Unknown

Pureclad Sled

Designed c. 1940s

Aluminum, enameled steel, vinyl fabric

15 x 55 x 18 in. (38.1 x 139.7 x 45.7 cm)

Manufacturer Unknown, aluminum produced by Reynolds Aluminum, Louisville, Kentucky

K2014.47

THE BELOVED WINTER PASTIME OF DOWNHILL SLEDDING HAS SEEN A WIDE variety of equipment, both specialized and improvised, from random sheets of metal, wood, or cardboard to plastic saucers to steerable, padded-out vehicles on runners such as this aluminum sled. Modernist designers, including Harold L. Van Doren and John Gordon Rideout, added sled design to their repertoire. The sled shown here is a curious and puzzling object of industrial design. Not much is known about it, but there is evidence that the sled was not factory-produced, and was more likely a one-off, perhaps even a prototype. The *Pureclad* 24S aluminum used to construct the sled is similar to that used in Buckminster Fuller's Dymaxion house of the 1940s. It is a high-strength aluminum alloy that was typically used in World War II aircraft. The construction of the sled indicates some element of manufacturing expertise—the rivets and the steering wheel are all well executed—but other elements such as the stitching in the vinyl upholstery and the finials at the ends of the sled are less well-made. The dents in the cone suggest a few crashes in snowbanks or trees. —PM

Harold L. Van Doren (1895–1957)
John Gordon Rideout (1898–1951)

Skippy Sno-Plane Sled

Designed c. 1933 (D91,796)

Painted plywood, enameled and chromium-plated steel, plastic

7⅜ x 46⅜ x 22½ in. (18.7 x 117.8 x 57.2 cm)

Produced by The American National Company, Toledo, Ohio

K2010.85

HAROLD L. VAN DOREN WAS A MAJOR FORCE IN DESIGN IN THE 1930S AND WAS CREDITED BY design historians Sheldon and Martha Cheney with introducing industrial design to midwestern America. They admired his artistic sensibility in designing objects that ranged from appliances to "exquisite little perfume bottles" to car tires, a gas pump, and children's cars.[8] The *Skippy Sno-Plane* sled was designed during his 1931–35 partnership with John G. Rideout. Van Doren credits the design to his partner Rideout, although both Van Doren and Rideout are listed on the design patent and the utility patent.[9] In his influential book *Industrial Design: A Practical Guide*, Van Doren used the *Sno-Plane* as one of his case studies and proclaimed that "Children Love Streamlining."[10] The name *Sno-Plane* is stenciled in black and red on the central wooden slat with seven arrows that resemble airport landing-strip guidelines, using the metaphor of flight so common in streamlining. The steering was meant to be an innovation, making the sled steer like a velocipede by means of a separate center runner, but it ultimately didn't work well and was abandoned in later designs. The sled sold well anyway because of its streamlined style, and it led to a series of children's vehicles for American National, including the *Skippy-Racer* scooter. –DAH

right Advertisement for *Skippy Sno-Plane* streamlined sleds, c. 1933.

Clarence J. Hickman (dates unknown)

True-Flex Skis and Poles

Designed c. 1947 (utility patent 2,560,693)

Aluminum, leather, steel

Skis: 74½ x 4⅞ x 3¼ in. (189.2 x 12.4 x 8.3 cm) each
Poles: 51 x 6¼ x 6¼ in. (129.5 x 15.9 x 15.9 cm) each

Produced by TEY Manufacturing Corp., Milford, Connecticut

K2010.134

ANCIENT PAINTINGS SUGGEST THAT HUMANS IN CHINA WERE SKIING MORE THAN TEN THOUSAND YEARS ago. The modern ski traces its origins to mid-nineteenth century Norway when the cambered ski was created. Older skis were very thick and rigid to prevent skiers from sinking into the snow, but cambered skis are arched upward for weight distribution and shock absorption. Like their predecessors, they were carved from hardwood, but they could be made thinner and lighter for ease of movement and controllability, which would naturally reduce the occurrence of certain types of injuries. The first solid aluminum skis were prototyped in 1928, and in 1948 these *True-Flex* laminated aluminum skis were introduced by TEY Manufacturing Corp. The previous decades had seen numerous improvements to ski safety, and these skis employed a new form of one of the most important: their soon-to-be-patented *True-Hold* bindings were a type of release binding that provided "a rigid steel heel strap for perfect control," and most importantly, "a safety toe iron which permits the boot to 'kick out' in a bad fall,"[11] preventing injuries caused by the twisting of ankles, knees, and hips when the feet are trapped in ski bindings. Various types of aluminum skis dominated the market for much of the mid-twentieth century, but by 1968, fiberglass skis had been introduced and quickly became the preference of serious and casual skiers alike. —PM

Designer Unknown

Archery Bow

Designed 1940–60

Aluminum, cord

59⅛ x 3⅛ x 1¼ in. (150.2 x 7.9 x 3.2 cm)

Manufacturer Unknown

K2013.40

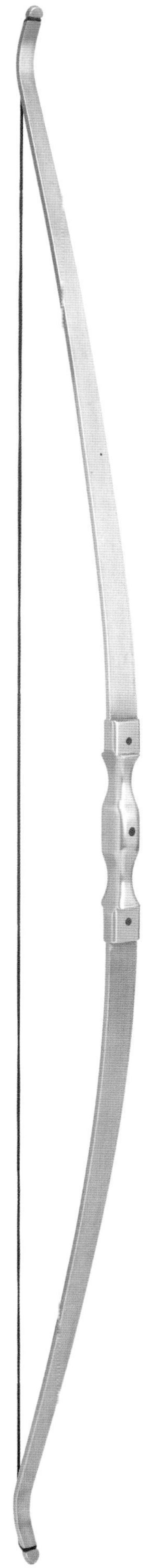

THE USE OF THE BOW AND ARROW CAN BE TRACED BACK MANY thousands of years to Kenya, and the oldest extant bows found in one piece were discovered in Denmark and date to 9,000 BCE. Archery bows traditionally have been made of wood for millennia, but steel versions began to appear in the 1880s. Aluminum bows began to appear in France by the 1940s, as did bows for target shooting that were made of Duralumin. Aluminum and magnesium compound bows—which have a mechanism to make pulling back the bowstring easier—rose in popularity with hunters and archers in the 1960s for their increased accuracy. The process of making metal bows improved over time as forging gave way to extrusion. Modern bows are made of other durable-yet-flexible materials such as carbon fiber and metal alloys. –PM

below

Olof A. Norlund (dates unknown)

Ice Creepers

Designed c. 1910 (utility patent 962,042)

Aluminum, steel

1⅞ x 5¼ x 8½ in.
(4.8 x 13.3 x 21.6 cm) each

Produced by Norlund's,
Williamsport, Pennsylvania

K2010.136

opposite

Milton M. Muncie (1908–1985)

Wading Sandals

Designed c. 1967 (utility patent 3,464,127)

Aluminum, leather

1⅜ x 4½ x 8⅝ in.
(3.5 x 11.6 x 21.9 cm) each

Manufacturer Unknown

K2010.135

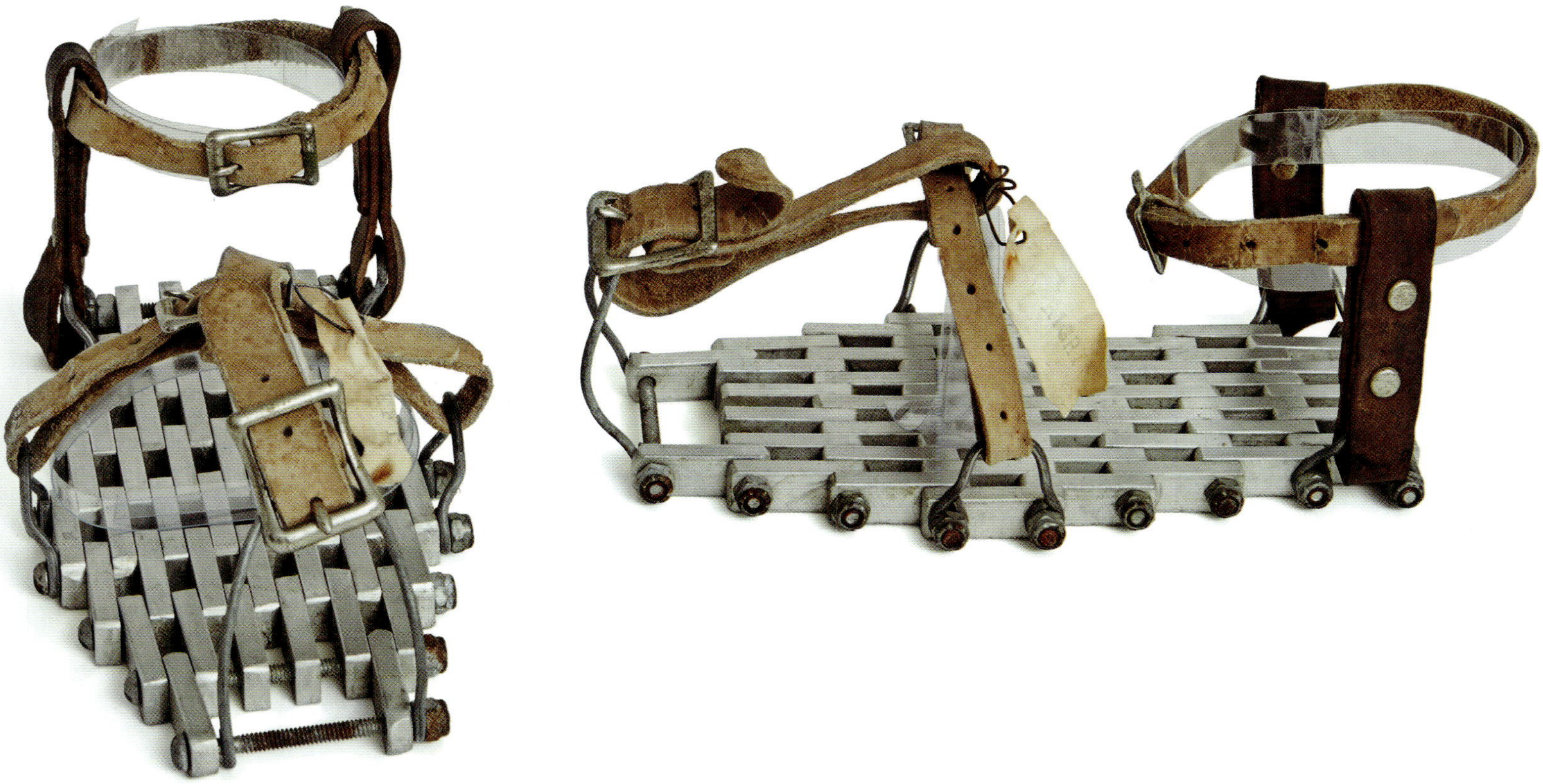

THE ORIGINS OF ICE CLEATS GO BACK TO THE VIKINGS ABOUT A THOUSAND years ago. It is not known exactly when ice cleats first appeared in modern times, but during the 1930s and 1940s, numerous patents were taken out for versions of ice cleats, or "ice creepers," as they were often called at the time. The sharp teeth on these cleats would allow anyone, such as soldiers or hunters, to walk on ice for long distances, especially on uneven terrain. According to their patent specifications, the cleated wading sandals by Milton M. Muncie "facilitate wading in fast water; walking over wet, slippery surfaces; and walking over rocky and mountainous terrain." –PM

Marc Newson (born 1963)
MN02 Bicycle
Designed 1998–99
38¼ x 66½ x 22½ in. (97.2 x 168.9 x 57.2 cm)
Aluminum, metal, rubber, plastic, vinyl
Produced by Biomega, Copenhagen, Denmark
K2013.128

THE DANISH START-UP COMPANY BIOMEGA, founded in 1998 by Jens Martin Skibsted and Elias Grove Nielsen, selected the Australian-born Marc Newson to design a simplified urban bike. Newson said he "looked for the simplest form that was able to do this, to connect the headset to either the bottom bracket or seat and then to the rear wheel."[12] The glow-in-the dark frame is molded in two halves from metal that is heated until it is softened but not liquefied, then bonded together with the highly durable epoxy Permabond SP110. Unsightly cables for the gears and brakes are mostly tucked away within the zigzag frame. The result is a sleek, strong, lightweight bike that is well-suited to city streets in both its straightforwardness and its in-built nighttime visibility. Newson designed another bike for the company, the *MN01 Extravaganza*, made of brushed aluminum. –PM

above View of the *MN02* bicycle's glow-in-the-dark frame.

Hajime Sorayama (born 1947)

AIBO Entertainment Robot with original box and accessories, Model No. ERS-110

Designed 1999

Plastic

10½ x 16¼ x 6 in. (26.7 x 41.3 x 15.2 cm)

Produced by Sony Corporation, Creative Center, Tokyo, Japan

K2010.7

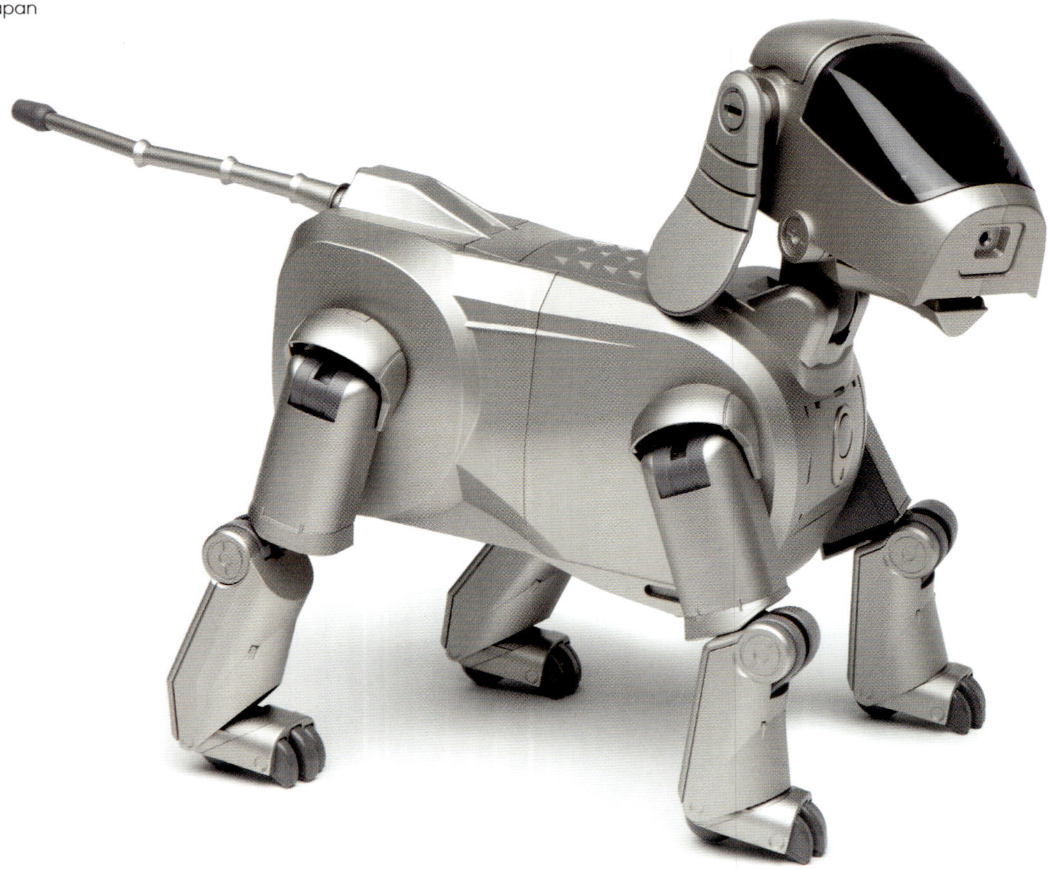

THE *AIBO* MODEL NO. ERS-110 WAS SONY'S FIRST AIBO ROBOT DOG in the form of a silver electronic pet, with a beagle-like appearance, and was first commercially released by Sony in 1999—with an initial production run of five thousand released in Japan and the United States. The name "AIBO" is a play on words: it is an abbreviation for Artificial Intelligence Robot and also connects to the Japanese word *aibou*, meaning "pal" or "partner." This robot dog is programed to exhibit many of the movements and behaviors of a real dog. It responds to visual, tactile, and auditory stimuli, as well as basic verbal commands. It also assimilates new information, "learning" from its environment and owners over time. Unlike its living counterparts, however, *AIBO* can help its owners keep track of appointments and even send them emails. —PM

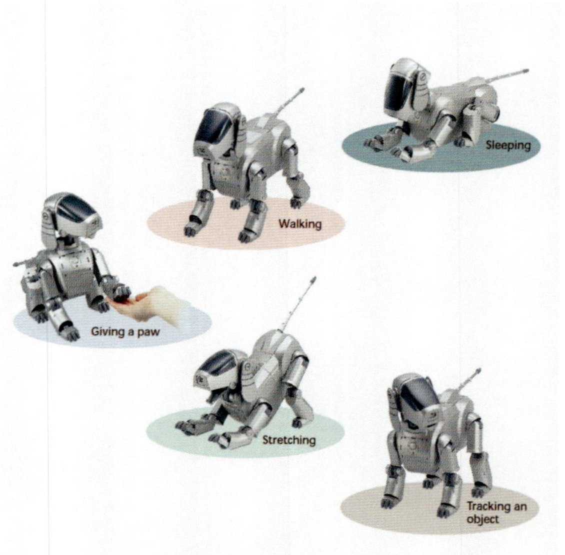

Right Illustration from the Sony *AIBO* ERS-110's user manual, showing the robot's capabilities, 1999.

Eero Aarnio (born 1932)
Two *Puppy* Children's Seats from the Me Too Collection
Designed 2005
Polyethylene
Red: 17⅝ x 22½ x 13⅜ in. (44.8 x 57.2 x 33.4 cm)
Green: 21¾ x 27 x 16½ in. (55.2 x 68.6 x 42 cm)
Produced by Magis, Torre di Mosto, Italy
K2008.64.1–2

IN 2001, THE INNOVATIVE PLASTICS COMPANY MAGIS DECIDED TO PRODUCE A COLLECTION of furnishings made especially for two- to five-year-olds that would be different from those made for adults; it did not want scaled-down grown-up furniture. Eero Aarnio's *Puppy* design, for Magis's Me Too collection of objects and furniture for children, is abstract and simplified in form. It can be identified as a dog from its proportions and frisky pose. The colorful animal figure has its roots in the tradition of zoomorphic objects, and the features of the puppy are reminiscent of those drawn by small children. Aarnio believes that industrial form comes from within, from the material, construction, and purpose of the product—where true form follows function. Functionality is the first aim. Making the product interesting, expressive, surprising, or playful involves the use of colors, textures, and the proportions of parts and details. The soft, round, and bright-colored nature of Aarnio's designs is characteristic of plastics and mold-cast products. –PM

Closca Design (founded 2013)
Cul-de-Sac Creative (founded 2008)
Fuga Collapsible Bicycle Helmet
Designed 2015
Plastic, polystyrene, nylon
Open, with brim attached:
4¾ x 11 x 8¼ in. (12.1 x 27.9 x 21 cm)
Produced by Closca Design, S.L., Valencia, Spain
K2016.19

Fuga bicycle helmet with removable visor, collapsed for carrying or storage, c. 2015.

IN MANY CITIES WEARING A HELMET WHILE RIDING A BIKE IS NOT LEGALLY REQUIRED, but a good helmet dramatically reduces the risk of head injury or death in the event of an accident. This patented foldable bike helmet made of plastic and polystyrene is a convenient solution aimed at the urban cyclist who wants to be both safe and stylish and have the convenience of an easy-to-carry helmet. Its top dome and two concentric rings can be popped open for use or pushed closed, reducing its height by half so as to easily slide it into a bag or backpack. In addition to this collapsible helmet, Closca has developed cycling apps such as RideU, an app to find rental bikes in more than 240 cities around the world. The *Fuga* helmet comes with a built-in near-field-communication (NFC) tag that is programmed with useful information such as emergency numbers and rental bike locations, which can be read with a smartphone. –PM

CLOSCA

TRANSPORTATION DESIGN IN THE TWENTIETH CENTURY

Russell Flinchum and Hampton C. Wayt

FIG 1
View of the west crossover of the London Underground's Central London Railway extension at Liverpool Street Station, 100 yards from the station, 1912.

FOR INDUSTRIALIZED NATIONS AROUND THE WORLD, THE ADVENT OF THE TWENTIETH CENTURY ushered in a new age of speedy and versatile transportation options that imparted, arguably more than any other design field, a symbiotic and quick-paced fluidity to human life that came to define the modern era. Mass transportation systems such as railroads, trolleys, and subways—which were first developed in the nineteenth century—connected and redefined urban areas as they expanded their reach through improved technology and increased demand. The underground rail system in London, also known as the Underground, or the Tube (fig. 1), was improved in 1905 through the use of electric traction, which helped facilitate the development of "Metroland" suburban housing centers around stations along the train lines, thus greatly expanding the city's metropolitan footprint.

Developing aviation technologies in these early years of the century revealed that man's centuries-long dream of flight was near at hand. Count Ferdinand von Zeppelin witnessed Germany's first successful flight of his LZ 1 airship of 1900 (fig. 2), and Alberto Santos-Dumont flew his airship around the Eiffel Tower in Paris in 1901. Both offered great promise for lighter-than-air travel, but ultimately the most important early aviation event was Orville and Wilbur Wright's first controlled flight of a heavier-than-air craft at Kitty Hawk, North Carolina, in 1903.

Advancements in personal transportation, too, offered new promises of mobility in the first decade of the twentieth century. The liberating effect of improved technologies such as pneumatic tires and paved roads, which resulted from the safety bicycle craze of the 1880s, literally paved the way for the burgeoning automobile market. In 1899, Camille Jenatzy traveled at 105.88 kilometers per hour (65.79 miles per hour) in his torpedo-shaped *La Jamais Contente* electric-motored race car in France (fig. 3). Henry Ford's implementation of the moving assembly line, which launched in 1913, offered cheap prices for his Model T car, allowing consumers of modest means to have access to reliable and swift mechanical transportation, forever changing the way people commute to work and live in their communities.

Around the same time as all of these developments, visions of future societies in which transportation was integral to architecture were birthed. One fantastical, popular example was *King's Dream of New York*, an illustration in a book published by Moses King in 1908 that

FIG 2
above left First ascent of Count von Zeppelin's *LZ-1*, Friedrichshafen, Germany, 1900.

FIG 3
above right Camille Jenatzy at the wheel of his electric car, *La Jamais Contente* ("Never Satisfied"), c. 1899.

imagined stories-high train lines, flying machines, and airships in constant movement amid a crowded metropolis of skyscrapers piercing the sky (fig. 4).

By the end of the first three decades, transportation's concern with increased speed and distance, improved reliability, standardized parts, ease of manufacturability, and other engineering attributes culminated in a number of notable feats. Charles Lindbergh's crossing of the Atlantic in his *Spirit of St. Louis* in 1927 set a new bar for airplane and pilot endurance. A year after Lindberg's historic adventure, the LZ 127 (the *Graf Zeppelin*) airship was making speedy trans-Atlantic voyages from Germany to North America, and in 1930 completed an around-the-world trip. In 1931, the German Dornier DO-X, the world's largest passenger airplane (and flying boat) up to that time, traveled the length of Africa and South America before being welcomed to New York City Harbor by thousands of admirers lining Manhattan's shore. Both forms of air travel would become symbolic in the following decade (although Zeppelin's approach was marginalized by the Hindenburg tragedy at Lakehurst, New Jersey, in 1937).

The financial crash of 1929 and the ensuing Depression, however, found many businesses shifting focus from scientific and technological advancements to the fanciful aesthetic ballyhoo of a nascent design field in an effort to attract customers. In an October 1940 retrospective of the 1930s, the magazine *Architectural Forum* designated America's previous ten years as the "design decade." Perhaps a more apt name would have been the *re*-design decade: due to the limited funds at hand, as well as the solid technological investments of the several preceding decades, many designs of the era amounted to little more than a sprucing up of the exterior of existing units without any change occurring in the underlying mechanisms.[1]

Much of this type of work entailed merely placing a new shroud over the old apparatus, as seen with Henry Dreyfuss's steam locomotive *Mercury*, made for the New York Central Railroad in 1936. Side-by-side comparisons of the *Mercury*, with a smoothly rounded bathtub carapace, and its predecessor reveal little more than a traditional steam locomotive that has received a new steel suit (fig. 5). Although such treatments (often termed "styling" rather than "design") were commercially successful, some designers and critics found these types of design approaches somewhat artificial in nature. But such eye-catching transformations

were more than a dolling up, through their signaling of "progress," a heavily promoted term both in America and Europe that decried the doldrums of the Depression by speaking of an unceasing societal forward momentum. Claims of progress also encouraged the idea that the overall quality of life, even during the Depression, was nevertheless better than it had appeared only a few decades earlier.[2]

The "streamline" craze of the 1930s—or the application of aerodynamic shapes to products through scientific (or more commonly, artistic) determination—is commonly credited to the popularity of airplanes during the era, although in many ways airships were just influential muses to designers, as witnessed in automobiles. In 1922, Zeppelin Company chief engineer Paul Jaray patented and built running prototypes of his design for a double-teardrop, dirigible-based automobile body that greatly reduced aerodynamic drag, even at speeds as low as thirty-five miles per hour. His modified Zeppelin form did not catch on initially, but with more powerful engines, smoother roads, aerodynamic body testing, and improved stamping technology in the early 1930s, automobile designers realized that this aerodynamically-efficient form was necessary for the increased speeds of the future. Even so, only a handful of production or serially-produced automobile designs, such as Hans Ledwinka's 1934 Tatra 77 limousine and the Chrysler Corporation's 1934 Airflow, featured coachwork that was closely related to Jaray's ideal; the rest of the industry interpreted the double teardrop as it saw fit.[3]

A much-publicized bright spot during the Depression came in the field of railroading, with the Pennsylvania Railroad (PRR) implementing massive electrification of the northeast

FIG 4
opposite left Postcard illustration of a futuristic city inspired by *King's Dream of New York*, published by Moses King, 1908.

FIG 5
opposite right *J-1c Hudson* locomotive (left) and Henry Dreyfuss's *Mercury* locomotive, New York Central Railroad, 1936.

FIG 6
right *SS Normandie*, New York Harbor, c. 1937.

corridor of the United States, which remains in use today. Complementing this impressive feat was industrial designer Donald R. Dohner's custom designed and iconic double-ended GG-1 Class locomotive featuring a smooth body with integrated hood and centrally located cabin providing better protection for the engineer. Not only was Dohner's work with the PRR's engineering department a rare instance during the Depression where a designer collaborated during (as opposed to after) engineering development, the unit (which industrial designer Raymond Loewy was later hired to create a paint scheme for, resulting in five stylish stripes) also happens to be one of the earliest known examples of the then-emerging trend of developing a 3D form sculpturally rather than orthographically—a process that has defined the appearance of nearly all product forms from the mid- to late-1930s onward.[4]

Since 1900, the great transatlantic ocean liners have been dominated by a French aesthetic that promoted luxurious and, for rich Americans living under Prohibition, alcohol-fueled lifestyles while treating the ship's interiors as floating galleries for decorative art. Designed under the direction of Roger-Henri Expert, the French-built *SS Normandie*—which at its core was an engineering tour de force that set a speed record for transatlantic crossing during its maiden voyage in 1935 (fig. 6)—featured a multitude of decorative works, including a colossal glass lighting installation by René Lalique and murals by Jean Dupas that were also made in glass but totally different in style. The teardrop shape of the three streamline smoke stacks rising above the main deck inspired silversmith-turned-designer Peter Müller-Munk's pointed *Normandie* pitcher, which utilized the sharp tapered edge to aid in pouring its contents (page 190).

FIG 7
Jeeps on parade at Fort Myer, Virginia, 1942.

The World War II years, beginning in 1939, and the resulting manufacture of paraphernalia for use in war produced a number of significant designs that appealed to a modernist mindset. This outlook promoted the idea that design without consideration of function should not be considered design, and that beauty, or at least a typeform, would emerge only by treating utility first. Representative of this way of thinking is Museum of Modern Art (MoMA) curator Arthur Drexler's critique of the World War II Jeep (fig. 7), which was hailed by MoMA in its 1951 *8 Automobiles* exhibition as "one of the few genuine expressions of machine art." Arguing for this philosophy, Drexler stated in the exhibition's accompanying catalogue, "The admirable Jeep seems to have the combined appeal of an intelligent dog and a perfect gadget." And this appeal was "so vast that this wonderful tool for transportation has won approval for much more than its practicality, though the engineers who perfected it worked without the concern for style with which other automobiles are designed."[5]

Various forms of escapism became popular during the war; designers were unable to produce for industry due to rationing and severe restrictions in place on the manufacture of many civilian "luxury" items (as defined by the War Office) ranging from toasters to automobiles. To help lift people's spirits, magazines regularly published futuristic visions of a victorious postwar world offering a cornucopia of consumer delights. Particularly popular among these scenes were fantastical automobile concepts (heavily associated with the war illustrations of industrial designer George W. Walker) sporting airplane-like "bubble top" canopies and cutting-edge plastic composite bodies. Immediately following the war, Allied victory spurred a technological utopianism in America that envisioned a society full of flying cars that could easily skip over gridlocked urban arteries. Translating these so-called Blue Sky designs (reputedly named after the endless blue horizons featured in the backgrounds of the illustrations) into successful commercial products proved tougher than anticipated, however. The prototypes of these few dual-purpose units that reached functionality, such as Robert Fulton's 1946 Airphibian or Moulton Taylor's 1949 Aerocar, were found to be neither good cars nor good airplanes.

The same ethos for technological marvel that produced flying cars transformed the hope-filled progress of the 1930s into prideful prowess after the war, in the 1940s and 1950s, again finding automobile designs sporting motifs taken from unrelated forms of transportation—especially jet technology. Perhaps the most representative of these is the 1951 Le Sabre "dream car" by legendary General Motors design chief Harley Earl, with its "NACA" air intakes, exhaust tube body shape, and taillights, fins, and other aeronautically inspired styling cues that in no way reflected the car's many actual engineering feats. It did, however, mirror the idea of "designing for mass acceptance" that was promoted by Dohner (who has since been deemed the father of American industrial design education by the Industrial Designers Society of America). This philosophy rejected Immanuel Kant's mantra of "art for art's sake," and upheld, by contrast, that a designer's primary role is to avoid personal taste biases and instead consider consumer psychology, appealing forms and colors, user convenience, and more, all in an effort to find those special qualities that will attract the broadest section of the general population.[6]

New materials and construction methods, often introduced as a result of wartime shortages, can be found in many postwar transportation designs. The original 1953 Chevrolet Corvette, also designed under Earl, did not feature the bubble top popularly imagined in the 1940s, although it did have a plastic body made out of fiberglass, each one individually crafted. Benjamin Bowden's organic, futuristic streamline prototype bicycle of 1946, called the Classic, originally featured an innovative and stylish pressed aluminum monocoque frame consisting of two large halves sandwiched together—as opposed to the typical multipiece welded tubular frame. It was shown that year in the *Britain Can Make It* exhibition, which was organized by the Design Council (then the Council of Industrial Design), but production of Bowden's machine would not come until the late 1950s and after a move to the United States, when the aluminum was changed to fiberglass and the bicycle was renamed the "Spacelander," after the space race that was raging at the time.

Aviation in the postwar years saw a dramatic change from propeller to jet propulsion. The Lockheed Constellation, with its flowing and porpoise-like shape, took aerodynamic, propeller-driven airplanes to their zenith in the early 1950s. The more efficient but far less elegant Boeing 707 was the first successful commercial jet and featured an impressive interior designed under the leadership of Frank Del Giudice for Walter Dorwin Teague. First manufactured in 1958, the 707 would easily overtake the Constellation and other propeller-driven competitors and ultimately became the prototype on which all planes that followed were based, as air travel transitioned from a rare event for the privileged few to an everyday occurrence for the masses.[7]

The democratization and expansion of air travel that began with the Constellation and solidified with the 707 took a heavy toll on travel by ocean liner. With passenger numbers dwindling, vessels were often converted to "cruise" ships that became floating hotels for vacationers traveling to exotic ports, rather than a means of transportation. The 1965 *SS Oceanic,* arguably the first of these new aquatic vacationlands, featured a "magrodome" sliding glass roof over a swimming pool on the Lido deck. It offered a pleasant swimming

FIG 8
above left TGV high-speed train, France, c. 1980s.

FIG 9
above right Concorde in flight, British Airways, c. 1990s.

FIG 10
opposite Promotional photograph of the Volkswagen *Beetle*, 1998.

environment, regardless of weather conditions (as first envisioned by Bel Geddes in his own ocean liner proposal of 1932).[8]

While the growing preference for air travel in the 1960s caused many countries around the globe to neglect or abandon their railroads, some chose to invest heavily in rail projects with great success. Japan's renowned high-speed and long-distance "bullet" lines, with development traced as far back as the 1930s, entered service in 1964. France's likewise-speedy intercity *Train à Grande Vitesse*, or TGV, debuted in 1981 (fig. 8), and through continued development reached a world speed record of 574.8 kilometers per hour (357.2 miles per hour) in 2007. The success of both systems has made them symbols of national pride, yet they remain at the heart of a reemerging interest in passenger rail travel, which began in the late twentieth century and continues into the twenty-first.

The dominance of air travel in the 1970s saw the 1976 debut of one of the most impressive airplanes before or since: the British-French designed Concorde Supersonic Transport passenger plane (fig. 9). It was Chuck Yeager who paved the way for this most impressive of airplanes, when he broke the sound barrier in the Bell X-1 in 1947. Including prototypes, only twenty Concordes were built, but they offered luxurious Mach II travel across the Atlantic (Air France's version was replete with flatware designed by Raymond Loewy) at more than twice the speed of their competitors. Travel by Concorde was an unqualified success until its nearly flawless record was shattered by a disastrous runway mishap in July 2000 that resulted in the removal of all Concordes from service.

After human space flight reached an apex in 1969 with the first moon landing, NASA began work on the Space Shuttle program. Unlike other types of manned space flights, the shuttles were designed to land on a runway like a commercial aircraft after reentry into the earth's atmosphere. Successfully launched in 1981, the space shuttles

teased humankind with the possibility of space travel, yet only reached low earth orbit. The shuttle program has been retired, but private companies such as Richard Branson's Virgin Galactic and Elon Musk's SpaceX continue to develop and test the potential for space travel for commercial purposes.

Throughout the 1960s, the waning of the technological utopianism that was embraced so emphatically by the public in the previous decade brought with it a rise in automobile designs that denoted certain lifestyles. Muscle cars of the mid- to late-1960s, named for their power and overtly muscle-like body designs (such as the ornithologically inspired 1969 Pontiac Firebird), were overpowered and under-braked automobiles with throaty exhaust that advertised raw power, youth, and speed. By contrast, Ferdinand Porsche's Volkswagen *Beetle*, a simple, durable, anachronistic holdover of early 1930s automobile design, launched counterculture into the mainstream with the company's "Think Small" and "Lemon" ad campaigns of 1959 and 1960.

The design by J Mays and Freeman Thomas for the reintroduced Volkswagen Beetle in 1998 (fig. 10) was more about nostalgia than lifestyle or technology, but it gave new life to one of the most enduring and endearingly iconic forms of transportation of the twentieth century. Moving into the next millennium, however, the pendulum continues to swing between technology and style with the increased integration of gadgetry and design in our lives, as seen with products such as the smartphone. Perhaps the embodiment of both aspects, at Tesla Motors (now Tesla, Inc.) the use of stylish forms to connote the advanced electrical drive of its twenty-first century automobiles speaks to the cutting-edge nature of science and mobility in an age when the driverless automobile seems just around the corner.

NOTES

1 Howard Myers, ed., "The Design Decade," *Architectural Forum* 73, no. 4 (October 1940).

2 In 1976, architectural historian Richard Pommer termed the placing of shrouds over preexisting mechanisms as "the skin game." See Richard Pommer, "Loewy and the Industrial Skin Game," *Art in America* 64 (March–April 1976), 46–47.

3 For Jaray's United States patent, see no. 1,631,269 (filed Aug. 19, 1922; granted June 7, 1927).

4 Donald Dohner's contribution to the GG1 locomotive, which was solely credit to Raymond Loewy for decades, was first revealed in 2009 (see Hampton C. Wayt, "Donald Dohner: The Man Who Designed Rivets," *Classic Trains* (Summer 2009), 30–35). Wayt's more recent, unpublished research also disproves nearly all of Loewy's claims surrounding his improvements to the GG1's design, including the use of welding instead of riveting for the shell.

5 Arthur Drexler, *8 Automobiles* (New York: Museum of Modern Art, 1951).

6 In 1998, the Industrial Designers Society of America (IDSA) awarded Donald Dohner a Personal Recognition Award designating him the "father" of American industrial design education, after Dohner's contributions in the field were revealed by designer-turned-historian Jim Lesko. See Lesko, "Industrial Design at Carnegie Institute of Technology, 1934–1967," *Journal of Design History* 10, no. 3 (1997), 269–92. For Dohner's philosophy on designing for mass acceptance, see Donald R. Dohner, "Designing for Mass Acceptance," *Industrial Arts* (UK) 1, no. 4 (Winter 1936), 253–56.

7 Frank Del Giudice, Burien, Washington, detailed his work on the 707 in a letter to Russell Flinchum, July 25, 1993.

8 Norman Bel Geddes, *Horizons* (New York: Little Brown, 1932), 36–43.

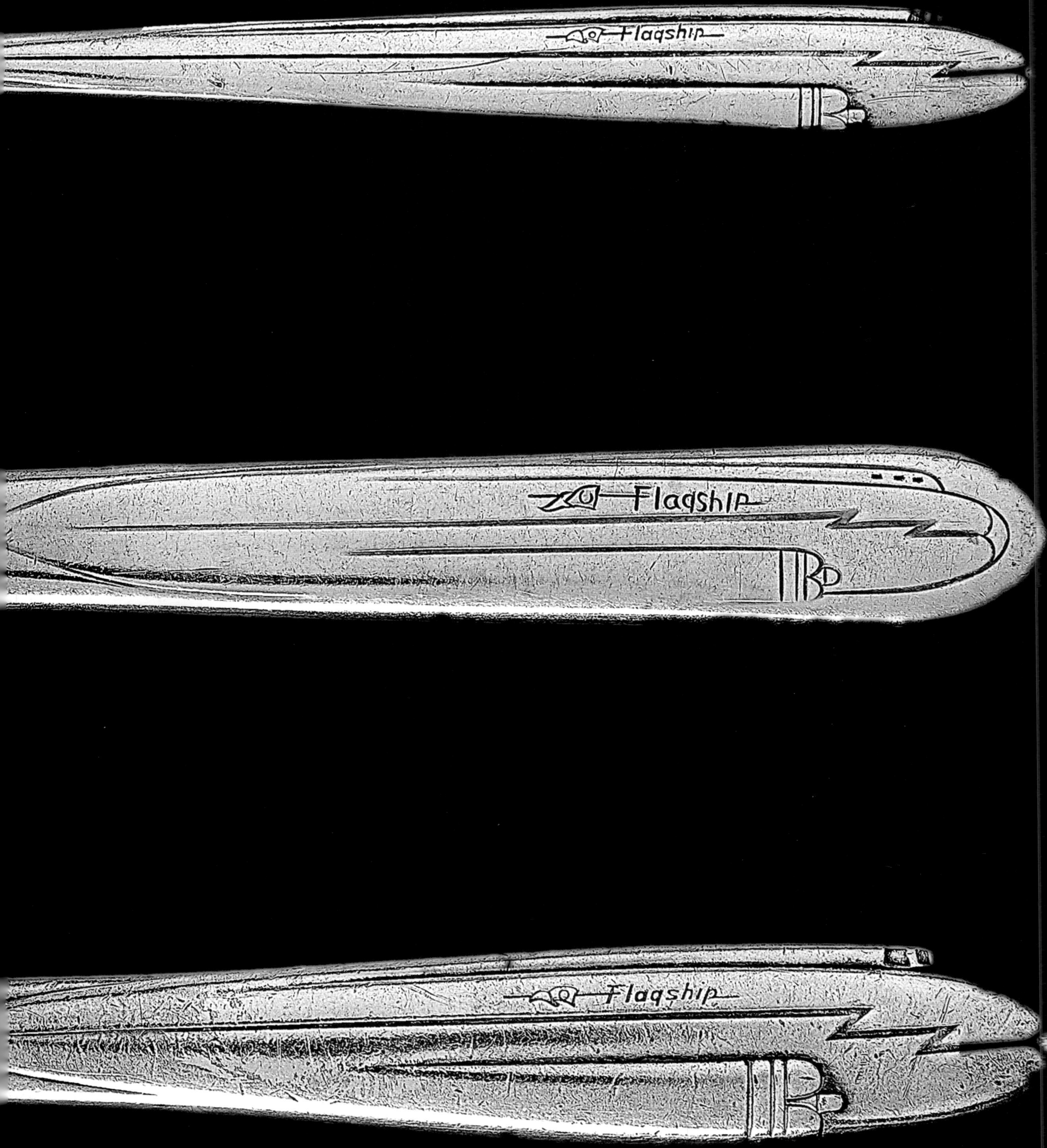
Flagship
Flagship
Flagship

11
objects

TRAVELING

WE SPEND MUCH OF OUR TIME TRAVELING, WHETHER COMMUTING to work, embarking on a vacation, or simply going to the grocery store. Mass travel is one of the defining characteristics of the modern age, and especially as the population has exploded over the last several decades, designers have had to address its material implications. From the objects of transportation themselves—automobiles, trains, ships, planes—to their interiors, to the items we take with us, the opportunities for design have been enormous. The Kravis collection includes many travel-related objects. One set of objects is related to air travel; another set addresses the issue of portability, even if it is limited to moving from room to room; and the last set of objects contains examples of actual transportation, from a Segway to a compact automobile.

Arne Jacobsen (1902–1971)

Giraffe Armchair from SAS Royal Hotel

Designed 1958–60

Laminated elm, upholstery

40¾ x 24 x 24½ in. (103.5 x 61 x 62.2 cm)

Produced by Fritz Hansen Eft., A/S, Copenhagen/Allerod, Denmark

K2011.167

View of SAS Royal Hotel dining room with the *Giraffe* armchairs, Copenhagen, Denmark, c. 1960.

IN THE POST-WORLD WAR II PERIOD, DENMARK EMERGED AS AN INTERNATIONAL CENTER FOR modern furniture design. Designers such as Finn Juhl and Hans Wegner achieved great fame for their works in the traditional medium of wood, while contemporaries like Poul Kjaerholm and Arne Jacobsen employed industrial materials such as steel and aluminum. Jacobsen emerged as a significant furniture designer in the 1950s. Three of his most acclaimed pieces were the *Ant* side chair (1951–52) and the *Swan* and *Egg* lounge chairs (both 1958). One of Jacobsen's most important architectural commissions was the Scandinavian Airlines System (SAS) Royal Hotel, which gave him the opportunity to design not only the building but also every element of the interiors. The *Giraffe* armchair was a custom piece for the dining room of the hotel and was never put into mass production. Only about 120 examples were made. It is a slightly idiosyncratic design for Jacobsen: the wood base recalls a form that he used for the *Grand Prix* side chair (1957), and his use of a wood trim on the edge of the shell is atypical. –RCM

Designer Unknown

Armchair for Train Lounge Car

Designed c. 1949

Aluminum, upholstery

33⅝ x 25⅝ x 31 in.
(85.4 x 65.1 x 78.7 cm)

Produced by Pullman Company,
Chicago, Illinois

K2013.8

ALTHOUGH MOST SEATS ON TRAINS ARE ATTACHED TO THE FLOOR AND TO ONE ANOTHER FOR stability, chairs for the lounge car, for much of the twentieth century, were different: they could be moved and arranged as suited the passenger and required special functional considerations. They had to be stable, hence the heavy structure, multiple supports, and sturdy base. The steel construction of Pullman Company's lounge car chair consists of parallel supports on the sides and three vertical supports at the rear that flare out at the base where they are attached to a base runner. Though freestanding, this chair could not be easily sent off balance. Upholstered back, seat, and arm cushions provided comfort on long train rides. The manufacturer's name, "The Pullman Co.," is etched in script on the inner side of the right front leg of this chair, and on a bar beneath the seat is stenciled "BILTABITO," which was the name of an observation/lounge/sleeper car with a seventeen-seat observation lounge, four drawing rooms, and one double bedroom.[1] –DAH

Advertisement for Pullman-Standard Car Manufacturing Company, showing a Pullman armchair in the observation car of the *City of Miami* train, 1941.

Edward J. Conroy (1912–1991)

Flatware from American Airlines Douglas DC-3 Flagship

Designed c. 1940 (D122, 113)

Silver-plated and stainless steel

Large knife: 7 in. (17.8 cm)

Produced by International Silver Company, Meriden, Connecticut

K2013.6.1

FIG. 1

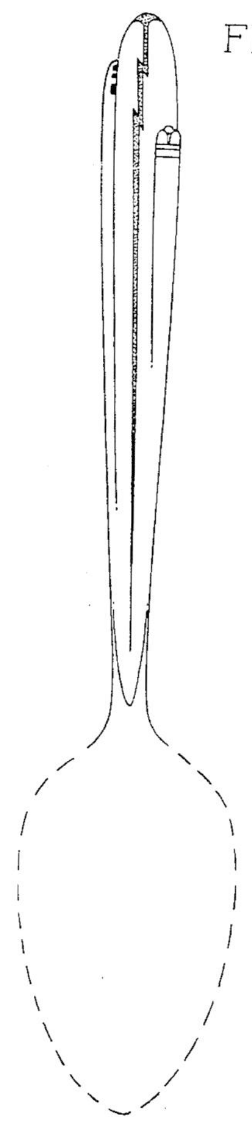

THE CUSTOM PLACE SETTING WAS A SIGN OF LUXURY THAT AMERICAN AIRLINES offered, beginning with their nonstop New York-to-Chicago service in 1936.[2] This flatware is from a place setting designed for the airline's Douglas DC-3 Flagship, which was launched that year. Just as the DC-3 was streamlined, so was this style carried out in the cutlery, which incorporated the famous aircraft body into the design of the handles, impressed on both sides. Other than the handles, the cutlery followed traditional flatware silver forms. –DAH

above left Postcard showing an American Airlines Douglas DC-3 Flagship, c. 1936.

above right Detail of U.S. design patent D122,113 for American Airlines flatware.

Flagship
Flagship
Flagship
Flagship

Marc Newson (born 1963)
A380 Inflight Service Teapot
Designed 2008
Porcelain
3 x 6 x 4⅝ in. (7.6 x 15.2 x 11.8 cm)
Produced in Thailand for Noritake, Nagoya, Japan, for Qantas Airlines, Winton, Australia
K2008.853.1

AUSTRALIAN BY BIRTH, MARC NEWSON HAS SPENT MOST OF HIS CAREER WORKING IN London and Paris. He was initially trained as a jeweler but has evolved into a true industrial designer. He first gained international fame with designs such as the *Lockheed* chaise lounge (1986–88) and the *Pod of Drawers* cabinet (1987). Over the last three decades, he has created elegant biomorphic designs for companies like Alessi, Cappellini, and Magis, but his work also includes designs for bicycles, cars, boats, and even airplanes. A comparison of Studio Joe Colombo's designs in 1970–72 for Alitalia Airlines (page 361) and Newson's work in 2008 for Qantas is most revealing about how modernism evolved over decades. Joe Colombo experienced an earlier, more austere form of modernism when the focus was on function and industrial materials like plastics—where less was more. Newson, on the other hand, was working in an opulent, if not somewhat decadent period, when more was more. His task thus was to design a large, sumptuous dinner service, but one that was still imbued with a minimalist aesthetic. The *A380 Inflight Service* teapot, made of white porcelain, was part of this service for the first class cabin of Qantas Airlines. Newson used similar bowl and handle forms for other pieces, such as coffee cups, creamers, and even gravy boats. Newson's *A380 Inflight Service* is a telling tale of two different times and cultures. –RCM

Studio Joe Colombo
Joe Colombo (born Cesare Colombo, 1930–1971)
Ambrogio Pozzi (born 1931)
Ignazia Favata (born 1943)

Linea 72 Cutlery
Designed 1970–72
Plastic
Knife: 6½ in. (16.5 cm)
Produced by SILPA, Perugia, Italy, for Alitalia Airlines
K2013.95

JOE COLOMBO WAS TRAINED AS A PAINTER AND A SCULPTOR AS WELL AS AN architect. He opened his design studio in 1962, and for a brief moment was one of the most creative and influential modernist designers in Italy until his death at the age of 41. Colombo was a prolific designer who worked in every medium. He employed the latest technology and materials in his work and created radically new interpretations of traditional forms, the chair in particular, where his designs seem more like sculpture than functional objects. Like many of his contemporaries, Colombo was fascinated with designing systems and environments. His fame was such that he was asked by Alitalia Airlines to design the *Linea 72* dining utensils for both first class and economy class cabins. Colombo responded to the demands of an airplane where space and movement within such a limited environment are critical factors. His designs for economy class were perhaps the most inventive: the various modular bowls fit into a sculptural dinner tray so they could not move. The cutlery—in white and clear plastic—had concave handles so that pieces could be stacked efficiently in a limited space. –RCM

above *Linea 72* economy class tray for Alitalia, showing Colombo designs, c. 1972.

left

Designer Unknown

Essco Iron

Designed 1940–1960

Chromium-plated steel, Bakelite

3⅞ x 6⅜ x 3⅛ in. (9.8 x 16.2 x 7.9 cm)

Produced by Southern Cross Porcelain Company, Melbourne, Australia

K2010.295

center

Designer Unknown

Steemco Steemette Travel Iron, Model No. 700

Designed c. 1950

Chromium-plated steel, Bakelite

4⅞ x 4⅜ x 6⅞ in. (12.4 x 11.1 x 17.5 cm)

Produced by Steam Electric Manufacturing Company, St. Louis, Missouri

K2010.280

right

Christopher Kerr (1899–1970)
Harold Thomas Holder (1902–1975)
Eric Lucas (dates unknown)
Norman Harry Lucas (1914–2000)

Smoothie Portable Electric Iron

Designed 1946

Bakelite, chromium-plated steel

2¼ x 4½ x 2 in. (5.7 x 11.4 x 5.1 cm)

Produced by Lucas Holder & Co., Coventry, England

K2010.299

ALTHOUGH THE BULKY *STEEMETTE* DOES NOT LOOK LIKE A CONVENIENT TRAVEL iron by our modern standards, its original box proudly states it as "America's First Steam Travel Iron." Its Bakelite handle, with a single flared support above the teardrop shaped body, curves back as if in a wind tunnel. According to the instructions, the handle was updated for this edition to be removable by a quarter turn to the right or left. The water was stored in a reservoir below the slanted flat deck. According to the accompanying instructions, "With your *Steemco* iron you actually iron on a cushion of fabric-vitalizing, germ-destroying *live* steam,"[3] a feature its predecessors did not have. The *Steemette* is an example of late streamlining, characterized by a move from the sculptural forms of earlier aerodynamic irons of the 1930s and 1940s to flatter designs of the 1950s. This American example is shown here with the British *Smoothie* and the Australian *Essco* travel irons. The *Smoothie* is a well-known gem—grips on the Bakelite sides replaced a bulky handle so it was streamlined, petite, and lightweight—perfect for travel. The *Essco* travel iron also makes use of Bakelite and streamlining with its forward-thrusting handle and tail in reddish brown. –DAH

John G. Thew (born 1927)

Turnpike Toll Gun with original instructions

Designed 1962 (utility patent 3,191,588)

Aluminum, enameled steel

3¾ x 6⅝ x 1½ in. (9.5 x 16.8 x 3.8 cm)

Produced by Lyman Metal Products, Norwalk, Connecticut

K2012.33a,c

ACCORDING TO THE DESIGNER, JOHN G. THEW, "THE TURNPIKE TOLL GUN SEEMED LIKE AN easy way to get quarters into the then existing toll baskets. Mr. Ray Lyman of Lyman Metal products produced and sold the toll gun and paid me a royalty. I made an extended line of prototypes of products for Lyman Metal Products in 1963–64."[4] Thew studied with his artist father, making lamps, trivets, and weathervanes. The toll gun's close resemblance to an actual gun is surprising, though it is more reminiscent of a gun that would have been carried by Flash Gordon. Indeed, there is a bit of fantasy in the scene depicted in the photo on the instructions, in which the driver has fired the gun from her convertible, and the coin flies through the air. The instructions for operation explain how to load the gun with coins by removing and replacing the magazine spring. A slot in the handle just behind the trigger shows how many coins are left. After the ball-tipped rod is pushed into the barrel until a click is heard, the internal spring is cocked and ready to propel the coins several feet to the toll basket. –DAH

TURNPIKE TOLL GUN

INSTRUCTIONS FOR THE OPERATION of your TURNPIKE TOLL GUN

1. Remove magazine spring from bottom of gun by unscrewing counterclockwise.
2. Load gun with coins (no dimes).
3. Replace magazine spring by depressing and screw in clockwise.
4. Cock gun by pushing rod with ball on end as far as it will go. A definite click will be heard. (The best way to do this is to grasp gun in one hand and push ball into palm of the other). Gun is now ready to be fired.

Note — To insure proper functioning, do not use bent or damaged coins.

LYMAN METAL PRODUCTS • Norwalk, Conn.

Operating instructions for the *Turnpike Toll Gun*.

Reinhold Weiss (born 1934)
Electric Hairdryer, Model No. HLD 231
Designed 1964
Enameled metal, plastic
1⅞ x 5¼ x 3¾ in. (4.8 x 13.3 x 9.5 cm)
Produced by Braun AG,
Frankfurt, Germany
K2008.392

BRAUN IS KNOWN FOR ITS SIMPLE, TIMELESS FUNCTIONAL DESIGNS IN THE 1950S AND 1960S. Reinhold Weiss worked for Braun from 1959 to 1966 as deputy director of the design department and helped to create many of the products and establish the strong brand identity for which Braun became famous throughout the world. A graduate of the Ulm School of Design, Weiss helped to forge the close relationship between the school and Braun. Braun's first hair dryer was this compact modern form for travel, which provided inspiration for countless other dryers that followed, including one by Dieter Rams for Braun in 1970. Its smooth cylinder encasing the motor served as a grip, and it was operated by a single on/off switch. Its simplicity, harmony, and clearly expressed function were typical of the Braun style. –DAH

Saburo Kobayashi (dates unknown)
Nobol Yabe (dates unknown)
Shunji Kashiyama (dates unknown)
Masao Kawamura (dates unknown)
Takahiro Fuse (dates unknown)
Yukinori Hirasawa (dates unknown)

Pocket Television, Model No. TV-10
Designed c. 1983 (utility patent 4,581,654)
Plastic, chromium-plated metal
3⅛ x 5⅛ x 1⅛ in. (7.9 x 13 x 2.9 cm)
Produced by Casio Computer Co., Ltd., Tokyo, Japan
K2008.751

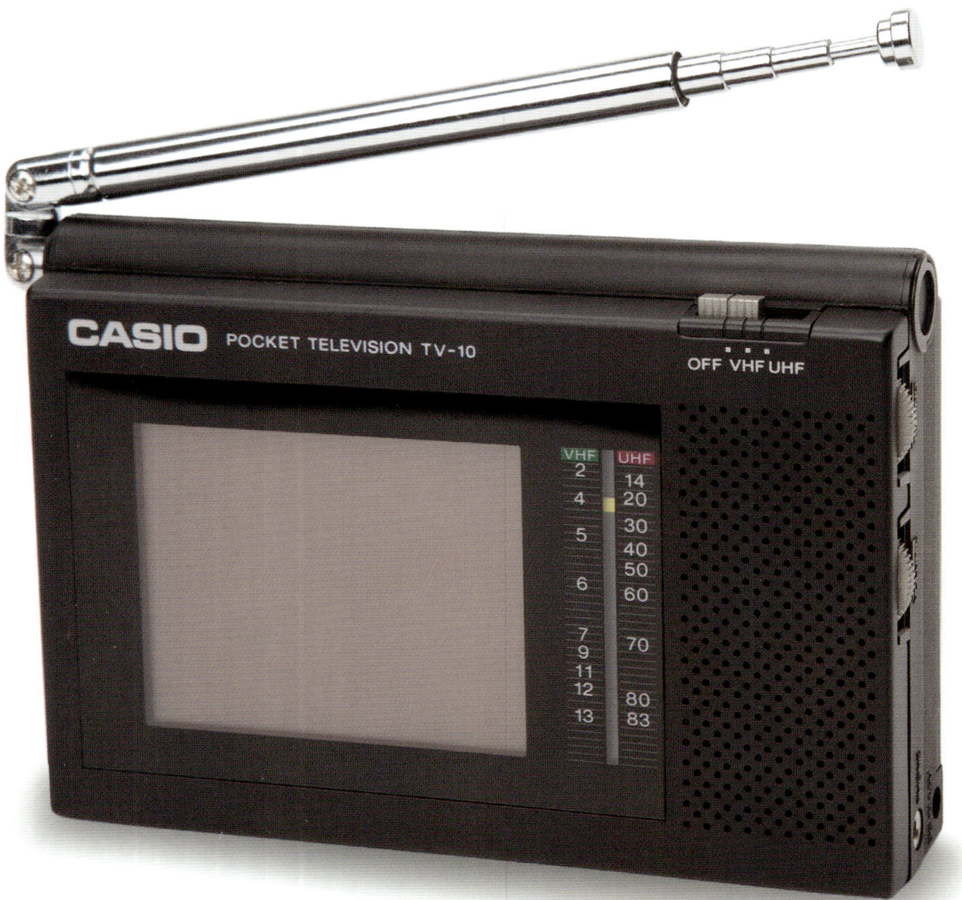

WITH ITS 2⅝-INCH (6.7-CM) SCREEN, THE BLACK-AND-WHITE CASIO TV-10 WAS CONSIDERED the world's smallest television when it was introduced in June 1983.[5] It was the first portable mass-produced TV with a liquid-crystal display (LCD), replacing the cathode-ray tube used by earlier pocket TVs.[6] The pocket-size TV was developed over many years, beginning with visionary concepts in the 1960s, when the idea was nearly unbelievable: "Flat TV picture tubes are really here," wrote Larry Steckler in a 1966 article for *Popular Mechanics*. "I've seen them in action. I've held them in my hands." The flat-tube sets were smaller and less complex, hence, "Truly miniature, book and pocket-size sets become feasible."[7] Major manufacturers developed designs for the new products, including Motorola, Sinclair, Sony, and Panasonic. Casio was founded under the name Kashio Seisakujo in 1946 by Kashio Tadao in Tokyo, and became Casio Computer Company in 1957.[8] In 1985, Casio released the TV 1000, the first color-LCD pocket TV.[9] The world of miniaturization in electronic products continues to evolve today. –DAH

Sam Hecht (born 1969)
Two *Jetlag* Travel Alarm Clocks
Designed 2007
Plastic, metal
⅞ x 3½ x 2⅜ in. (2.2 x 8.9 x 6 cm) each
Produced by I.D.E.A. International,
China, for Takumi, Japan
K2008.1020.1-2

THESE SLEEK, COMPACT TRAVEL ALARM CLOCKS ARE EACH ACCOMPANIED BY FOLDOUT instructions. Flat and rectangular, the miniaturized forms are reminiscent of the austere designs by Dieter Rams for Braun. I.D.E.A. International is a Chinese manufacturer and specialty retailer of lifestyle products, such as clocks, appliances, cosmetics, and watches, available in shops and online. Its various brands include Takumi, which sold these clocks. The *Jetlag's* sole purpose as an alarm clock is revealed on the clock's screen: the clock time and the alarm time are both on continuous display. With a footprint barely larger than a credit card, the clock fits easily into a small suitcase, or even a coat pocket. –DAH

Dean L. Kamen (born 1951)

Segway HT i167, Titanium

Designed c. 2001

Plastic, steel, rubber

Lowest setting: 42¼ x 25 x 19½ in.
(107.31 x 63.5 x 49.53 cm)

Produced by Segway Inc.,
Bedford, New Hampshire

K2002.1

THE *SEGWAY HT i167* (CODE NAME "GINGER") IS A PERSONAL SELF-BALANCING, zero-emissions electric vehicle. The *HT*—short for *Human Transporter* and later changed to *PT* for *Personal Transporter*—was first announced in December 2001 and released to the public in November 2002. Its rechargeable and highly efficient battery could travel up to twelve miles on a single charge. The *HT*'s acceleration, stopping, and turning functions all respond to the subtle movements of the rider's body: a slight lean forward to advance, a slight lean backward for reverse, returning fully upright to stop, and so forth. This mode of operation without pedals, levers, or steering wheel can take some getting used to, but is not too difficult to master. According to the Segway website, multiple sensors register the tilt of the machine for dynamic stabilization, which "balances the rider by applying forces to the ground in the direction of the lean."[10] The development costs for the revolutionary *HT* exceeded $100 million, with the mission of reducing the necessity of combustion engines for travel. With a maximum speed of 12.5 miles per hour (average human walking speed is just over 3 miles per hour), this transporter could be perfect for short journeys that are not practical by foot. –PM

smart Division, Daimler AG

2008 *smart fortwo Passion Coupé*

Designed 2007

Steel frame with thermoplastic
body panels, safety glass

106 x 61½ x 60½ in. (269.2 x 156.2 x 153.7 cm)

Manufactured by Daimler AG,
Stuttgart, Germany

K2008.1078

THIS TWO-SEATER COMPACT *SMART FORTWO* WAS DEVELOPED WITH CITY DRIVERS IN MIND. First introduced in 1998, this version debuted at the Bologna Motor Show in 2006, although this is a later model from 2008. Gerhard Steinle—credited as the lead designer of the *smart* cars—first developed the prototype car during the early 1990s for Mercedes-Benz ("smart" derives from a joint venture between Swatch and Mercedes and was an acronym for Swatch Mercedes *art*). The *smart fortwo* was officially launched in the United States in 2008, and marketed as the most efficient fully gasoline-powered car for sale in the country, although it was outperformed by certain gas-powered models from Mitsubishi and Scion, as well as by some hybrids. The car design had some attractive features such as lightweight recycled-plastic body panels that would not dent, and could be exchanged for a different color. —PM

ENDNOTES

01 HOMEMAKING

1 Eliot F. Noyes, *Organic Design in Home Furnishings* (New York: Museum of Modern Art, 1941, repr. 1969), 4.

2 From a statement made by Girard in 1966, reproduced in Leslie Piña, *Alexander Girard Designs for Herman Miller*, 2nd ed. (Atglen, PA: Schiffer, 2002), 130.

3 Jonathan Olivares in Mateo Kries and Jochen Eisenbrand, eds., *Alexander Girard: A Designer's Universe* (Weil am Rhein, Germany: Vitra Design Museum, 2016), 359.

4 Budapest Poster Gallery, "Bottlik, József (1897–1984)," accessed July 5, 2017, budapestposter.com/artists/bottlik-jozsef.

5 Radiomuseum, "History of the Manufacturer Orion; Budapest," accessed December 25, 2016, www.radiomuseum.org/dsp_hersteller_detail.cfm?company_id=690.

6 KMP Furniture website, www.kmpfurniture.com/designer-news/ingo-maurer---the-light-bulb_112.html, accessed October 16, 2017.

7 Interview of Gianni Colombo by Vittorio Fagone, in Ignazia Favata, *Joe Colombo and Italian Design of the Sixties* (Cambridge, MA: MIT Press, 1988), 120.

8 Tina Mount, "Modernism, Potted," *Los Angeles Times*, May 27, 2004.

9 Museum of Modern Art press release no. 51-312-20a, March 25, 1951, p. 1. MoMA Press Release Archive.

10 Art Directory, "Christian Dell (Hanau 1893–Wiesbaden 1974)," accessed January 7, 2017, www.dell-christian.com/.

11 Charlotte Fiell and Peter Fiell, *1000 Lights* Vol. 1, *1879–1959* (Köln: Taschen, 2005), 261.

02 HOUSEKEEPING

1 United States Patent and Trademark Office (USPTO), design patent application for a "Washing Machine Agitator" filed by James B. Kirby, assignor to the Apex Electrical Manufacturing Company, November 15, 1937, granted December 20, 1938: D112,658.

2 USPTO, utility patent application for a "Washing Machine" filed by James B. Kirby, assignor to the Apex Electrical Manufacturing Company, December 18, 1937, granted March 18, 1941: 2,235,452 (specs p. 2, lines 27–35).

3 Apex Electrical Manufacturing Company, "Only Apex Gives You the Spiral Dasher," advertisement, printed in the *Saturday Evening Post* (May 10, 1947): 115.

4 Knapp-Monarch Company, "Knapp-Monarch Electrical Appliances: 45 Perfect Servants at Your Finger Tips," marketing brochure, c. 1939.

5 USPTO, utility patent application for a "Sadiron Rest" filed by William E. Gundelfinger, assignor to the Knapp-Monarch Co., September 12, 1938, granted April 21, 1941: 2,239,171 (specs, p. 1). USPTO, design patent application for a "Sadiron or the Like" filed by William E. Gundelfinger, assignor to the Knapp-Monarch Co., June 12, 1939, granted August 15, 1939: D116,161.

6 Sheldon Cheney and Martha Cheney, *Art and the Machine* (New York: Whittlesey House, 1936, repr. by Acanthus Press, 1992), 43.

7 USPTO, utility patent application for a "Dust Collecting Apparatus" filed by Robert A. Allen, August 6, 1908, granted March 7, 1911 (985,765).

8 Carroll Gantz, *The Vacuum Cleaner: A History* (Jefferson, NC: McFarland, 2012), 55–56.

9 Advertisement, "Mais oui! Je sors! J'ai enfin des loisirs depuis que j'ai mon TORNADO (Oh yes! I'm going out! I finally have free time since I have my TORNADO)," published for Tornado France, 1950s.

10 Pia Wallén's website, www.piawallen.se/collaborations/electrolux, accessed October 16, 2017.

11 Ole Jensen, quoted in www.normann-copenhagen.com/product/product-collections/washing-up-bowl, accessed October 30, 2017.

03 CONTROLLING CLIMATE

1 USPTO, five design patent applications for a "Control Instrument" filed by Henry Dreyfuss, assignor to Minneapolis-Honeywell Regulator Company, June 19, 1942, granted December 14, 1943: D136,848–D136,852.

2 Advertisement for the *Honeywell Round*, 1953, published in *LIFE*, November 9, 1953.

3 Victoria and Albert Museum Collections, "Electric Fan Heater, Christian Barman," accessed January 8, 2017, collections.vam.ac.uk/item/O22784/electric-fan-heater-barman-christian.

4 Oxford Reference, Christian Barman biography, accessed January 8, 2017, oxfordreference.com/view/10.1093/oi/authority.20110803095447676.

5 Victoria and Albert Museum Collections, "Electric Fan Heater, Christian Barman."

6 Laurel Ciohon, "75 Years of History in Baileys Harbor," *Door County Today*, October 8, 2013, accessed November 3, 2016, http://doorcounty.wehaaserver.com/article-20361-75-Years-of-History-in-Baileys-Harbor.html.

7 Ralph Ormiston obituary, *The Herald Scotland*, October 30, 2007.

8 Richard Chamberlain, Alan Peat, Geoffrey Rayner, et al., *Austerity to Affluence: British Art and Design, 1945–1962* (London: Merrell Holberton, 1997), 96.

9 David A. Hanks and Anne Hoy, *American Streamlined Design: The World of Tomorrow* (Paris: Flammarion; Montreal: Stewart Program for Modern Design, 2005), 177–78. USPTO, design patent application for an "Electric Fan or Similar Article" filed by Robert Heller, assignor to the A. C. Gilbert Company, January 13, 1937, granted April 27, 1937: D104,259.

10 "Products and Practice: 'Airflow Electric Fan,'" *The Architectural Forum* 66, no. 3 (March 1937): 200.

11 USPTO, utility patent application for an "Air Circulator" filed by Max Weber, March 22, 1937, granted July 12, 1938: 2,123,448 (specs p. 1, lines 20–21, 23–25).

12 Mel Byars, *The Design Encyclopedia* (London: Laurence King, 2004), 578.

13 DeVilbiss Company, *DeVilbiss Electric Steam Vaporizer No. 148*, instruction manual, c. 1943, back cover.

14 Jeffrey L. Meikle, "The Midwestern City and the American Industrial Design Aesthetic," in Dennis P. Doorden, et al., *The Alliance of Art and Industry: Toledo Designs for a Modern America* (Toledo: Toledo Museum of Art, 2002), 100.

15 "Notes on Finishing Phenonlic Plastics," *Durez Plastics News: Notes on Product Improvements* 14, no. 3 (March 1947): 3.

16 USPTO, utility patent application for a "Vaporizer" filed by Joy B. Schmitt, assignor to the DeVilbiss Company, April 5, 1943, granted November 21, 1944: 2,363,365.

04 MEASURING

1 USPTO, design patent application for a "Scale" filed by Arnold M. and Melvin D. Thompson, March 3, 1948, granted November 8, 1949: D155,944.

2 *Machine Art*, exh. cat. (New York: Museum of Modern Art, 1934), cat. no. 348.

3 Russel Wright, "Unintentional Modern," *Arts & Decoration* 40, no. 3 (January 1934): 51.

4 Nicholas Fox Weber, *Patron Saints: Five Rebels Who Opened America to a New Art, 1928–1943* (New York: Alfred A. Knopf, 1992), 198.

5 Juliet Kinchin, "Machine Art: Elements of a New Beauty," in David A. Hanks, ed., *Partners in Design: Alfred H. Barr Jr. and Philip Johnson* (New York: The Monacelli Press; Montreal: The Stewart Program for Modern Design, 2015), 153.

6 Zeiss website, "Milestones in the Company's History," accessed February 21, 2017, www.zeiss.com/corporate/int/history/company-history/milestones.html#inpagetabs-2.

7 Phyllis Ross, *Gilbert Rohde: Modern Design for Modern Living* (New Haven, CT: Yale University Press, 2009), 89–92.

8 Aluminum Company of America, "It's a Lot of Greek," *Aluminum News-Letter* (June 1937): 4.

9 Tilmann Buddensieg and Henning Rogge, *Industriekultur: Peter Behrens und die AEG* (Berlin: Gebr. Mann Verlag, 1979), 181–82.

10 "Futuristic Clocks Put on Exhibition; One Resembles Half of Billiard Ball, Another a Child's Wood Building Block," *The New York Times*, October 14, 1949.

05 PREPARING & STORING FOOD

1 See Penny Sparke, "Cookware to Cocktail Shakers: The Domestication of Aluminum in the United States, 1900–1939," in Sarah Nichols, ed., *Aluminum by Design* (New York: Harry N. Abrams, 2000), 112–39.

2 World Cat Identities, Online Computer Library Center, "Ames, Mary Ellis," accessed October 9, 2016, www.worldcat.org/identities/np-ames,%20mary%20ellis/.

3 Cook's Info, "Mary Ellis Ames," accessed October 10, 2016, www.cooksinfo.com/mary-ellis-ames.

4 John Stuart Gordon, *A Modern World: American Design from the Yale University Art Gallery, 1920–1950* (New Haven, CT: Yale University Art Gallery and Yale University Press, 2011), 200.

5 Russell Flinchum, *Henry Dreyfuss, Industrial Designer: The Man in the Brown Suit* (New York: Cooper-Hewitt, National Design Museum, Smithsonian Institution, and Rizzoli, 1997), 50–51.

6 USPTO, utility patent application for a "Baking Utensil" filed by Raymond L. Barton, January 19, 1935, granted October 26, 1937: 2,096,726 (specs, col. 1, lines 11–17).

7 Hanks and Hoy, *American Streamlined Design*, 122.

8 The approximate design date for this toaster was taken from utility patent 2,271,485, assigned to Sunbeam for the heating element invented by Ludvik J. Koci, the drawings for which clearly show the outline of this unusual toaster casing.

9 "Sunbeam Toaster," *Durez Plastics News: Notes on Product Improvements* 13, no. 11 (November 1946): 1.

10 Jasper Morrison, quoted in Marcel Wanders, *International Design Yearbook 20* (New York: Abbeville Press, 2005), 47.

11 Hanks and Hoy, *American Streamlined Design*, 105.

12 Stefano Giovannoni, in an interview with *designboom*, November 18, 2014. www.designboom.com/design/stefano-giovannoni-interview-11-18-2014/, accessed October 16, 2017.

13 USPTO, design patent application for a "Saucepan" filed by Paul V. Gardner, assignor to Corning Glass Works, November 22, 1939, granted January 9, 1940: D118,445. Utility patent application for a "Vessel Assembly" filed by Deo R. Mattoon, assignor to Corning Glass Works, July 13, 1944, granted May 18, 1948: 2,441,892.

14 Elizabeth Gordon, "Fine Art for 39¢," *House Beautiful* (October 1947).

15 Museum of Modern Art press release no. 11, February 11, 1955, p. 1. MoMA Press Release Archive.

16 USPTO, utility patent application for a "Bread Server or Analogous Seal Tight Container" filed by Earl S. Tupper, May 8, 1950, granted November 30, 1954: 2,695,645 (col. 1, lines 17, 18, 43).

17 Hanks and Hoy, *American Streamlined Design*, 93.

18 Ibid., 263.

19 USPTO, design patent application for a "Combined Pitcher and Cover Therefor" filed by Ralph E. Kruck, assignor to Westinghouse Electric & Manufacturing Company, January 10, 1939, granted July 25, 1939: D115,833.

20 Raymond Loewy, *Industrial Design* (Woodstock, NY: Overlook Press, 1979), 98.

06 EATING & DRINKING

1 Advertisement for the American Thermos Bottle Company, "Drink with the Hollywood Set!" c. 1937.

2 Advertisement for the American Thermos Bottle Company, "Now you can sell 'The Hollywood Set' as low as $3.95," c. 1937. Thanks to Laura Smith, Archivist in the Archives & Special Collections at the Thomas J. Dodd Research Center, University of Connecticut Library, for providing extensive information from the archives regarding the American Thermos Bottle Company. E-mail to Jonathan Lans of the Kravis Design Center, June 13, 2017.

3 Frederica Todd Harlow, in Martin Eidelberg, ed., *Design 1935–1965: What Modern Was; Selections from the Liliane and David M. Stewart Collection*, 2nd ed. (New York: Abrams, 2001), 132.

4 "Jens Quistgaard, 88, Designer of Popular Tableware, Is Dead," *New York Times*, February 2, 2008.

5 "Adventure in Aluminum," *Industrial Design* 7, no. 5 (May 1960): 54.

6 Industrial Designers Society of America website, "Russel Wright, FIDSA," 2016, accessed December 27, 2016, www.idsa.org/content/russel-wright-fidsa.

7 William J. Hennessey, *Russel Wright: American Designer* (Cambridge, MA: MIT Press, 1983), 21.

8 M.S., "Raymor: Over 25 Years of Pioneering Have Made More Than a Good Business of Good Design," *Industrial Design* 1, no. 1 (February 1954): 124–27.

9 Revere Copper and Brass Company sales catalogue, 1935.

10 Gordon, *American Design from the Yale University Art Gallery*, 138.

11 Jewel Stern, in Rachel Delphia and Jewel Stern, *Silver to Steel: The Modern Designs* (Pittsburgh: Carnegie Museum of Art, 2015), 48; 178, no. 156.

12 Ibid., 178, no. 157.

13 International Silver Company, advertisement in *House & Garden*, November 1928.

14 Jewel Stern, *Modernism in American Silver*, exh. cat. (Dallas: Dallas Museum of Art, 2005), 358, cat. nos. 124–126. In today's equivalent, the cost would be approximately $1,070, $500, and $640, respectively.

15 USPTO, two design patent applications for a "Table Set" filed by Jean G. Theobald, assignor to the International Silver Company, July 17, 1928, granted October 29, 1928: D76,564 and D76,565.

16 Stern, *Modernism in American Silver*, 35, 335.

17 Obituary for Peter Schlumbohm, *The New York Times*, November 7, 1962, 39.

18 Ralph Caplan, "Cross Section," *Industrial Design* 9, no. 12 (December 1962): 121.

19 USPTO, design patent application for an "Insulated Container for Bottles" filed by Peter Schlumbohm, April 23, 1945, granted November 6, 1945: D142,795.

20 USPTO, utility patent application for a "Cocktail Shaker" filed by Peter Schlumbohm, July 6, 1946, granted December 5, 1950: 2,532,373 (specs, col. 1, lines 2–3 and 10–11).

21 USPTO, design patent application for a "Combined Flask and Handle for a Coffeemaker or the Like" filed by Peter Schlumbohm, January 12, 1943, granted May 23, 1944: D137,943.

22 "Peter Schlumbohm," *Industrial Design* 9, no. 12 (September 1962): 14.

23 Walter Dorwin Teague, "Plastics and Design" (condensed from an address before the National Conference on Plastics, December 12, 1939), *Architectural Forum* 72 (February 1940): 94.

24 For more information about the Schreckengost family, see "The Life," in Henry Adams, *Viktor Schreckengost and 20th-Century Design* (Cleveland: Cleveland Museum of Art, 2000), 4–29.

25 Flinchum, *Man in the Brown Suit*, 112.

26 Society of Industrial Designers, *U.S. Industrial Design, 1949–50* (New York: Studio Publications, 1949), 25.

27 "Housewares Parade," *House & Garden* 77, no. 4 (April 1940): 40.

28 Mary Wright and Russel Wright, *Guide to Easier Living* (New York: Simon and Schuster, 1950), 28.

29 Stephen Heyman, "Scholten & Baijings's Bright Ideas," *The New York Times Style Magazine*, August 21, 2015, accessed November 27, 2017, www.nytimes.com/2015/03/21/t-magazine/scholten-baijings-design-home.html

30 Gordon, *American Design from the Yale University Art Gallery*, 164.

31 Leslie Piña, *Fostoria: Designer George Sakier* (Atglen, PA: Schiffer, 2007), 28–29.

32 Cheney and Cheney, *Art and the Machine*, 77.

33 Ibid., 80.

34 *Machine Art*, exh. cat. (New York, Museum of Modern Art, 1934), n.p., cat. nos. 185–86, 243.

35 See Adams, *Viktor Schreckengost and 20th-Century Design*, 115–16.

36 USPTO, design patent application for a "Plate" filed by Frank H. Sebring Jr. and Donald Schreckengost "said Schreckengost assignor to Sebring," October 17, 1934, granted January 1, 1935 (D94,245).

37 USPTO, design patent application for a "Cup" filed by Vincent Broomhall and Herbert A. Smith, assignors to the Salem China Company, March 4, 1935, granted September 3, 1935: D96,722; design patent application for a "Cup" filed by Vincent Broomhall and Herbert A. Smith, assignors to the Salem China Company, March 4, 1935, granted September 3, 1935: D96,722.

38 Glass Etch and Pattern Gallery, "Bartlett-Collins Company History," April 28, 2007, accessed November 8, 2016, chataboutdg.com/gallery/details.php?image_id=5928.

39 Dianna Everett, "Glass Manufacturing," The Encyclopedia of Oklahoma History and Culture, accessed October 29, 2016, www.okhistory.org/publications/enc/entry.php?entry=GL002.

40 *Shiro Kuramata 1934–1991*, exh. cat. (Tokyo: Hara Museum of Contemporary Art, 1996), 133.

41 Ibid.

42 Nicholai Wiig Hansen, quoted in Olivia Martin, "Design Idea of the Week: A Fool Proof Flask," *Dwell*, April 12, 2013. www.dwell.com/article/design-idea-of-the-week-a-fool-proof-flask-6dc3a6a3.

43 Massimo Vignelli, *Vignelli from A to Z* (Mulgrave, Australia: Images, 2007), 133.

44 David A. Hanks, ed., *The Century of Modern Design: Selections from the Liliane and David M. Stewart Collection* (Paris: Flammarion; Montreal: Stewart Program for Modern Design, 2010), 288.

45 Amber Bravo, interview with Steve Frykholm, "Herman Miller's Poster Child," *WHY Magazine*, undated, accessed October 15, 2016, www.hermanmiller.com/why/herman-millers-poster-child.html.

07 LISTENING

1 USPTO, design patent application for "Radio Receiver or Similar Article" filed by Daisuke Kajiwara on October 12, 1971, granted November 20, 1973: D229,347.

2 Panasonic, "Panasonic's Craziest Color Radio and What to Do with It," *Toot-A-Loop* advertisement, c. 1969–72.

3 Panasonic sales brochure clipping, c. 1969–72.

4 USPTO, utility patent application for a "Telephone Handset" filed by Knut Hugo Blomberg and Ralph Åke Gösta Lysell, October 15, 1941, granted April 22, 1947 (2,419,388); and utility patent application for a "Telephone Set" filed by Knut Hugo Blomberg, December 30, 1942, granted August 6, 1946 (2,405,543).

5 Flinchum, *Man in the Brown Suit*, 97.

6 "Ekco Radio" brochure, produced for E. K. Cole, Ekco Works, Southend, England, c. 1934.

7 Danielle Shapiro, *John Vassos: Industrial Design for Modern Life* (Minneapolis: University of Minnesota Press, 2016), 98–102.

8 Gordon, *American Design from the Yale University Art Gallery*, 129

9 Shapiro, *John Vassos*, 82–84.

10 Hanks and Hoy, *American Streamlined Design*, 173.

11 "'Radio Nurse' Watches Child," *Mechanix Illustrated* 20, no. 3 (July 1938): 73.

12 *Phaidon Design Classics*, Vol. 2, *334–666* (New York: Phaidon Press, 2006), object no. 535.

13 Bev Roberts, "David Ogle DSC, MBE (1922–1962): Sad Career of a Creative Designer," *Inopian*, January 21, 2015, accessed November 5, 2016, www.inopian.com/david-ogle-dsc-mbe-1922-1962-sad-career-of-a-creative-designer. Note that there are discrepancies across sources for Ogle's birth year. England & Wales, Civil Registration Birth Index, 1916–2005, via Ancestry.com, confirms birth in June 1921.

14 RCA Manufacturing Company, *RCA Broadcast Station Equipment, 1941*, 5, 9, 10; and *RCA Broadcast Station Equipment Price List, February 1, 1940.*

15 Sony Corporation advertisement for Sony TranSound XF-5000, 1981.

16 Press release, "Jawbone® Introduces JAMBOX™: The World's First Intelligent Wireless Speaker and Speakerphone," November 4, 2010. http://content.jawbone.com/static/www/pdf/press-releases/pr-110410.pdf.

17 Alex Steinweiss in a 1990 interview, quoted in Steven Heller, "Alex Steinweiss, Originator of Artistic Album Covers, Dies at 94," *The New York Times*, July 19, 2011.

08 VIEWING

1 Eric Pace, "Edwin H. Land Is Dead at 81; Inventor of Polaroid Camera," *The New York Times*, March 2, 1991.

2 Alice Rawsthorn, "How Dieter Rams Made Braun Look Cool," *The New York Times*, June 6, 2011.

3 "Bush TV22 Television 1950," accessed December 23, 2016, bakelite_world_2001.tripod.com/itsbakeliteyouknow/id48.html.

4 Bernd Polster, *Braun: Fifty Years of Design and Innovation* (Stuttgart: Edition Axel Menges, 2009), 501.

5 Ibid., 80.

6 "Early Technology: Televia," accessed December 23, 2016, www.earlytech.com/earlytech/item?id=187.

7 Museum of Modern Art press release,"Marco Zanuso and Richard Sapper: Selections from the Design Collection," August 1993. MoMA Press Release Archives.

8 Ibid.

9 Revere Camera Company, *Owners Operating Manual: Revere 16, 16mm Projector*, 1948, 3.

10 Goldberg Brothers website, "Our History," accessed December 11, 2016, goldbergbrothers.com/about.

11 Paillard Inc., *H16 Reflex* user manual, c. 1956.

09 WORKING

1 From the Collection of the Deutsche Museum, via Google Arts & Culture, accessed December 29, 2016, www.google.com/culturalinstitute/beta/asset/z%C3%BCndapp-zr-automatic-128-b-sewing-machine/0wFD3zqBxdAfQg?hl=en.

2 Clarke website, accessed October 28, 2016, aboutus: https://www.clarkeus.com/aboutus/ourhistory.aspx.

3 *Vibro-Tool* advertisement, *Popular Mechanics* 84, no. 4 (October 1945): 145.

4 Clarence Monroe Burton, William Stocking, and Gordon K. Miller, eds., *The City of Detroit, Michigan, 1701–1922* (Detroit: S. J. Clarke Publishing, 1922), 961.

5 USPTO, utility patent application for an "Electrically Driven Tool" filed by Samuel Duncan Black and Alonzo Galloway Decker, assignors to the Black & Decker Manufacturing Company, December 4, 1914, granted November 6, 1917: 1,245,860 (specs, p. 1, lines 37–39).

6 *Black & Decker ¼" Drill* advertisement, *Popular Mechanics* 64, no. 4 (October 1935): 131A.

7 USPTO, utility patent application for "Eye Goggles" filed by Harvey S. Cover, February 13, 1922, granted October 31, 1922: 1,433,676 (specs, p. 1, lines 11–15).

8 John Kovach, "Inventor of 'Dust Protector,' 'Eye Guards' Had It Covered," *South Bend Tribune*, June 9, 2008.

9 USPTO, design patent applications for an "Eye Shield" filed by Joseph E. DuBois, February 20, 1943, granted September 21, 1943: D136,378 and D136,379.

10 USPTO, design patent application for a "Power Mower" filed by Clifford H. Flanigan, October 14, 1948, granted June 6, 1950: D158,847.

11 USPTO, utility patent application for an "Ornamental Lawn Sprinkler" filed by Joseph Lessler, granted December 19, 1871: 121,949 (specs, p. 1, col. 2).

12 Lance Wyman, interview by Alexandra Lange, "Talking Pictures," *WHY Magazine*, undated, c. 2014, https://www.hermanmiller.com/stories/why-magazine/talking-pictures/, accessed October 16, 2017.

13 Vincent Scully, quoted in Stanley Abercrombie, *George Nelson: The Design of Modern Design* (Cambridge, MA: MIT Press, 2000), 177.

14 Howard Miller promotional copy, c. 1965, quoted in Stephanie Murg, "Safety First: George Nelson Explains It All!" *Adweek*, March 14, 2012, accessed October 16, 2017, http://www.adweek.com/digital/george-nelson-explains-it-all/.

15 Ibid.

16 Eliot Noyes, quoted in "Good Design Is Good Business," *IBM100: Icons of Progress*, IBM website, accessed October 16, 2017, http://www-03.ibm.com/ibm/history/ibm100/us/en/icons/gooddesign/.

17 USPTO, utility patent application for a "Stapling Machine" filed by Herbert W. Marano assignor to the Wilson-Jones Company, July 7, 1949, granted June 8, 1954: 2,680,245.

18 Society of Industrial Designers, *U.S. Industrial Design*, 98.

19 USPTO, design patent application for a "Strip Server" filed by Egmont Arens, assignor to the Nashua Gummed and Coated Paper Company, January 20, 1942, granted June 16, 1942: D132,802.

20 Gordon, *American Design from the Yale University Art Gallery*, 332.

21 USPTO, utility patent application for an "Adding Machine" filed by Thomas O. Mehan on June 17, 1937, granted March 24, 1942: 2,277,498.

22 Marianne Brandt, "Letter to the Younger Generation," in Eckhard Neumann, ed., *Bauhaus and Bauhaus People: Personal Opinions and Recollections of Former Bauhaus Members and Their Contemporaries* (New York: Van Nostrand Reinhold, 1970), 98.

23 Hanks and Hoy, *American Streamlined Design*, 74.

24 USPTO, design patent application for a "Design for a Pencil Sharpener" filed by Edward C. Hoffmann, April 13, 1939, granted December 5, 1939 (D117,901).

10 PLAYING

1 Juliet Kinchin, "De Stijl, Children, and Constructivist Play," in Juliet Kinchin and Aidan O'Connor, *Century of the Child: Growing by Design 1900–2000* (New York: Museum of Modern Art, 2012), 70–71.

2 Ibid., 71.

3 Henry Wiencek, *The World of LEGO Toys* (New York: Abrams, 1987), 50.

4 Ibid., 48.

5 Date of model from Daniel Lipkowitz, *The Lego Book* (London: Dorling Kindersley, 2012), 170. Thanks to Gabe Dougherty for bringing this to our attention.

6 Museum of Modern Art press release, "The Museum of Modern Art Christmas Shop Opens Fifth Season," November 13, 1970. MoMA Press Release Archives.

7 Instruction sheet for *Vasarely Planetary Folklore Participations No. 1*, produced by Éditions William Wise, 1969.

8 Cheney and Cheney, *Art and the Machine*, 84.

9 USPTO, design patent application for a "Sled" filed by Harold L. Van Doren and John G. Rideout, assignors to the American National Company, January 15, 1934, granted March 20, 1934: D91,796. Utility patent application for a "Sled" filed by Harold L. Van Doren and John G. Rideout, assignors to The American National Company, January 29, 1934, granted May 26, 1936: 2,041,982.

10 Harold Van Doren, *Industrial Design: A Practical Guide* (New York: McGraw-Hill, 1940), plate 32.

11 Advertisement for TEY Manufacturing Corp., "New Tracks in Skis," published in *Ski Magazine* (November 1, 1948), reprinted in *New England Ski Museum Newsletter*, no. 48 (Autumn 1999): 2.

12 Marc Newson, quoted in "Superformer," *Wired* (October 1, 2000).

11 TRAVELING

1 Classic Trains website, comment from an anonymous user passenger fan, accessed February 20, 2017, http://cs.trains.com/ctr/f/3/t/48375.aspx?page=32.

2 Hanks and Hoy, *American Streamlined Design*, 220.

3 *Directions for Use: The Steemco Steemette*, produced for Steam Electric Manufacturing Company, St. Louis, Missouri, c. 1950.

4 John G. Thew to Jonathan Lans of the Kravis Design Center, e-mail, July 5, 2017. USPTO, utility patent application for a "Coin Gun" filed by John G. Thew, assignor to Raphael Lyman, October 10, 1962, granted June 29, 1965: 3,191,588 (specs, col. 1, lines 37–39).

5 Casio website, company history, June 1983 timeline entry, accessed February 8, 2017, world.casio.com/corporate/history/chapter02.

6 Guenthoer.de website, "The Short 'History' of Pocket-TV: A Chronology on the Development of Pocket-size Televisions," accessed February 8, 2017, www.guenthoer.de/e-history.htm.

7 Larry Steckler, "Fantastic New Portable Color TV," *Popular Mechanics* (February 1966): 91, 93.

8 Casio website, company history, June 1983 timeline entry.

9 Guenthoer.de website, "The Short 'History' of Pocket-TV."

10 Segway website, "How does the PT move? How does it balance?" http://www.segway.com/about/our-story.

SELECTED BIBLIOGRAPHY

Albrecht, D. et al., *Russel Wright: Creating American Lifestyle.* New York: Abrams, 2001.

Aldersey-Williams, H. *World Design: Nationalism and Globalism in Design.* New York: Rizzoli, 1992.

Antonelli, P. *Mutant Materials in Contemporary Design.* New York: Museum of Modern Art, 1995.

Antonelli, P. *Objects of Design from the Museum of Modern Art.* New York: Museum of Modern Art, 2003.

Attfield, J. *Wild Things: The Material Culture of Everyday Life.* New York: Bloomsbury, 2000.

Baudrillard, J. *The System of Objects.* London: Verso, 1996.

Bel Geddes, N. *Horizons.* New York: Dover, 1977.

Bush, D. *The Streamlined Decade.* New York: George Braziller, 1975.

Cheney, S., and M. Cheney. *Art and the Machine.* New York: Whittlesey House, 1936.

Delphia, R., and J. Stern. *Silver to Steel: The Modern Designs of Peter Müller-Munk.* Pittsburgh: Carnegie Museum of Art, 2015.

De Zurko, E. R. *Origins of Functionalist Theory.* New York: Columbia University Press, 1957.

Doblin, J. *One Hundred Great Product Designs.* New York: Van Nostrand Reinhold, 1970.

Doordan, D. P., et al. *The Alliance of Art and Industry: Toledo Designs for a Modern America.* Toledo: Toledo Museum of Art, 2002.

Dormer, P. *The Meanings of Modern Design: Towards the Twenty-First Century.* London: Thames and Hudson, 1990.

Dreyfuss, H. *Designing for People.* New York: Simon and Schuster, 1955.

DuBois, J. H. *Plastics.* Chicago: American Technical Society, 1946.

Eidelberg, M., ed., et al. *Design 1935–1965: What Modern Was.* 2nd ed. New York: Abrams, 1991.

Fiell, C., and P. Fiell. *Industrial Design A–Z.* Köln: Taschen, 2006.

Flinchum, R. *Henry Dreyfuss, Industrial Designer: The Man in the Brown Suit.* New York: Cooper-Hewitt, National Design Museum, Smithsonian Institution, and Rizzoli, 1997.

Forty, A. *Objects of Desire: Design and Society 1750–1980.* London: Thames and Hudson, 1986.

Fraser, W. H. *The Coming of the Mass Market, 1850–1914.* London: Macmillan, 1981.

Frederick, C. *Household Engineering and Scientific Management in the Home.* Chicago: American School of Home Economics, 1919.

Gordon, J. S. *A Modern World American Design from the Yale University Art Gallery, 1920–1950.* New Haven, CT: Yale University Art Gallery and Yale University Press, 2011.

Greif, M. *Depression Modern: The Thirties Style in America.* New York: Universe Books, 1975.

Halen, W. *Christopher Dresser.* London: Phaidon Press, 1990.

Heskett, J. *Industrial Design.* New York: Oxford University Press, 1980.

Heskett, J. *Toothpicks and Logos: Design in Everyday Life.* New York: Oxford University Press, 2002.

Hiesinger, K., and G. Marcus. *Landmarks of Twentieth-Century Design: An Illustrated Handbook.* New York: Abbeville Press, 1993.

Julier, G. *The Culture of Design.* London: Sage, 2000.

Kaufmann, E., Jr. *Introduction to Modern Design.* New York: Museum of Modern Art by Arno Press, 1969.

Kirkham, P., ed. *The Gendered Object.* Manchester, United Kingdom: Manchester University Press, 1996.

Loewy, R. *Never Leave Well Enough Alone.* New York: Simon and Schuster, 1951.

Marchand, R. *Advertising the American Dream: Making Way for Modernity, 1920–1940.* Berkeley: University of California Press, 1985.

McCarty, C. *Information Art: Diagramming Microchips.* New York: Museum of Modern Art, 1990.

McCracken, G. *Culture and Consumption: New Approaches to the Symbolic Character of Consumer Goods and Activities.* Bloomington: Indiana University Press, 1988.

Naylor, G. *The Arts and Crafts Movement: A Study of its Sources, Ideals and Influence on Design Theory.* London: Studio Vista, 1971.

Pevsner, N. *Pioneers of Modern Design: From William Morris to Walter Gropius.* Harmondsworth, United Kingdom: Penguin, 1960.

Phaidon Design Classics, Vol. 1–3, *001–999.* New York: Phaidon Press, 2006.

Polster, B. *Braun: Fifty Years of Design and Innovation.* Stuttgart: Edition Axel Menges, 2009.

Pulos, A. J. *The American Design Adventure, 1940–1975.* Cambridge, MA: MIT Press, 1988.

Pulos, A. J. *American Design Ethic: A History of Industrial Design to 1940.* Cambridge, MA: MIT Press, 1983.

Shapiro, D. *John Vassos: Industrial Design for Modern Life.* Minneapolis: University of Minnesota Press, 2016.

Sparke, P. *Japanese Design.* London: Michael Joseph, 1987.

Sparke, P. *The Plastics Age: From Bakelite to Beanbags and Beyond.* Woodstock, NY: Overlook Press, 1993.

Sudjic, D. *The Language of Things: Design, Luxury, Fashion, Art: How We Are Seduced by the Objects Around Us.* London: Penguin, 2011.

Teague, W. D. *Design This Day: The Technique of Order in the Machine Age.* London: Studio Publications, 1946.

Van Doren, H. *Industrial Design: A Practical Guide.* New York: McGraw-Hill 1940.

Woodham, J. *Twentieth-Century Design.* New York: Oxford University Press, 1997

Wright, M., and R. Wright. *Guide to Easier Living.* New York: Simon and Schuster, 1950.

CONTRIBUTORS

PENNY SPARKE is a British writer and academic specializing in the history of design. She has been teaching at universities in the United Kingdom since 1972, and she is currently a professor of design history and the director of the Modern Interiors Research Centre at Kingston University in London. Sparke has written fifteen books about twentieth-century design. She has also curated many exhibitions, including *The Plastics Age* at the Victoria and Albert Museum in London.

RUSSELL FLINCHUM is associate professor in the department of graphic design and industrial design at North Carolina State University's College of Design. He received his doctorate from City University of New York's Graduate Center in 1998, after guest curating an exhibition and writing a book on industrial designer Henry Dreyfuss for Cooper Hewitt, Smithsonian Design Museum. He wrote *American Design* for the Museum of Modern Art, drawing on more than ten years of experience as a guide to the architecture and design collections. He began teaching at North Carolina State in 2014 and lives in Raleigh.

DAVID A. HANKS is the curator for the Kravis Design Center and has assisted George Kravis in the growth and development of his collection. Hanks has held curatorial positions at the Art Institute of Chicago and the Philadelphia Museum of Art, and has been curator for the Stewart Program for Modern Design in Montreal since 2000. Traveling exhibitions he has curated for the Stewart Program include *American Streamlined Design: The World of Tomorrow* and *Partners in Design: Alfred H. Barr Jr. and Philip Johnson.*

JULIET KINCHIN is a curator, university professor, and writer on aspects of nineteenth- and twentieth-century design and material culture, with a longstanding interest in the social and political contexts of modern design. She joined the Museum of Modern Art, New York, as a curator in the department of architecture and design in 2008, and has since organized exhibitions including *Counter Space: Design and the Modern Kitchen; Century of the Child: Growing by Design 1900–2000; What Was Good Design? MoMA's Message 1939–55; The New Typography, Designing Modern Women 1900–2000;* and *How Should We Live? Propositions for the Modern Interior.*

CARA MCCARTY is a curator, lecturer, and writer on modern and contemporary design, as well as curatorial director at the Cooper Hewitt, Smithsonian Design Museum, where she leads the museum's curatorial vision and exhibition planning. She was actively involved in the 2014 transformation of Cooper Hewitt into a twenty-first-century museum. Previously she was curator and head of the department of decorative arts and design at the Saint Louis Art Museum and held curatorial positions in the department of architecture and design at the Museum of Modern Art. Her numerous exhibitions and accompanying publications include *Access + Ability; Tools: Extending Our Reach; Structure and Surface: Contemporary Japanese Textiles; Masks: Faces of Culture; Information Art: Diagramming Microchips;* and *Designs for Independent Living.* She received a bachelor's degree from Stanford University and was a Loeb Fellow at Harvard's Graduate School of Design.

CATHERINE MCDERMOTT has for more than twenty years curated and published in the area of contemporary design as a professor at the Kingston School of Art, one of the United Kingdom's leading design schools. In 2015 she was awarded a National Teaching Fellowship. In 2016 she won the prestigious Sir Misha Black Award for innovation in design and pioneering the ways that design is curated and communicated. The program she founded in 2000 to offer a master's degree in curating contemporary design, a collaboration between Kingston University and the Design Museum in London, has inspired a generation of talented new curators around the world. Her practice focuses on interdisciplinary curating, which includes pioneering new sites of practice and exploring cultural dialogue and emerging roles for curators. She is currently leading projects that support emerging designers in Chile and China.

PAT KIRKHAM is professor of design history at Kingston University, London (with a mandate to bring together studies in design and film) and professor emerita at Bard Graduate Center. The first woman to attain a first-class honors degree in history at the University of Leeds, she obtained her doctorate from the University of London. A founding member of the Design History Society and a founding editorial board member of the *Journal of Design History*, Kirkham has published and lectured widely on design, gender, and film. She has written scholarly studies of designers, design reformers, and design partnerships (for example, Charles and Ray Eames, Saul and Elaine Bass, Eva Zeisel, and Harry Peach) and directed major research, exhibition, and publishing projects such as *Women Designers in the USA, 1900–2000: Diversity and Difference* and *History of Design: Decorative Arts and Material Culture, 1400–2000*.

PAUL MAKOVSKY is an editor, writer, and curator, as well as the vice president of design at *Metropolis* magazine, where he has worked since 1999. As an expert in architecture, technology, and design at all levels, he has curated and collaborated on exhibitions about contemporary American graphic design, emerging Norwegian product design, and Canadian painting and design. He also co-curated *Knoll Textiles, 1945–2010* at the Bard Graduate Center. A graduate of McGill University and the University of Toronto, he is currently working on a book about the life and work of Florence Knoll Bassett, as well as researching a book on modernist food photography.

R. CRAIG MILLER has been one of the leading advocates for modern and contemporary design of his generation, through his work at the Metropolitan Museum of Art, Denver Art Museum, and Indianapolis Museum of Art. He built major collections for each institution, helping to define a new, more inclusive concept of modern design. During the course of his career, he has also organized numerous exhibitions and written and contributed to dozens of publications, as well as giving lectures in Asia, Europe, and North America. He is currently a visiting scholar at the Philadelphia Museum of Art and continues to work as a curator and consultant.

TOM TREDWAY is an assistant professor in the department of design at California State University, Long Beach. His published works include contributions to *Biography, Identity and the Modern Interior; Eva Zeisel: Life, Design, and Beauty;* and *History of Design: Decorative Arts and Material Culture, 1400–2000*.

HAMPTON C. WAYT is a twentieth-century design historian and historiographer who specializes in industrial and transportation design. He has been a leading force in the preservation of design history, having recovered dozens of designers' archives throughout the course of his career, and his scholarly interests include the interrelationship of art, technology, and culture; the design process and design education; and the cultivation of designers' legacies. He received his master's degree from the Bard Graduate Center, New York, in 2012.

KATE CLARK is an editor and researcher who has worked for David A. Hanks & Associates since 2001. She has assisted Hanks on numerous exhibitions and such books as *American Streamlined Design: The World of Tomorrow; The Century for Modern Design;* and *Partners in Design: Alfred H. Barr Jr. and Philip Johnson*.

ALANA EMBRY has been the collections manager/registrar for the Kravis Design Center in Tulsa since its founding in 2015. She holds master's degrees in museum studies from the University of Oklahoma and in art history from the University of Missouri. Her work for the Kravis Design Center includes inventory, research, and documentation of the collection, coordinating loans and gifts to other institutions, and visitor education. Embry taught in the art department at the University of Arkansas-Fort Smith from 2010 to 2017.

JONATHAN LANS has been the preparator for the Kravis Design Center in Tulsa since it was founded in 2015. His roles in the collection include organizing, researching, and documenting the objects, assisting with photography shoots, building stands and storage units, and giving tours of the collection. Lans is originally from Brisbane, Australia, where he began his career teaching English and Indonesian; he has taught in both elementary and secondary schools in Australia, Russia, Japan, and India.

ACKNOWLEDGMENTS

WITHOUT THE KRAVIS DESIGN CENTER and its dedicated, capable staff, this publication would not have been possible. In particular, Alana Embry, registrar and collections manager, took on, with energy and enthusiasm, the gathering of all curatorial data; coordinated and supervised the new photography of the objects; and oversaw the enormous task of obtaining all contextual photos and permissions. Preparator Jonathan Lans shared in this work with dedication and precision, carrying out essential research to locate patent drawings and biographical information about the designers. Rick Phillips, administrative assistant to George Kravis, made sure that all work went smoothly and that all necessary support was made available.

At Rizzoli, Margaret Chace and David Morton were excited about the project since it was first proposed in 2014. The task of editing and coordinating the text by various authors was the responsibility of Andrea Danese, who managed this enormous undertaking with determination. The editing and research were shared by Kate Clark of David A. Hanks & Associates, who was also responsible for coordinating many details of the publication with a commitment to quality and an unwavering attention to detail that was remarkable. Megan Conway was responsible for copyediting the text.

We thank Sarah Gifford for a book design that is contemporary yet timeless, reflecting the Kravis design philosophy. Her talent, professionalism, and patience were invaluable.

British design historian Penny Sparke was responsible for the concept of the book, its thematic organization, and the insightful introduction. She was undaunted by the complex task at hand. We are particularly grateful to the scholars and curators who contributed text to this publication. To summarize the history of industrial design and make it accessible to both student and general reader was challenging. The authors of the descriptions accompanying each object are acknowledged by their initials following each entry. The research undertaken for this publication has brought fresh insights to many familiar designs, and uncovered new discoveries about many objects that have never been published until now.

Extensive new photography was carried out especially for this publication. We were fortunate to have the services of photographers John Faier of Chicago and Shane Culpepper of Tulsa for many of the objects featured in this volume. Also included are photographs by Adam Murphy and Edward Robison. Contextual photographs were provided by numerous museums and archival sources; a complete list is in the photo credits. We are grateful to the following individuals who went above and beyond the call to assist us: Janice Hussain of the Cooper Hewitt, Smithsonian Design Institution; Amy Auscherman of Herman Miller Archives; Claudine Nicol of the Montreal Museum of Fine Arts; Ginger Grilli of Ohio's Salem Historical Society; Jacklyn Hoyt, of the Special Collections Research Center at Syracuse University Libraries; and Matthias Pühl of Vitra Design Museum.

Research and assistance from many institutions were essential to the success of this project. We are indebted to the following individuals, who were particularly helpful in research: collection specialist Paul Galloway of the Museum of Modern Art; curatorial consultant R. Craig Miller; design historian Phyllis Ross; author Jewel Stern; inventor John G. Thew; and archivist Laura Smith of the University of Connecticut Library.

We would also like to extend our thanks to individuals who supported our project with their advice and specialized expertise: Angéline Dazé, Harry Greenberger, Michael Jefferson, Andrea Monfried, and Jan Spak.

Above all, we are grateful to George Kravis for supporting the scholarship that has made this publication possible. We acknowledge his vision and generosity in sharing his knowledge of design and his collection with the authors.

David A. Hanks
Curator, Kravis Design Center

INDEX

Note: Page numbers in italic type indicate illustrations.

D

E

F

G

H

I

J

K

L

M

N

O

P

PHOTO CREDITS

Courtesy of Apple Corp. via Getty Images: 250 (left)

Iwan Baan: 218, 219

Florence Knoll Bassett portfolio of photographs and articles. Florence Knoll Bassett papers, 1932–2000. Archives of American Art, Smithsonian Institution: 13

Bauhaus-Archiv Berlin, © 2017 Artists Rights Society (ARS), New York/VG Bild-Kunst, Bonn: 18 (right); Photo: Lucia Moholy, 281 (left); Photo by Werner Zimmermann, 281 (right)

bpk Bildagentur/Staatsgalerie, Stuttgart, Germany/Willi Baumeister/Art Resource, NY, © 2017 Artists Rights Society (ARS), New York/VG Bild-Kunst, Bonn: 214 (right)

Photo courtesy of Black+Blum Ltd.: 11

Photo by Margaret Bourke-White/Time & Life Pictures/Getty Images, Photo © Estate of Margaret Bourke-White/Licensed by VAGA, New York, NY: 2, 3

© BRAUN P&G: 271

Marcel Breuer Papers, Special Collections Research Center, Syracuse University Libraries: 19

Courtesy of Cappellini: 33

Foundation Achille Castiglioni: 247

Photo from the Cleveland Press Collection, Michael Schwartz Library, Cleveland State University, Photographer Robert C. Hoffner, Viktor Schreckengost™ with permission from American da Vinci, LLC: 330

Courtesy of Closca: 342, 343

Photo: Matt Flynn © Cooper Hewitt, Smithsonian Design Museum: 47; 58; 60; 68; 75; 94 (left); 140; 164; 196; 227; 230; 236; 246; 274; 275; 320; © Eames Office, LLC, 52; Gift of OXO, 143 (left); Gift of Industrial Designers Society of America, 153; © The Edith Lutyens and Norman Bel Geddes Foundation, Inc., 189; © Estate of Margarete Marks. All Rights Reserved, DACS/ARS 2017, 202; Photo: Ellen McDermott © Cooper Hewitt, Smithsonian Design Museum: 30 (left); Photo © Cooper Hewitt, Smithsonian Design Museum © Eames Office, LLC, 91 (right)

Shane Culpepper, Tulsa, OK: 6; 59; 66–67; 70–71; 78–79; 80 (top); 81; 82–83; 86 (left); 89; 102; 103; 104; 106; 107; 110; 111; 112; 114; 119; 120; 125; 132 (left); 136; 137; 138; 139; 150; 151; 155; 156; 158; 159; 160 (top); 161; 162; 165; 166; 167; 174; 176; 178; 179 (left); 186–87; 188; 194; 197; 198–99; 200; 209; 210; 211 (bottom left and bottom right); 222; 224; 225; 229; 237 (top); 240; 241; 243; 245; 248; 249; 250 (right); 251; 252; 255; 261; 264; 265; 266; 267; 268; 269; 270; 272; 273; 276 (left); 279; 290; 291; 292; 293; 294; 295; 296–97; 298; 299; 303; 305; 306; 308–309; 310; 311; 317; 321; 324; 325; 328; 329; 332; 333 (top); 335; 338–39; 339 (right); 340 (top); 341; 356; 360; 361 (bottom); 362–63; 364 (left); 366; 367; 368; © 2017 Artists Rights Society (ARS), New York/ProLitteris, Zurich, 71 (lamp at far right); © Dyson, 84; © 2017 Artists Rights Society (ARS), New York/VISDA, 88; © 2017 The Josef and Anni Albers Foundation/Artists Rights Society (ARS), New York, 254; Viktor Schreckengost™ with permission from American da Vinci, LLC, 331

Design Museum Denmark, Photographer Pernille Klemp: 22

Image © Electrolux, © 2017 Artists Rights Society (ARS), New York: 85

John Faier Photography: 4; 9 (left and right); 10; 61; 62; 63; 65; 69; 72; 77; 96; 100; 108–09; 115; 116; 122; 123; 124; 126; 128; 134; 148; 152; 169; 171 (top); 172–73; 180; 182–83; 185; 201; 203 (top); 204; 205; 207; 208; 220; 223; 226; 238–239; 244; 253 (top); 256; 258–59; 262; 277; 284; 304; 307; 315; 322; 334; 352; 359; 365; © 2018 Girard Studio, LLC, 44, 55; © 2017 Artists Rights Society (ARS), New York, 64; © 2017 Artists Rights Society (ARS), New York/VISDA, 87; © 2011 Artists Rights Society (ARS), New York/KUVASTO, Helsinki, 191; © 2017 The Isamu Noguchi Foundation and Garden Museum, New York/Artists Rights Society (ARS), New York, 235; © 2017 Artists Rights Society (ARS), New York/ADAGP, Paris, 313, 318, 327

Ignazia Favata/Studio Joe Colombo: 361 (top)

Courtesy of Russell Flinchum: 346 (right)

Georg Fuchs and F.H. Newbery, *Arts Decoratifs Modernes: Exposition de Turin*, 1902, page 88: 213 (left)

Courtesy of fuseproject: 146

Hagley Museum and Library, Raymond Loewy Collection (2004.255), © Loewy Estate: 26 (top and bottom)

Archives, David A. Hanks & Associates: 175, 186 (bottom)

Harry Ransom Center, The University of Texas at Austin, © The Edith Lutyens and Norman Bel Geddes Foundation, Inc.: 27

Courtesy of Herman Miller Archives: 145 (left), 211 (top)

HfG-Archiv Ulm, Photo: Wolfgang Siol: 282

Photo: Frederick Hollyer © National Portrait Gallery, London: 17 (left)

Photo by Hulton Archive/Getty Images: 344

Courtesy of the Indianapolis Museum of Art: 193 (bottom)

Courtesy Intel Corp.: 92

Courtesy of Marcin Jakubowski, CC-BY-SA 4.0 International: 144

Courtesy of Jawbone: 253 (bottom)

Kartell: 86 (right)

Kravis Design Center: 16; 18 (left); 57 (bottom); 74; 76 (bottom); 79 (top); 80 (bottom); 98; 101; 141 (left); 141 (right); 154; 160 (bottom); 163; 168; 179 (right); 181; 212; 228 233; 234; 237 (bottom); 251; 260; 276 (right); 278; 312; 333 (bottom); 346 (left); 357; 358 (left); 364 (right); By permission of Oxford University Press, 41 (left); © Herwin Schaefer, 41 (right); © 2017 Artists Rights Society (ARS), New York/ADAGP, Paris, 326; © Sony Corporation, 340 (bottom)

Image courtesy of http://www.laurelhollowpark.net/: 203 (bottom right)

Courtesy of Layer Design: 94 (right)

Library of Congress Prints and Photographs Division, Washington, D.C.: Reproduction number LC-USZ62-19261, 25; Reproduction Number LC-USZ62-39060, © The Edith Lutyens and Norman Bel Geddes Foundation, Inc., 216 (right); Reproduction number LC-USZ62-100556 345 (left); FSA/OWI Collection, C-USE6-D-009321, 348

Photo by Tony Linck/The LIFE Picture Collection/Getty Images: 195

Photograph by Los Angeles Modern Auctions (LAMA): 50, 133, 242

Photo courtesy of Magis: 31 (right)

Photograph © MAK–Austrian Museum of Applied Arts/Contemporary Art: 17 (right)

Oleksiy Maksymenko/Alamy Stock Photo, © Mercedes-Benz: 369

Foto Marburg/Art Resource, NY: 213 (right)

Image copyright © The Metropolitan Museum of Art. Image source: Art Resource, NY: © 2017 Artists Rights Society (ARS), New York / ADAGP, Paris, 31 (left); © Shao Fan (b. 1964), 37

© Ministère de la Culture-Médiathèque du Patrimoine/Henri Baranger/Dist. RMN-Grand Palais/Art Resource, NY, © 2017 Artists Rights Society (ARS), New York/ADAGP, Paris: 216 (left)

Courtesy MIT Museum, © 2017 Artists Rights Society (ARS), New York/KUVASTO, Helsinki: 21

Photo Montreal Museum of Fine Arts, Christine Guest: 34 (left), 36

Adam Murphy Photography: 99, 105, 157, 190, 206, 232, 286–87, 288, 336, 337

The Museum of the City of New York/Art Resource, NY: 347

Digital Image © The Museum of Modern Art/Licensed by SCALA/Art Resource, NY: 28; 40; 53; 142 (right); 143 (right); Alfred H. Barr, Jr. Papers, © 2017 Artists Rights Society (ARS), New York/VG Bild-Kunst, Bonn, Art © Estate of Alexander Rodchenko/RAO, Moscow/VAGA, New York: 214 (left); Photograph by Wurts Brothers, 215

Repro-photo: René-Gabriel Ojéda, © RMN-Grand Palais / Art Resource, NY: 345 (right)

Verner Panton Design © Panton Design, Basel: 46

Photo © Philips Company Archives, © F.L.C./ADAGP, Paris/Artists Rights Society (ARS), New York 2017: 217 (right)

Photo: Jean-Claude Planchet © CNAC/MNAM/Dist. RMN-Grand Palais/Art Resource, NY, © 2017 Artists Rights Society (ARS), New York/ADAGP, Paris: 319

Courtesy of Race: 217 (left)

Radisson Blu Royal Hotel, Copenhagen: 355

Photograph by Edward C. Robison III: 76 (top), 302, 316 (top)

Image courtesy of Sotheby's, © Sotheby's, Inc. 2012: © Gehry Design, LLC, 51 (top); © 2017 Artists Rights Society (ARS), New York/VG Bild-Kunst, Bonn, 170; © Sotheby's, Inc. 2016: © 2017 The Isamu Noguchi Foundation and Garden Museum, New York/Artists Rights Society (ARS), New York, 118 (right); 192; 193 (top)

© Philippe Starck: 31 (center)

Stewart Program for Modern Design: 316 (bottom)

Stiftung Deutsches Technikmuseum Berlin, Foto: Historisches Archiv: 289

STOCKFOLIO®/Alamy Stock Photo: 350 (left)

© 2017 H. Stoll AG & Co. KG, www.stoll.com: 93

Jeremy Sutton-Hibbert/Alamy Stock Photo: 32

© Thonet GmbH: 91 (left)

Archivio Fotografico © La Triennale di Milano: 30 (right)

United States Patent and Trademark Office, www.uspto.gov: 113, 118 (left), 121, 127, 145 (right), 301, 358 (right), endpapers (details of patent drawings)

© University of Applied Arts Vienna, Victor J. Papanek Foundation. Reproduced with permission from University of Applied Arts Vienna: 142 (left)

University of California, Berkeley Library: 141 (center)

Vessel USA Inc., www.architecturalpottery.com: 66 (top)

Image courtesy of Vitra Design Museum: 51 (bottom); Alexander Girard Estate, photograph by Charles Eames, © Eames Office, LLC, © 2018 Girard Studio, LLC, 54

© Volkswagen of America Inc.: 351

Photo by Francis Ware: 34 (right)

© West Dunbartonshire Council Libraries and Cultural Services: 14

Nigel Wilkins/Alamy Stock Photo: 350 (right)

Brownie Wise Papers, Archives Center, National Museum of American History, Smithsonian Institution: 171 (bottom)

Courtesy of Wright: 56; 57 (top); 129; 130; 132 (right); 184; 300; 354; © 2017 Artists Rights Society (ARS), New York/KUVASTO, Helsinki, 48; © Eames Office, LLC, 49; © 2017 Artists Rights Society (ARS), New York/ProLitteris, Zurich, 131; © 2017 Artists Rights Society (ARS), New York/VG Bild-Kunst, Bonn 314

Russel Wright Papers, Special Collections Research Center, Syracuse University Libraries: 198 (left)

First published in the United States of America in 2018 by

Rizzoli Electa
A Division of Rizzoli International Publications, Inc.
300 Park Avenue South
New York, NY 10010
www.rizzoliusa.com

in association with

Kravis Design Center
PO Box 52506
Tulsa, Oklahoma 74152
www.kravisdesigncenter.org

Library of Congress Control Number: 2017962718

ISBN: 978-0-8478-6240-5

FOR THE KRAVIS DESIGN CENTER
Alana Embry, Collections Manager/Registrar
Jonathan Lans, Preparator
David Hanks, Curator
Kate Clark, Editor

FOR RIZZOLI ELECTA
Charles Miers, Publisher
Margaret Rennolds Chace, Associate Publisher
Andrea Danese, Editor

Design by Sarah Gifford

2018 2019 2020 2021 / 10 9 8 7 6 5 4 3 2 1

PRINTED IN CHINA

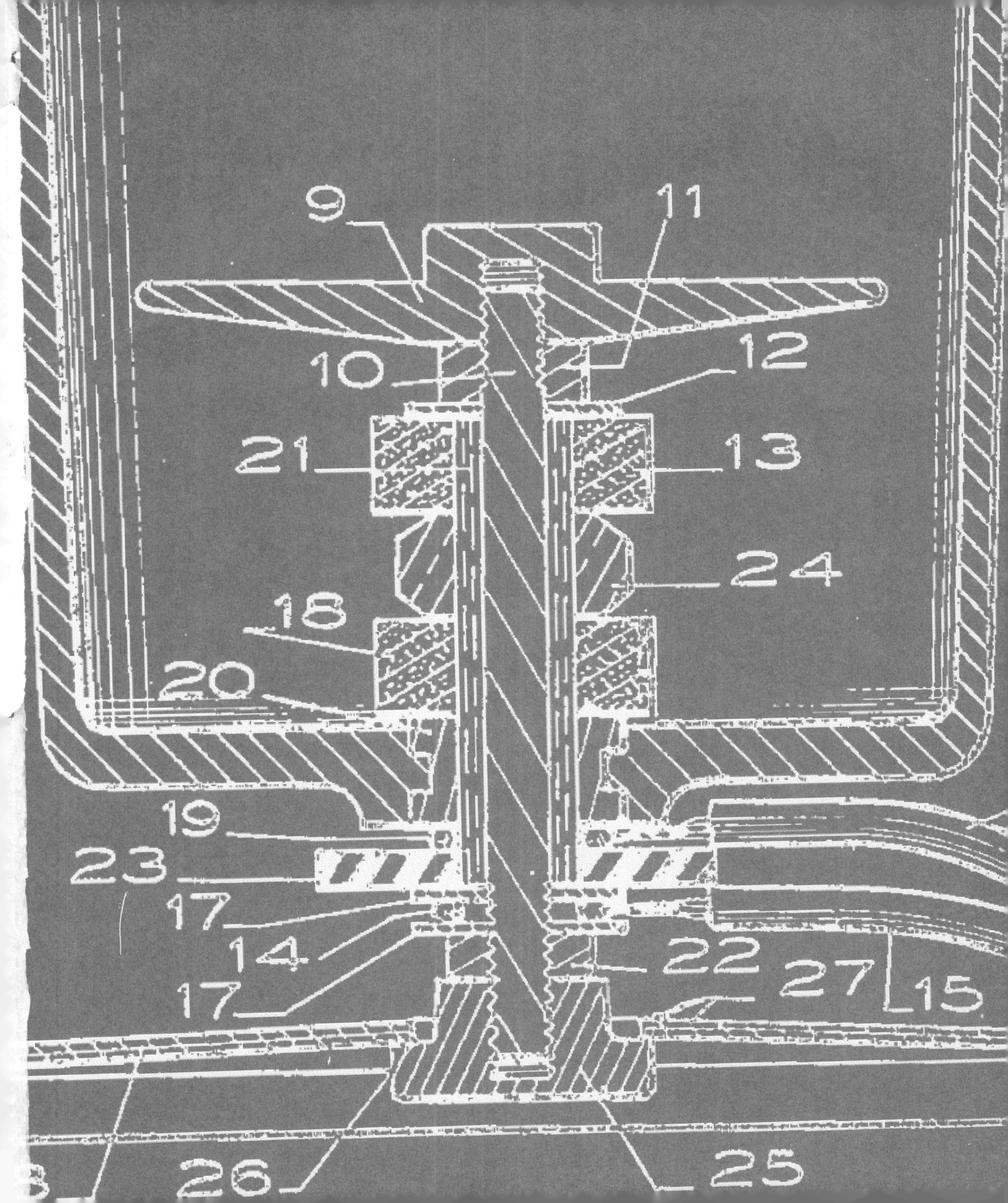

9
11
10
12
21
13
24
18
20
19
23
17
14
22
17
27
15
26
25